topics in
ocean engineering

volume 3

volume 3

TOPICS IN OCEAN ENGINEERING

charles l. bretschneider, editor

chairman, department of ocean engineering
university of hawaii

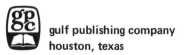

gulf publishing company
houston, texas

Topics in Ocean Engineering

Library of Congress Catalog Card Number 78-87230

ISBN 0-87201-600-5

Topics in Ocean Engineering, Volume 3 is part of the University of Hawaii's Sea Grant Program sponsored by the National Oceanic and Atmospheric Administration. It is based on a series of seminars given by experts in various disciplines of ocean engineering. The purpose of these seminars is to acquaint marine scientists, practicing engineers, laymen or technicians and graduate students involved with exploiting the sea with the "state of the art," which is changing rapidly because of the ever increasing knowledge of the marine environment and its effects on the procedures and techniques being used to solve marine engineering problems.

We consider ocean engineering at the University of Hawaii in three broad categories: Port and Coastal Engineering, Offshore and Continental Shelf Engineering and Deep Ocean Engineering.

In *Topics in Ocean Engineering*, we attempt to cover the above broad categories. We also attempt to select experts from various parts of the United States, as well as from outside the country.

We do not intend to have the speakers necessarily present results of new research. Instead, we wish to make use of their knowledge and approaches they have gained in applying their own research to solving engineering problems peculiar to the marine environment.

Charles L. Bretschneider, Professor
Department of Ocean Engineering
University of Hawaii

introduction

contents

topics in
ocean engineering

volume 3

part 1: john lyman

The Practice of Hyperbaric Chemistry

A native of Berkeley, California, John Lyman received his bachelor's degree in chemistry from the University there in 1936. His graduate work in chemical oceanography was carried out at Scripps Institution of Oceanography, leading to M.S. and Ph.D. degrees from U.C.L.A. in 1951 and 1957 respectively. During World War II he served on active duty in the United States Naval Reserve, from which he was recently retired with the rank of Captain.

His employment history includes service with the Union Oil Company of California as an analytical chemist, and as an oceanographer in administrative positions with the Hydrographic Office of the Navy, the National Science Foundation, the Bureau of Commercial Fisheries, and the Smithsonian Institution. He has also served as a consultant to the Atomic Energy Commission, The Office of Naval Research, the Environmental Science Services Administration, the U.S. Army Corps of Engineers, and the United Nations.

From 1968 to 1973 he was Professor of Oceanography at the University of North Carolina, Chapel Hill, and at North Carolina State University, Raleigh. He also served as Marine Science Coordinator for the university system of North Carolina and Principal Investigator of the North Carolina Institutional Sea Grant.

1

the practice
of hyperbaric chemistry

What do we mean by "hyperbaric chemistry"? We mean carrying out laboratory procedures in an environment where the ambient pressure is significantly higher than 1 bar.*

The environment where such an operation is most likely to take place is an actual or simulated underwater habitat. There is growing realization among ocean scientists that the solutions of many problems concerning the outer continental shelves and their living creatures are beyond the reach of devices, however sophisticated they may have become, that must be suspended from surface ships, and that the current generation of submersibles, while extremely useful in reconnaissance operations, lacks many degrees of accessibility to the actual environment.

The solution many investigators are turning to, therefore, is to leave the surface and operate in the ocean itself, not from behind the armored hull and viewing ports of the submersible, but in a pressurized underwater habitat. "Bounce" dives, where the SCUBA-equipped investigator penetrates the

hyperbaric ocean for a short period and then returns to the surface, have their uses, but the progressive loading of the blood with atmospheric gases, particularly nitrogen, as the time at depth is increased, imposes a heavy penalty in decompression time. Table 1-1, an extract from the U.S. Navy diving tables, illustrates the great penalty resulting from increased increments of time at depth, or of depth.

Thus, as was pointed out by Gillen (1), in bounce diving the working ability of each individual is severely limited, because he either stays only a short time on the bottom to avoid prolonged decompression, or he pays the price of an extensive decompression schedule.

A solution to this difficulty has been found in maintaining the individual at the pressure he wishes to work at for a period of days or even weeks. The chamber or capsule in which he lives is commonly known as a "habitat." At the end of the experiment or expedition the divers, in or out of the habitat, are returned to the surface through a single decompression schedule and thus have sacrificed a minimum of effective working time. For studying the behavior or migration of marine organisms (particularly where observations may be required throughout the entire diel cycle), for conducting marine archaeological observations, for detailed seafloor mapping and sampling, and for most industrial applications, such a procedure seems indispensable.

---

*One bar, one million dynes per square cm, is the c.g.s. pressure unit. It is nearly the same as the standard atmosphere, 76 cm of mercury, which is 1.0133 bars or 14.7 psi, and it is also 1020 grams per square cm. For oceanographic use, the decibar (one-tenth of the bar) is a convenient unit, since in sea water the depth in meters and the pressure in decibars are for many purposes equivalent. In English units, the pressure in psi is roughly half the depth in feet.

3

Table1-1. Decompression Time Required (Min.)				
Time on Bottom (Min.)	160 ft.	170 ft.	180 ft.	190 ft.
10	4	5	6	7
20	16	22	26	31
40	71	81	92	102
60	132	152	167	182

In recent years we have seen a variety of saturation dives, such as the U.S. Navy's Sealab I and Sealab II and the Interior Department's Project Tektite, to name only three of the most familiar operations. Now that the capability of such systems has been well demonstrated, we can expect to see scientific or commercial missions as the main object of future operations in the ocean with pressurized habitats. Of concern in this chapter are the effects of pressure and other aspects of the habitat atmosphere on carrying out laboratory procedures in chemistry.

The habitat atmosphere will vary as the depth of operation is varied. In the first 100 feet or so, commonly either ordinary air or an oxygen-nitrogen mixture with a partial pressure of oxygen of two decibars (the same partial pressure as in air at sea level) is supplied. In Tektite I, for example, which was carried out at a depth of 50 feet, the habitat was supplied with 92% nitrogen and 8% oxygen (2). The air supply for the SCUBA gear with which the participants made excursions into the waters of Great Lameshur Bay, however, was ordinary atmospheric air. The change in breathing gas composition is made because a high partial pressure of oxygen can be toxic. A great advantage of using oxygen-poor mixtures in habitats, in addition, is the freedom from combustion hazards they provide. The tragic death of three astronauts at Cape Kennedy in January, 1967, in an accidental fire in an oxygen-rich atmosphere has provided an example that will not soon be forgotten.

Thus, any procedure involving a flame must be abandoned or modified underwater. Just as in professional baseball, tobacco-chewing is substituted for smoking. And the familiar bunsen burner must everywhere be replaced by the hot plate. The chemist will have to foresee well in advance all his requirements for bent and fire-polished glass tubing or depend on ordering these items from the support facility above sea level. Flame spectroscopy or atomic absorption analysis seems to be completely ruled out as a laboratory procedure.

Below a depth of about 150 feet, nitrogen in turn is found to be dangerous. A nitrogen partial pressure much greater than about 5 bars produces neurological symptoms in human subjects that are commonly described as "narcosis." Hence, in prolonged bounce diving or in saturation diving below this level, the gas mixture is changed to helium-oxygen. Cost factors generally make it mandatory that such a gas be recycled. Thus, the chemist has all but lost the utility of the "hood." It is indeed possible to design the habitat's ventilation system so that the outlet from the living spaces to the breathing-mixture purification system is in fact the laboratory hood, but extreme care must be taken that laboratory procedures do not produce volatile contaminants at levels that exceed the ability of the purification system to remove them. For example, particular attention must be given to the possibility that evolution of a volatile strong acid like hydrogen chloride does not overpower the absorption system for the weak acid carbon dioxide, which of course is being continuously produced in significant quantities by the divers themselves, for carbon dioxide even at relatively low levels is extremely toxic.

Fortunately, a great deal of information on the purification of breathing mixtures is available as a result of U.S. Navy investigations of conditions in Polaris-carrying submarines. That this technology has become available to civilian enterprises is well exemplified by the successful two-month drift of the *Ben Franklin* in the Gulf Stream from Florida to New England in 1969. One specific aspect of atmospheric purification will be discussed later in this chapter.

Helium-oxygen mixtures have been successfully employed in simulated dives to 1000-1100 feet. At these levels, human subjects begin to exhibit muscular spasms or tremors, which many investigators believe to be a specific effect of helium. Some attention therefore is currently being directed to the possibility of using hydrogen-oxygen mixtures for support of operations deeper than 1000 feet. Since at that depth the mixture need only be one part of oxygen to 149 parts of hydrogen (by volume), there is no explosion hazard; yet the necessity of careful control of the mixture during initial compression and decompression and the awesome specter of what might happen should equipment failure result in the creation of an oxygen-hydrogen mixture of combustible proportions under elevated pressure combine to assure that development in this area will be gradual and cautious.

Besides the flameless aspect of this new underwater environment and the inability henceforth to control his contamination of laboratory air merely by opening a window or flicking the switch of a hood, the chemist has two more problems to contend with under the sea. One is the general effect of the ambient high pressure, and the other is the specific effect of operating in a high partial pressure of gas of a low molecular weight.

These two problems are interrelated and cannot easily be discussed separately. Their significance can be introduced by relating a couple of anecdotes. The first concerns a hydronaut on the Navy's Sealab II operation, which was conducted in a habitat at a pressure of 7 atmospheres. One of the participants brought with him an imported underwater camera, a source of considerable pride, which was specifically designed to be watertight at any depth to which a man could take it. But when he went to open it, he found he was unable to do so. A pressure of over 100 psi on the few square inches of camera back that had to be folded out produced a load of over half a ton, quite overpowering the strength of 10 fingers! Undaunted, our hero then decided at least to inspect his supply of film, only to discover that each little can was now squashed flat!

The other incident occurred in December, 1968, when a simulated dive to 1000 feet was conducted in Duke University's hyperbaric facility at Durham, N.C. The participants subsisted on food prepared outside the chamber and passed to them through an airlock. Because of the time lag in adjusting to pressure, their coffee was put in styrofoam cups to keep it warm. But to the great dismay of all concerned, the first brim-full cups quickly compressed to about three-fifths of their original dimensions, and the resulting mess can be imagined!

Now it happened that one of the participants was celebrating his birthday during the test, and he was also fond of cheesecake. So his wife baked his favorite dish and brought it down to the laboratory for it to be passed through the airlock. Remembering the styrofoam coffee cups and the leathery disks that attempts to cook breakfast pancakes had produced in Sealab II, the Duke technicians predicted dire consequences as the cheesecake was equilibrated to pressure. But the good lady's optimism proved justified, and the cake remained just as fluffy at 30 atmospheres as it had been at one.

Now, why did the styrofoam collapse but not the cheesecake? The answer lies in the nature of the two materials. Styrofoam is relatively impermeable to gases (it does in fact tend to regain some of its original volume after a few hours), but the helium was able to pass readily into the pores of the cheesecake, so that there was a negligible pressure gradient between its exterior and its interior.

The obvious lesson of the camera incident to the chemist is that he will have to arrange his reagents in such a way that their stoppers do not become gastight. Glass-stoppered bottles appear to be completely out of the question, both because of the difficulty in withdrawing the stoppers in the pressurized habitat and the risk of implosion as a result of the pressure loading. Ordinary rubber stoppers maintain sufficient dimensional stability to retain their usefulness. Piercing each one with a fine hypodermic needle appears to provide a means of pressure equilibration, but the needle must be left in place throughout the operation at depth and the subsequent decompression to 1 atmosphere. Bottles containing volatile reagents should be jettisoned in accordance with a predetermined schedule before commencing decompression, for otherwise they will jet vapors of their contents into the laboratory atmosphere. The greater the free air space in the reagent bottle, the greater this effect will be.

Similar stopper venting is required for any samples that have been collected at depth (biological as well as chemical samples) since a screw-cap jar, for example, sealed at 1000 feet, is the equivalent of a hand grenade at the surface.

The cheesecake incident recalls a property of helium that becomes of great significance in hyperbaric operations—its small molecular size.* Helium thus has a great ability to penetrate tiny orifices, and it penetrates many barriers impassable to ordinary air. Television tubes will be ruined by hyperbaric helium; the only solution is to keep them in an armored housing at 1 atmosphere pressure. Ordinary thermal lagging with closed pores, like styrofoam, is squashed into a small volume by high-pressure gases, although there is some recovery of shape with time. One solution that has been tried is to use an open-pore polyurethane artificial sponge, but this substance in turn is so permeable to helium that it becomes inefficient. It appears that more work needs to be done on the design of heat-shielding materials for use in high-pressure light-gas atmospheres. Perhaps a bulk of hollow glass spherules in a glass wool matrix might provide both dimensional stability and a reasonable degree of resistance to heat transfer.

The question of heat brings up two more problems of concern to the chemist. One is the elevation of boiling point at elevated pressures. At 30 bars, water boils at 230°C, which will melt the solder out of a tin can. Greater pressures are likely to involve temperatures that will soften the glass before the water boils. Ordinary laboratory distillation procedures appear to be ruled out in pressurized laboratories. The denser atmosphere will be found to conduct heat much more efficiently

*Since the hydrogen molecule weighs only half as much as the helium molecule, any effect due to the small helium molecule will be twice as significant in the case of hydrogen. As mentioned, it is likely that hydrogen will become the inert gas for hyperbaric operations below 1000 feet.

than ordinary laboratory air. Hot plates have to be two or three times the ordinary wattage to provide the same heating to the sample since the heat content and conductivity of the surrounding atmosphere are so much greater.

Even the ordinary light bulb proves unworkable at pressure, generally imploding at 100 psi (3). A solution has been found by providing a cluster of 12 12-volt auto tail-light bulbs in series. The whole cluster fails when one bulb fails, but it is easy to spot the culprit and replace it.

Another effect of high pressure is the low compressibility of the gases. A vacuum pump will run as though it were pumping water. High-pressure gases also tend to form layers, with the lightest components on top, so that an efficient vertical circulation system is required to minimize this effect. Gas chromatography cannot be performed in a high-pressure environment.

One toxic volatile component that existing life-support systems are unable to remove from a closed atmosphere is mercury vapor. Thus, many of the familiar devices of a surface laboratory, such as McLeod gages, manometers, Van Slyke apparatus, and the like, are strictly taboo in a pressurized habitat. Opinions differ as to the use of mercury-in-glass thermometers. Some practitioners would bar them in favor of thermisters, which seem to be unaffected by pressure and light gases. Others will accept such thermometers if they are water-jacketed or otherwise fitted to prevent air contamination in case of breakage but point out that the pressure effect on a mercury-in-glass thermometer must be taken into account. Oceanographers are already familiar with this effect and in fact use it as a method of determining depth in the sea but may forget that even the ordinary laboratory thermometer will have its reading increased about 1°C for 100 decibars.

Still another effect of the denser atmosphere will be the increased effect of buoyancy on weighings. Fortunately, the use of helium at greater depths tends to compensate to some extent for this effect, yet helium at 30 atmospheres is more than four times as heavy as air at 1 atmosphere. Both the layering effect already mentioned and the possible rise and fall of the tide (easily amounting to a pressure fluctuation of one-third of an atmosphere, the equivalent of a 10-foot tide) will introduce considerable uncertainties into the ambient value of air density and suggest that the analyst concerned with weighing water or biological samples and striving for high precision would do well to arm himself with a calibrated set of weights constructed from some inert material of a density around 1.03, thereby eliminating most of the uncertainty.

Finally, specific pressure effects on chemical reactions or physical properties must be considered. Under pressure, thiosulfate solutions evolve hydrogen sulfide. Standard buffer solutions shift pH, and the magnitude of the shift varies with their nature; borates are different from phosphates, for example. Some recent measurements of the electrical conductivity of sea water show a relatively large and nonlinear response of conductivity to pressure, indicating that salinometers depending on electrical conductivity may not be capable of operating in a pressurized habitat with the precision they deliver at sea level (4).

These considerations all suggest that much of the practice of chemistry in hyperbaric situations will have to depend on the old-fashioned "wet chemical" techniques, with volumetric processes forming the mainstay of laboratory procedures in submerged habitats.

Acknowledgments

Fruitful discussions with H. A. Saltzman of Duke University Medical Center and Ralph Brauer and J. Morgan Wells of the Wrightsville Marine Bio-Medical Laboratory are gratefully acknowledged.

References

1. Gillen, William H. 1969. Saturation exposures, advantages and limitations. In *Topics in Ocean Engineering*, vol. 1. C.I. Bretschneider, ed. Houston: Gulf Publishing, 153-159.
2. Quinn, Alison. 1969. Project Tektite I—four-man underwater laboratory. *Undersea Technology*, vol. 10, Feb., 48-49.
3. Linderoth, L.S., Jr. 1968. *The design of hypo-hyperbaric chambers for medical centers.* ASME Preprint 69-DE-6.
4. Mays, Michael. 1968. *The effect of pressure on the ionic conductance through the upper 2000 meters of the ocean's water column.* U.S. Naval Postgraduate School M.S. Thesis.

part 2: salvatore comitini

Development, Rational Use and Management of the Marine Environment

Salvatore Comitini is an associate professor in the Department of Economics and a staff member at the Economic Research Center of the University of Hawaii. He holds B.S. (1951) and M.S. (1955) degrees from the University of Alabama and a Ph.D. (1960) from the University of Washington.

His honors include: Who's Who in the West, American Men of Science, Pi Gamma Mu and fellowships and awards from the National Science Foundation, the Mershon Center for Education in National Security and the Ford Foundation.

Dr. Comitini has served as an analytical statistician in economics for the Bureau of Commercial Fisheries, U.S. Department of the Interior in Washington (1960-1961) and as a consultant in such projects as:

NORFISH Project, Center for Quantitative Science, University of Washington, Seattle

Indo-Pacific Fisheries Council Working Party of Experts on Economic and Social Aspects of National Fisheries Service, NOAA, U.S. Department of Commerce, College Park, Md.

Institute of Social, Economic and Government Research, University of Alaska

Fisheries Division, Organization for Economic Cooperation and Development, Paris, France

Fisheries Department, Food and Agriculture
Organization of the U.N., Rome, Italy

Comitini served as a member of the advisory committee to the Governor's Task Force on Oceanography for preparation of their report to the governor, "Hawaii and the Sea." He also served as executive director of a task group submitting a report on "Hawaii's Role in the Pacific Basin" in 1969.

Among his papers are

"Licensing and Efficiency: An Empirical Study of the Japanese Tuna Fishing Industry," *Malayan Economic Review* (1971) with David S. Hwang.
"Development of Marine Resources and Regional Economic Growth in Alaska," *Science in Alaska* (1970).
"The Influence of Management Regimes on the Availability of Capital for Fishery Development," FAO (1969), *FAO Fish. Rep.*
"An Economic Report of World Trends in Fisheries Development with Broad Implications for the Indo-Pacific Region," FAO (1970), *IPFC Occasional Papers* (1971).
"Exploiting Marine Resources," *The Encyclopedia of Marine Resources* (1969).
"Economic and Legal Aspects of Japanese Fisheries Regulation and Control," *Washington Law Review* (1967).
"A Study of Production and Factor Shares in the Halibut Fishing Industry," *Journal of Political Economy* (1967) with David S. Hwang.
"Marine Resources Exploitation and Management in the Economic Development of Japan," *Economic Development and Cultural Change* (1966).

2

development, rational use and management of the marine environment

Introduction

The factors promoting regional economic growth, the goals of regional economic development and the criteria for judging the rate of growth are all interrelated. Basically, they are (1) an increase in real income per capita, and (2) an increase in real output per unit of input. The former measures economic welfare, or well-being of the populace. The latter measures productivity of economic resources (e.g., labor and capital), or the allocative efficiency of the economy. In addition to these primary goals of economic growth, others can be identified: (3) public welfare—which involves considerations dealing with income distribution, health and education, (4) national security—which involves questions concerning strategic materials, skills and the defense posture, and (5) the balance of payments—which identifies those factors which promote growth through exports of primary or manufactured goods or through interregional flows of investment capital.

Given goals (1) and (2), there can occur conflicts with goals (3), (4) or (5) if there are different functional relationships which tend to maximize the various goals separately. For example, as real income per capita increases (through whatever source), private and public demands for uses of marine resources—commercial fishing, natural resources extraction, transportation, communication, recreation and waste disposal—all increase, thus affecting costs of operating in the marine environment and hence the overall efficiency and welfare of the regional economy. The basic question, then, is: What should be optimized in the process of developing marine resources for economic growth? Then, the related question: What criteria should be used to make rational judgments as to priorities among alternative uses?

The Marine Environment—Development Opportunities and Externalities

Since marine resources are, almost without exception, public resources and therefore fall within the public domain, it is generally considered that the best criteria for judging alternatives in promoting or reducing various marine activities is some form of benefit-cost analysis (1). This procedure tends to approximate the calculation of profit opportunities in private-enterprise activities. However, given their characteristic as "common-property" or "open-access" resources, there are the important

11

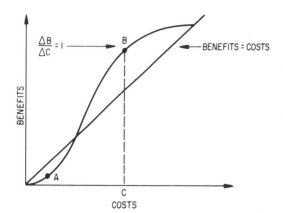

Figure 2-1. Benefit-cost analysis.

questions of when and how much public intervention is desirable in affecting the development of marine resources.

The Marine Science Commission, completing its two-year study in 1969, suggested, for example, that the Great Lakes could serve as a laboratory in establishing national water quality standards for the marine environment (2). Concerning the desired level of water quality in this or any other body of water, the report states that the objective should be to optimize the *net* economic benefits of enhancement, i.e., after taking into account the accrued costs of attaining any particular level. The character of the Lakes is such as to make interdependency between uses inevitable. Thus, it is felt that government intervention not only can improve efficiency in terms of market value of uses but can also take into account extra-market values. Moreover, a public body can more accurately gauge the significance of the various interdependencies and, through public policy, promote efficiency through multipurpose use of the water resource.

The type of benefit-cost analysis required to determine the proper level of water quality to promote is illustrated in Figure 2-1. At low levels of enhancement, the incremental costs of improved water quality will be greater than the value of associated benefits. At higher levels, the value of benefits will rise—incremental benefits

equaling incremental costs at A. At still higher levels, the value of benefits will continue to rise, but the ratio will slowly decrease until incremental benefits equal incremental costs again at B. To go beyond point B would mean that incremental costs will exceed incremental benefits. Thus, the optimum level of water quality is implied directly from the derived optimum cost level at C. Benefits derived from improved water quality which can be quantified include. enhancement of land values; reduced cost of water treatment for domestic, municipal and industrial supplies; enhancement of the fishery resource, both commercial and sport; enhancement of water-based recreation activities and increased potential for water-based recreation opportunities; minimized potential public health hazard; improved aesthetic appeal of the marine environment and attraction of tourists.

Even if the objective of water quality enhancement were simply to make the greatest possible contribution to the income of the coastal region, there still is the question of comparing the contribution made by the program in different years (3). To arrive at some number for the overall contribution of the program requires the summing up of the contributions to regional income over the course of the program's life. Each year's income, however, must be given an appropriate weight which reflects its relative value as compared to the income generation of another year. According to traditional capital theory, income to be received in future years should be weighted by an appropriate interest or discount rate reflecting both the productivity of an investment and the time preference of society to sacrifice present consumption for the greater future consumption implied by that investment. The process of linking future value flows (income) with present value stocks (capital) is to find the *present*, or *discounted value* of some future income stream and convert it into its equivalent capital value (4). The present value of a sum, b_n, due n years hence at interest rate i, is found by solving

$$B_O = b_n (1 + i)^{-n}$$

The present value of an entire series of sums, b_k, available at different points of time in the future is found by

$$B(n) = \sum_{k=1}^{n} b_k (1 + i)^{-k}$$

If the yearly income stream is constant, then,

$$B(n) = b \sum_{k=1}^{n} (1 + i)^{-k} = \frac{b}{i} [1 - (1 + i)^{-n}].$$

As n becomes very large, this expression takes on the simple form:

$$B = \lim_{n \to \infty} B(n) = \lim_{n \to \infty} \frac{b}{i} [1 - (1 + i)^{-n}] = \frac{b}{i}.$$

Because the nature of capital markets in a free enterprise economy evaluates investments based upon *individuals'* calculations, the interest rate which arises does not register *collective* considerations of investment. In effect, then, there is a divergence between *private* and *social* rates of discount which causes difficulties in evaluating public investment projects which necessarily must take social benefits and costs into account. Thus, in formulating and evaluating public marine resources development plans, no market interest rate is directly suitable as a discount rate. The basic difficulty in evaluating the social desirability of public investment projects is that to some degree this will involve displacement of private investments which have been evaluated at a different (private) discount rate. To decide whether the shift of resources from private to public activities is economically efficient, the evaluation of the time stream of net benefits of both projects must use the same rate. Only if the present value of benefits per dollar spent in the public sector is greater than that in the private sector—utilizing the social rate of interest in both computations—is the shift of resources socially desirable.

To illustrate the economic implications of this divergence between the social time preference rate and private rates, suppose a proposed public project in the coastal zone, e.g., water quality enhancement, has a capital cost of $100 million and produces a stream of benefits over time which, discounted at an assumed social discount rate of 2.5%, has a present value of $150 million. This gives a benefit-cost ratio for the project of 1.5:1. Suppose, however, that the marginal stream of benefits from private investment in the coastal zone, e.g., dredging and filling of bays and estuaries for industrial, housing and urban development, yields $0.05 of income benefits on each dollar of investment per year in perpetuity. Thus, private investment of $100 million in the coastal zone, discounted at the same social rate of 2.5%, produces benefits the present value of which is $200 million. If the public investment would displace the private investment if undertaken, it compels coastal regions to forego potential income which is socially valued at $200 million. Thus the "opportunity cost" (or foregone opportunities) of the public investment is not the "money" cost of $100 million but, rather, the $200 million of benefits which would have accrued from alternative investments. Therefore, if these are competing activities in the marine environment, the cutoff benefit-cost ratio is 2:1, rather than 1:1, to correct for the market's undervaluation of the social benefits of investment.

For the special case in which the benefit streams from the public investment are constant over a uniform economic life, a "synthetic" discount rate can be devised which takes into account the opportunity cost of the displaced investment as well as its social desirability. Use of this synthetic rate is equivalent to using the social discount rate and the appropriately higher benefit-cost cutoff ratio; thus the rule approving public investments with benefit-cost ratios exceeding unity can be applied. There are two methods of expressing this criterion. First, if the opportunity cost of the proposed public investment is $a per dollar of outlay while the (constant) annual benefit is $b per dollar of incremental outlay, the criterion for approving the outlay is:

$$\frac{b}{i} \ [1 - (1+i)^{-n}] \geqslant a,$$

where i is the social discount rate and n is the life of the project. Second, since the opportunity cost per dollar of outlay is a = r/i, where r represents the rate of yield in perpetuity on an alternative investment, the criterion can be written:

$$b \geqslant \frac{r}{1-(1+i)^{-n}}$$

or

$$\frac{b}{j} \ [1 - (1+j)^{-n}] \geqslant 1$$

where j is the discount rate such that

$$\frac{1-(1+j)^{-n}}{j} = \frac{1-(1+i)^{-n}}{r}$$

The rate j is the discount rate which synthesizes both the opportunity costs of the public outlay and social time preference. For the example given above, taking n = 40,

$$\frac{1-(1+j)^{-n}}{j} = \frac{(0.62757)}{0.05} = 12.551$$

and referring to any standard compilation of interest tables, j is found to be approximately 7.5%.

Where there are large-scale economies from certain marine activities, the inherent social costs resulting from competing uses of the marine environment can dissipate the rewards to the coastal region from private innovation and new investment (5). These social costs are what economists call *external economies*, or *external diseconomies* of production, or simply *externalities* (6). These externalities represent divergences between private and social costs and returns. Where individual self interest and the social interest do not coincide, economic activity can

lead to a misallocation of resources since the market price system does not account for the social costs and benefits of private operations. If, for example, external economies (in the form of lower costs of operations) accrue to activities B and C from activity A, private enterprise may normally produce a less than optimum amount of activity A since its contribution to economic welfare is not compensated. Alternatively, if activities B and C incur external diseconomies (in the form of higher costs of operations) from activity A, private enterprise may normally produce an overoptimal amount because part of the cost is borne by activities B and C and external to activity A. Nonrecognition of these external diseconomies by private operators would imply that they are "overvaluing" their expected rate of return since they have not taken the external, or social costs of their operations into account in making the investment decision. Thus capital invested is probably higher and the rate of development of marine resources is probably faster than it otherwise would be.

The consequences to the coastal region economy of not taking these external costs into consideration when making investment plans is vividly illustrated by the economic cost of the Santa Barbara oil spill, as shown in Table 2-1 (7). The most important category of external costs from offshore oil production operations represented the costs of clean-up and property restoration from the beginning of the spill in January, 1969, to December, 1970. To the extent that the oil spill necessitated using labor and material resources with alternative uses for cleaning the water and beaches and property restoration it represents a social cost. Altogether, including the costs incurred by the Union Oil combine, the U.S. Department of the Interior, the State of California, and the City and County of Santa Barbara, this amounted to well over $11 million! It was concluded from study of tourist statistics that any adverse (social cost) effects on tourism in the Santa Barbara area were offset by gains in adjoining or nearby counties. Therefore, there were no *net* social costs involved but only

Table 2-1

Estimates of the Economic Cost of the Santa Barbara Oil Spill*

Union Oil Company on behalf of itself and three partners: Gulf, Mobil and Texaco		
A. Beach cleanup	$4,887,000	
B. Oil well control efforts	3,600,000	
C. Oil collection efforts	2,000,000	$10,487,000
U.S. Department of the Interior		382,000
State of California		200,000
County of Santa Barbara		57,200
City of Santa Barbara		negligible
Damage to tourism		negligible
Damage to commercial fishing industry		804,250
Property value loss		1,197,000
Fish life damage		negligible
Bird life damage		7,400
Seal and sea lion damage		negligible
Intertidal plant and animal damage:	low estimate	1,000
	high estimate	25,000
Value of lost oil		130,000
Recreational value lost		3,150,000
	Low estimate	$16,415,850
	High estimate	$16,439,850

*From Walter J. Mead and Philip E. Sorenson, "The Economic Cost of the Santa Barbara Oil Spill." a paper prepared for the Santa Barbara Oil Symposium, University of California, Santa Barbara, December 16-18, 1970.

private costs borne by motel-hotel and restaurant owners near the beach in the city of Santa Barbara. The adverse social effects on the commercial fishing industry stemmed mainly from loss of income to boat owners and crew and reduction in the fish catch, not offset by increases elsewhere, resulting from the fleet tie-up in Santa Barbara harbor at the height of the spill and property damage to vessels. There was an adverse economic effect on the values of beach front property causing owners to suffer a loss in expected income (real or psychic) during the time of the publicity and after-effects of the Santa Barbara Channel oil spill.

Although the social costs from damage to the marine environment were probably greater than zero, because of the high degree of uncertainty concerning biological damage to plant and animal life, no reasonable (nonarbitrary) cost estimates could be made. The immediate effects based on surveys and studies of the biota of the channel appeared to be negligible. However, the long-run effects on the ecology of the marine habitat in terms of stress factors, mortality and disruption are unknown at this early date. The loss of oil during the spill represents a social cost due to the fact that the economy has less oil than could have been recovered in the absence of the spill. The loss of recreational value of the Santa Barbara coastal area was reflected in the decline in beach visits by residents and tourists to the extent that these were not offset elsewhere. The social value of lost recreation opportunities was probably higher to coastal area residents than to those living farther inland. Also, the damage to birds and wildlife and to the ecology of the region probably had an adverse effect on the aesthetic values imputed by area residents and users.

Offshore and Coastal Zone Development

A rational management system for development of the offshore and the coastal zone provides only a framework within which regional growth may take place. The full potential of marine activities—natural resources extraction, wastes management, transportation, recreation, conservation, and their respective interactions—will be realized only when science and technology are coupled with imagination and sound management to make existing uses more efficient and to introduce new beneficial uses.

As the coastal region becomes increasingly urbanized, attention will need to be directed to projects and possibilities for relieving pressures on shoreline space and reducing the risk of disastrous accidents and storm damage.* Systems are currently being developed by the oil industry for underwater storage of crude oil and petroleum

*Cf. National Council on Marine Resources and Engineering Development, *Marine Science Affairs—A Year of Plans and Progress* (Washington: U.S. Government Printing Office, 1968), pp. 61ff.; Battelle Memorial Institute, *Development Potential of U.S. Continental Shelves* (Washington: U.S. Government Printing Office, 1966), pp. III-1ff.; Commission on Marine Science, Engineering and Resources, *Our Nation and the Sea: A Plan for National Action* (Washington: U.S. Government Printing Office, 1969), pp. 49-81.

products; and there would appear to be an equally feasible potential for storing other bulky or dangerous products. These offshore and underwater cargo facilities should provide attractive alternatives to expensive dredging of channels for new, deep-draft ocean vessels. Moreover, development of completely submersed transport systems will create a need for new, revolutionary types of docking facilities in the not too distant future.

The feasibility of constructing large-scale underwater nuclear power facilities is also being studied with the purpose and aim of reducing accident possibilities, avoiding excessive heating of coastal waters (thermal pollution), and possibly enriching coastal waters by creating upwelling of nutrients. The opportunities for shifting transportation, storage and power generation functions offshore are considered sufficiently near at hand, and even compelling to the national interest, to warrant specific attention by the federal, state and local governments.

In most urbanized marine communities, outdoor recreation is a rush to the water for swimming, surfing, skin and scuba diving, pleasure boating, and sport fishing. The opportunities for development of the marine environment coupled with the public's demand for marine recreation require that governments recognize the need to include these recreational opportunities when applying cost-benefit criteria to particular marine projects, and additionally to those dealing with urban renewal, model cities, and land and water conservation.

Regional Economic Growth and Management of Marine Resources

The two important aspects of marine resources, from an economic point of view, are (1) their common property characteristic and (2) the externalities which are generated by their exploitation and development. Both these factors tend to affect, significantly, the prospects for coastal region development. Take the case of common property, or open-access marine resources. The absence of some form of property right vested in their utilization—their lack of ownership—has been hypothesized (8, 9), and generally confirmed (10, 11), to lead to excessive entry of capital and labor inputs (or, simply, "effort") in the sense of higher than necessary costs of operation. The rapidly diminishing returns as a consequence of this phenomenon result in low productivity, low monetary returns and a general unwillingness to invest in new technology and long-term profit opportunities.

These results are most dramatically exemplified in the marine fisheries. Marine mineral resources, on the other hand, are more easily capable of delineation; thus, exploitation rights to specific deposits are usually granted to private operators on an exclusive use basis. Whereas the difficulties in the fisheries are more closely related to open-access, those in the marine mineral industries—e.g., oil and gas—are of a regulatory and administrative nature and essentially concern the optimal leasing policy which would promote efficient development and yet not be detrimental to the public interest (10, note 1).

Where the underlying management regime or regulatory structure differs—e.g., between a controlled-access and an open-access resource—as Scott has shown in his model of resource allocation between the fisheries and agriculture sectors of an economy (12), there results a misallocation of capital and labor resources and a consequent decline in the welfare of the community, *even though* these inputs are earning their opportunity incomes. The loss to the community results from the loss of the economic surplus (or economic rent) through excessive capacity, fishing effort and overexploitation of the fishery resources—i.e., *incremental* output effort is less than *incremental* cost. In addition, as Smith has recently shown (13), the biological objective of maximizing the sustainable yield from a given fish stock can be very unstable over time, thus resulting in unstable returns and expectations and higher than normal risks of capital investment.

The additional externalities generated by open-access means there may not appear to be a stable long-run equilibrium rate of return in the fisheries compared to other marine resources, e.g., oil and minerals where exploitation rights are leased to private operators. Thus, the community would probably opt for encouraging the development of oil, gas and minerals, rather than fisheries, due to the fact that the latter are typically depressed, in a low-level equilibrium state, and attract low-opportunity-cost factors of production in line with the expected norm of low returns.

Under the basic regime within which marine fisheries operate there would seem to be little hope for their contributing significantly to regional economic growth. Essentially, under the present institutional structure, fishermen are caught in a "low-level equilibrium trap." Because of unlimited entry of effort, there occurs an instantaneous supply shift to meet any shift in demand so that prices are normally depressed. Even when the total catch of a given fishery is kept constant through biological management, the technological restrictions imposed on fishing ability and efficiency mean that costs would rise until all fishermen are operating at the break-even point. There is thus no incentive to innovate or even invest in new, modern vessels and equipment of advanced design.

The rational investment decision whereby the expected rate of return on capital exceeds the rate of interest (or opportunity costs of investable funds) clearly suggests that unless the productivity of capital is better in the fisheries than in alternative investment opportunities not much capital will flow in. Therefore, a management regime which prevents "unlimited" entry of capital and gear into the fisheries by favorably affecting the rate of return, paradoxically, would induce *more* capital investment, especially of the long-range, stable kind.

An increase in profit expectations can come about through increases in demand or decreases in cost. The existence of these sources of profits tends to act as an incentive to new capital investment in developing the fisheries of the oceans. With regard to innovative potential as between an unlimited versus a limited entry system, it would probably be greater under unlimited entry *in the short run*. This is to take advantage of opportunities for quick profits before they are dissipated through rapid adoption and ease of entry by competitors. It implies, however, that these innovations most probably take the form of gear improvements and fish-finding equipment—i.e., those involving relatively quick turnover of invested capital. Under a limited entry system, a licensed operator would most likely be interested in maximizing his *long-run* profits, thus undertaking *larger* capital investments in resource improvements, vessel design, gear efficiency, etc. Prevention of excessive entry—which otherwise dissipates the economic rent resulting from entrepreneurial ability—tends to encourage this type of capital investment.

Concluding Remarks

In fostering regional economic growth through development of marine resources, in view of the foregoing discussion dealing with open-access and externalities, the crucial considerations hinge on the formulation of a management regime which is conducive to growth. Clearly, free and open access to exploitation and utilization of marine resources results in the dissipation of the aggregate economic surplus from the resources and external diseconomies, to the detriment of all engaged and to the community in general. The advantage of a limited access regime is that the public authority can lease exploitation rights through a process which results in lowest cost producers obtaining the development rights. This system minimizes the total costs of production of marine products, promotes efficiency through innovation and adoption of advanced technology, and maximizes the economic return or surplus from the marine resources which can then be used to promote growth and development in the region. The economic rent collected by the public authority—through lease fees (bonuses), royalties or taxes on output—can be used for public investment in social overhead capital, i.e., health, education, transportation and communication, and other types of basic infrastructure which are not typical avenues of private investment and yet are required to sustain a viable, growing economy.

References

1. *Selecting policies for the development of marine resources* (working paper prepared by Resources for the Future, Inc., at the request of the Commission of Marine Science, Engineering and Resources, March 6, 1968), CLEARINGHOUSE for Federal and Scientific and Technical Information, Springfield, Va., pp. 14ff.

2. Panel Reports of the Commission on Marine Science, Engineering and Resources, Vol. 2, *Industry and technology—keys to oceanic development* Washington: Government Printing Office, pp. vi-129ff.

3. *Guidelines for estimating the benefits of public expenditures*, Hearings before the Subcommittee on Economy in Government of the Joint Economic Committee, Congress of the United States, 91st Cong., 1st Sess., May 12 and 14, 1969. Washington: U.S. Government Printing Office, 1969, Appendix I.

4. Davidson, R. K., V. L. Smith, and J. W. Wiley. 1962. *Economics: an analytical approach* Homewood, Ill.: Richard D. Irwin, Inc., pp. 103ff.

5. Comitini, Salvatore. 1969. Exploitation of marine resources. In *the encyclopedia of marine resources* New York: Van Nostrand Reinhold, pp. 211-212.

6. Baumol, William J. 1965. *Economic theory and operations analysis.* Englewood Cliffs, N.J.: Prentice-Hall, pp. 368-371.

7. Mead, Walter J. and Philip E. Sorenson. 1970. *The economic cost of the Santa Barbara oil spill.* A paper prepared for the Santa Barbara Oil Symposium, University of California, Santa Barbara, December 16-18, pp. 12ff.

8. Gordon, H. Scott. 1954. The economic theory of a common-property resource: the fishery. *Journal of Political Economy* (April), pp. 124-142.

9. Pontecorvo, Giulio. 1967. Optimization and taxation in an open-access resource: the Fishery. In *Extractive resources and taxation*, Mason Gaffney, ed. Madison: University of Wisconsin Press, pp. 157-167.

10. Crutchfield, James A. and Arnold Zellner.

1963. *Economic aspects of the Pacific Halibut Fishery*. Washington: U.S. Government Printing Office, pp. 41ff.

11. _____ and Giulio Pontecorvo. 1969. Baltimore: The Johns Hopkins Press, pp. 147-168.

12. Scott, Anthony. 1957. Optimal utilization and the control of fisheries. In *The economics of Fisheries*, R. Turvey and J. Wiseman, eds. Rome: FAO, pp. 42-56.

13. Smith, Vernon L. 1968. Economics of production from natural resources. *American Economic Review* (June), pp. 409-431.

part 3: richard paul shaw

Dr. Richard P. Shaw, professor in the Faculty of Engineering and Applied Science at the State University of New York at Buffalo, began his teaching career at Brooklyn Polytechnic in 1956, moving to the Pratt Institute (1957-1962) and then to SUNYAB (1962). In addition he has been a consultant to several engineering firms and has participated in several summer institutes, the latest of which was the NSF Special Program in Ocean Engineering at the University of Rhode Island 1971. Dr. Shaw was on leave during 1969-1970 under an ESSA-NRC post-doctoral research associateship with the Joint Tsunami Research Effort at the University of Hawaii where he was also a visiting colleague in the Department of Oceanography.

His educational background includes a B.S. in applied mathematics at Brooklyn Polytechnic (1954), M.S. in engineering mechanics (1955) and Ph.D. in applied mechanics (1960) from Columbia University where he was a Guggenheim Fellow in the Institute of Flight Structures and an NSF Fellow.

His main areas of research deal with wave propagation and he has published over 30 papers on acoustic wave scattering, elastic and viscoelastic wave propagation, harbor resonance, water waves, acoustic-elastic structure interaction as well as on oceanography, high pressure test chambers, etc. He is a member of the American Society of Mechanical Engineers, the Acoustical Society of America, the Marine Technology Society, the American Academy of Mechanics and the Society for Engineering Science.

3

experimental investigations of structure for deep sea applications

Designs of structures for deep-sea applications are significantly more difficult than corresponding structural design on dry land for a number of reasons. The "local" environment is corrosive to many structural materials, exerts an ambient hydrostatic pressure which may account for a considerable portion of the load-carrying capability of the structure and exerts other forces due to local current motion, response to structural motions, etc. While none of these problems is unknown in land-based designs, they are in general far more severe in the ocean environment. They are also compounded by the fact that *in-situ* testing is almost impossible to carry out with any degree of control over most environmental parameters. This in turn leads to the necessity of having laboratory "ocean environment" simulators in which some of these structural problems may be studied.

Before getting into details, let us define the general area we wish to consider. The majority of oceanographic structures will be found in depths not greater than 1000 feet; this corresponds to maximum hydrostatic pressures of approximately 500 psi. However, some applications such as deep-sea coring devices, deep-sea exploration vehicles (such as the bathyscaphe Trieste I), etc., will occur at very great depths (the maximum recorded manned dive was to 35,800 feet in the Trieste) and will

therefore be subjected to pressures up to 18,000 psi. The term "deep-sea" application then covers a rather broad spectrum; it has been used to describe all applications at depths greater than the continental shelf area, approximately 600 feet.

The functions of particular structures also vary greatly but can be divided roughly into two categories—shelter, i.e., a structure such as a habitat or submarine which protects personnel and/or instrumentation from the local environment, and support, i.e., a structure such as an undersea well truss which carries a useful load. On land, the support function of structures is predominant since effects of gravity are significant while the local environment is reasonably benign; at depth these roles are somewhat reversed since the local environment is exceedingly "hostile," while gravitational forces are counterbalanced to some degree by buoyancy. Nevertheless, these functions are not distinct and even predominantly protective shelters such as habitats must also carry loading other than the local hydrostatic pressure. The design of such structures then will require methods of analysis which may be somewhat more complicated than corresponding analysis of land-based structures.

For example, the finite deformation theory of shells under multiple forms of loading depends on

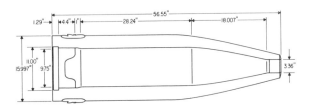

Figure 3-1. Basic dimensions of 16-inch high-capacity projectile Mk-13 Mod. 2.

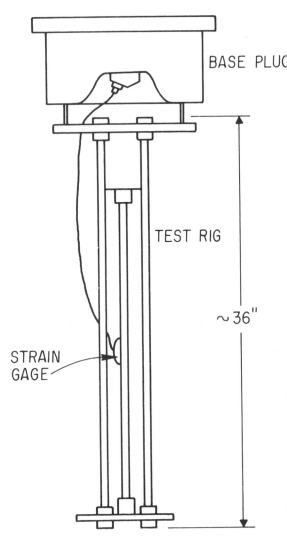

Figure 3.2. Support jig for structural testing within pressure chamber.

the sequence in which the loadings are applied; the end result will be different for a structure loaded at the surface and then brought to depth than for the reverse case. Similarly, the yield strength of ductile materials under combined stresses has been predicted reasonably well by octahedral shear stress theory; this too requires a particular form of loading in order to be applied to a triaxial state (1). This form of loading, in which the ratio of principal stresses remains constant, is even less likely to be found in deep-sea structural applications than in land-based structures.

A second categorization of structural function, which has a strong bearing on the types of material used, relates to mobility. A structure such as a submersible which must move at depth faces much more severe material restrictions—i.e., it is limited to high strength to weight materials—than would a fixed-site structure. This material weight restriction coupled with the severe corrosion problem caused by the environment has caused interest to turn from the usual metallic materials which dominate surface construction to nonmetallics such as glass, glass fiber reinforced plastics, concrete, ceramics, etc. (2, 3, 4). However, the behavior of these new materials under high hydrostatic pressure in a water environment is in no manner as well understood as the behavior of more common structural materials on land. Part of any experimental investigation preliminary to widespread use of these materials for deep-sea structures would be the determination of their material properties as a function of hydrostatic pressure (5).

Experimental Test Facilities

While some corrosion tests have been carried out *in situ* (e.g., Reference 6) those tests which require deliberate variation of the environmental parameters such as pressure are probably best carried out in the laboratory. A survey of existing high-pressure test facilities has recently been

Figure 3-3. *a.* Support jib in place being lowered into pressure chamber. *b.* Experimental facility.

made (7). Almost all of these facilities were used only to apply a hydrostatic pressure environment to examine structural integrity and not to apply additional nonhydrostatic loading to determine structural response. The remainder of this discussion will concentrate on a chamber designed and constructed at the State University of New York at Buffalo specifically for this purpose. Problems of corrosion were not considered directly since fresh water was used as the environmental medium to protect the test chamber itself. The discussion will therefore be restricted to structural loading effects alone, although material tests similar to those made in Reference 5 could be carried out as well.

SUNYAB High-Pressure Test Chamber

The SUNYAB chamber was constructed from a 16-inch naval rifle shell, making a fairly common form of pressure test chamber (8). The major modifications in this chamber were the inclusion of a hydraulically operated actuator within the test chamber which could apply axial forces to a speci-

men and a jig attached to the base plug of the chamber in which structural test specimens may be mounted (Figures 3-1, 3-2, 3-3). Construction details are given in Reference 9.

One of the experiments carried out in the test chamber was an investigation of the effect of hydrostatic pressure on the axial load-carrying capacity of a column. According to simple theory[10] there should be no effect of hydrostatic pressure on the theoretical buckling load of a perfectly straight uniform column. Some later investigations[11], however, predicted significant changes in the theoretical buckling load for hydrostatic pressure values of the order of the shear modulus of the material tested. While shear moduli for metals such as aluminum and steel are of the order 10^7 psi and therefore out of the range required for oceanographic purposes, some of the nonmetallic materials considered for deep-sea use do fall into this category, i.e., have shear moduli of order 10^4 psi.

The first set of experiments in the SUNYAB facility tested solid rectangular columns of steel and aluminum at hydrostatic pressures up to 5000 psi. Both the ambient fluid and the hydraulic actu-

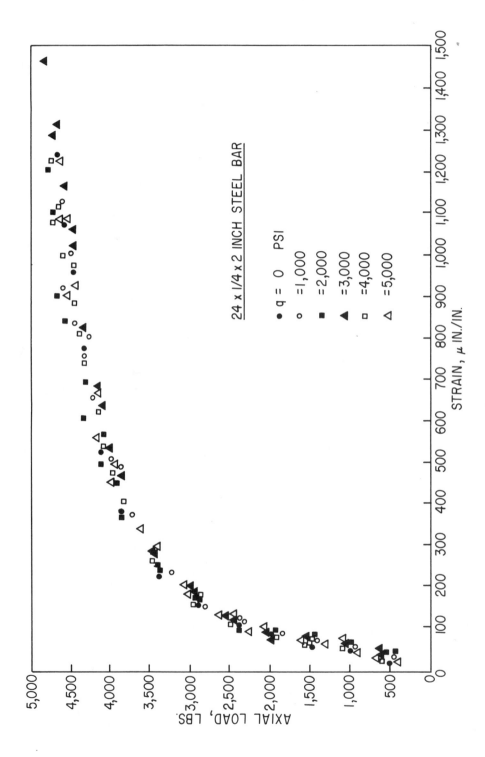

Figure 3-4. Results of SUNYAB experiments testing rectangular columns of steel at hydrostatic pressures up to 5000 psi.

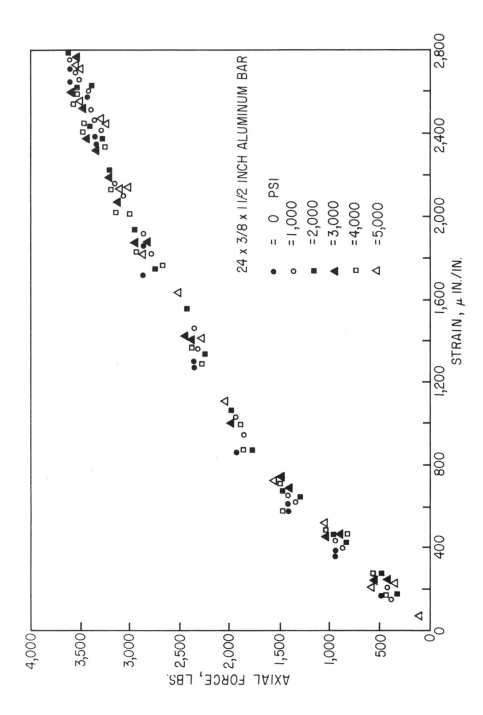

Figure 3-5. Results of SUNYAB experiments testing rectangular columns of aluminum at hydrostatic pressures up to 4000 psi.

ator were raised to the same pressure initially. Then the pressure in the actuator was increased gradually, thereby applying an additional axial load on the test column which was mounted in the test support jig and strains recorded from strain gauges cemented to the faces of the column. Results are shown in Figures 3-4 and 3-5. As predicted by both elementary theory and the later variational approach, at these levels there was no effect of hydrostatic pressure on the load-carrying capacity of these metal columns. This result was also found by a different experiment (1) which was carried out in a purely hydrostatic environment.

Problems for Further Study

The SUNYAB facility has been used for other tests as well (e.g., Reference 12), but a great deal of work remains to be done. Much of this work is connected with basic material properties research similar to that of Reference 5 but with larger scale specimens and with correlation of theoretical studies of yield strengths such as those based on the octahedral shear stress theory with actual tri-axial measurements (most experimental work has been done with biaxial stress behavior). Other material problems of interest are in the area of hydrostatic pressure effects on fatigue life.

Finally, the testing of structural models such as plastic coring tubes can be carried out directly under controlled environmental conditions. In fact, one of the purposes of this seminar is to encourage modification (at relatively small cost) of present hydrostatic pressure testing facilities to include this option of nonhydrostatic loading to carry out these experiments which the author feels are necessary before deep-sea structural design can be considered a straightforward engineering problem.

References

1. Breckenridge, R. and H. Haynes. 1967. Behavior of structural elements in the deep ocean. *Proc. of the Conf. on Civil Eng. in the Oceans* (ASCE 1968).

2. Bernstein, H. 1965. Materials for deep submergence capsules. *Ocean Science and Ocean Engineering*. Trans. Joint Conf. MTS and ASLO, pp. 583-600.

3. Levenetz, B. 1965. Potentials of nonmetallics for use as structural materials for deep submergence vehicles. *Ocean Science and Ocean Engineering*. Trans. Joint Conf. MTS and ASLO, pp. 601-622.

4. MacKenzie, J.D. 1969. Ceramics in ocean engineering. *Ocean Engineering* 1(5):555-571.

5. Uy, J.C., D.R. McCann and P.R. Hettwer. 1969. The tensile properties of polysulfone under a high hydrostatic pressure. *Ocean Engineering* 1(5):57-574.

6. Burkart, W.F. 1965. Deep ocean engineering at Budocks. *Ocean Science and Ocean Engineering*. Trans. Joint Conf. MTS and ASLO, pp. 171-187.

7. Keller, K.N. 1968. High pressure test chambers—state of the art. 68-WA/UNT-8 ASME Winter Annual Meeting.

8. Gray, K.C. and J.D. Stachiw. 1965. The conversion of 16-in. Projectiles to pressure vessels. Tech Note N-755, U.S. Civil Engineering Laboratory, Port Hueneme, California.

9. Shaw, R.P. and J. Sezna. 1969. An experimental facility for testing loaded structures in a high hydrostatic pressure environment and an application to axially loaded columns. DISR Rept. 50, State University of New York at Buffalo; *Ocean Engineering* 2(2)(Apr. 1971):57-74.

10. Peterson, J.P. 1963. Axially loaded columns subjected to lateral pressure. *AIAA J.* 1(6):1458-1459.

11. Kerr, A.D. and S. Tang. 1966. The effect of lateral hydrostatic pressure on the instability of elastic solids, particularly beams and plates. *J. App. Mech.* 33:617-622.

12. Shaw, R.P., W-N Dong and G. Gilley. 1970. An experimental investigation of a weak link for a deep moored instrument cable. *Mar. Tech.* pp. 1139-1145.

4

long period forced harbor oscillations

Fundamental Problem[1, 2]

Harbors may be considered as mechanical systems with inertial, dissipative and restoring forces. They are in nature acted upon by external forces which may give rise to greatly magnified responses within the harbor, e.g., resonances. To describe some of the techniques used to predict the harbor behavior under various inputs, we must clearly define the problem at hand:

1. *Harbor*: a basin of water, usually of nonuniform depth, with some internal structuring to support berthing of various-size ships (large freighters in a commercial harbor to pleasure craft in a marina) and some natural or artificial boundaries. These can be anything from gently sloping sand beaches to vertical concrete walls. By definition, a harbor must be connected in some manner to a larger body of water such as the open sea to be useful.

2. *Long-Period Waves*: while the response of harbors to a wide range of forcing functions is of interest, the usual harbor response studies divide into short-period waves (sea and swell) from 4-20 seconds and long-period waves from 2 minutes and up. (The intermediate range involves some features of both and is correspondingly more difficult to treat.) The response to these two classes of waves is distinctly different. For example, short-period waves are stopped by breakwaters and are readily absorbed on sloping beaches, while long-period waves are transmitted through most breakwaters with little energy loss and are almost totally reflected, even from gently sloping beaches. This discussion shall concentrate on long-period waves, although much will be applicable to waves of any period.

3. *Oscillations*: a periodic motion of a system under a periodic driving force (forced oscillations) or motions, which may die out, from some initially disturbed state (free oscillations). We are interested in forced oscillations and in particular those forcing frequencies which cause the maximum response, (resonance). For a harbor open to the sea where periodic waves are incident on the coastline, energy from the sea enters the harbor either directly or by diffraction from the entrance corners, is reflected from the walls and is partially trapped within the harbor basin. The mechanisms which prevent the energy in the basin from building up indefinitely are radiation from the harbor entrance back into the sea, frictional effects on the bottom and sides

29

of the basin, wave breaking and absorption on the beach. For long waves, the last two loss mechanisms are negligible compared to the first two.

4. *Seiche*: a related problem which occurs in completely closed basins due to tidal effects, seismic disturbances, atmospheric changes, etc. This problem may be studied by techniques very similar to those used in studying harbor oscillations. Note: some authors refer to harbor resonance problems as harbor seiching. Here the two terms are distinguished, harbor resonance applying to forced oscillations of open harbors and seiching to oscillations of closed basins.

Finally, we may discuss the practical problems caused by harbor oscillations. Basically, we recognize that all harbors have some resonant frequencies. We wish to "detune" the harbor design from those input frequencies found predominant in nature at the actual location of a proposed harbor or to discover the parameters requiring change in an existing harbor to avoid this problem if it has been experienced. This requires some manner of determining the resonant frequencies and the changes that can be made in them by altering the parameters of the system. This in turn can be done either theoretically or by some model experiment. The latter can be either a scale model (usually distorted) of the harbor in a water tank or some physical analog to this system. The corresponding problem of the input frequency spectrum for an actual harbor is not discussed here, although obviously this knowledge is required in order to "detune" the harbor from its surroundings.

Oscillatory Systems

Before discussing harbor problems, let us review quickly the concept of mechanical resonances. Consider a single degree of freedom, spring-mass-dashpot system which states that the total force, which is the forcing function, $F(t)$, plus the dissipative force opposing the velocity, $-C\overset{\circ}{X}$, plus the restoring spring force, $-KX$, must equal the mass times the acceleration, $M\overset{\circ\circ}{X}$. Then,

$$\overset{\circ\circ}{X} + (C/M)\,\overset{\circ}{X} + (K/M)\,X = F(t)/M.$$

Consider a periodic forcing function, $F(t) = KA\,e^{i\omega t}$, with A as a reference level for the displacement X; that is, under a static force of KA, the static displacement would be A. Then x must also be periodic, $x = Be^{i\omega t}$.

We may solve for the amplification factor, B/A:

$$(B/A) = \frac{K/M}{\dfrac{K}{M} - \omega^2 + \dfrac{i\omega C}{M}}$$

$$= \frac{1}{1 - \dfrac{\omega^2}{\omega_0^{\,2}} + \dfrac{i\omega c}{K}}$$

where $\omega_0 = (K/M)^{1/2}$, the natural frequency, has been introduced. For a nondissipative system, $C = 0$ and the amplification factor becomes infinite as ω approaches ω_0. This is called resonance and ω_0 the resonance frequency of the system. For dissipative systems, the amplification factor is not infinite but does reach a maximum at a resonant frequency which is not the same as the natural frequency, although for slightly damped systems the difference is small.

$$|B/A|_{max} = [(C^2/KM) - (C^4/4K^2M^2)]^{-1/2}. \tag{4-1}$$

$$\omega^2_{res} = \omega_0^{\,2}\left\{1 - \frac{C^2}{2KM}\right\} \tag{4-2}$$

These results are sketched in Figure 4-1. We see that dissipation reduces the magnitude of the amplification and lowers (slightly) the resonant frequency.

Water Wave Theory[1, 3]

We assume an inviscid, incompressible and irrotational fluid with small velocities ($\bar{u}$) and surface heights (ζ) (i.e., linear theory). Then continu-

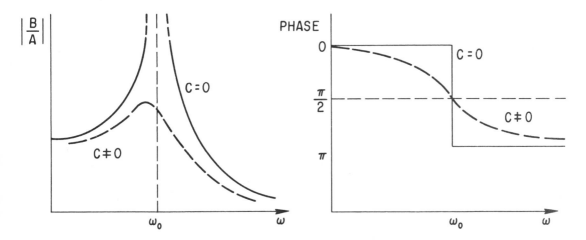

Figure 4-1. Schematic illustration of equations 4-1 and 4-2.

ity requires $\nabla \cdot \bar{u} = 0$, and irrotationality requires $\nabla \times \bar{u} = 0$ or $\bar{u} = \nabla \Phi$ where Φ is a velocity potential, leaving $\nabla^2 \Phi = 0$. We further assume in all cases discussed here that depth is constant. This allows a separation of variables:

$$\phi(x,y,z,t) = f(x,y,t) \cdot Z(z)$$

where

$$Z''(z) - k^2 Z(z) = 0 \text{ and}$$

$$\nabla_o^2 f + k^2 f = 0, \tag{4-3}$$

where ∇_0^2 is the two-dimensional Laplacian operator.

The boundary condition on the bottom requires $\partial Z/\partial z = 0$ on $z = -h$ and the corresponding solution for $Z(z)$ is $Z(z) = A \cosh k(z+h)$. On the free surface, $z = 0$, we have two linearized boundary conditions; one dynamic: $z = -(1/g)(\partial\Phi/\partial t)$ and one kinematic: $\partial\zeta/\partial t = \partial\Phi/\partial z$ where $\zeta(x,y,t)$ represents surface height. If we further assume simple harmonic motion, we find that the first condition gives $\zeta(x,y,t) = -(i\omega/g) A \cosh kh$ $f(x,y)e^{i\omega t}$ and the second gives the dispersion relationship

$$\omega^2 = gk \tanh kh \tag{4-4}$$

where f is to be determined from the 2D Helmholtz equation, equation (4-3); k is then the wave number and is related to ω and h by equation (4-4).

If we consider a closed rectangular basin we can determine $f(x,y)$ by a further separation of variables:

$$f(x,y) = X(x) \cdot Y(y)$$
$$X'' + p^2 X = 0$$
$$Y'' + q^2 Y = 0$$

where $k^2 = p^2 + q^2$. The boundaries are assumed to be rigid walls which require the normal derivative of f to vanish on the boundaries $x = o$, $x = L$, $y = o$, $y = B$; therefore, $p = m\pi/L$, $q = n\pi/B$ and $f(x,y) = A_1 \cos px \cos qy$; $k_{mn} = [(m\pi/L)^2 + (n\pi/B)^2]^{1/2} = \omega_{mn}/c$ where ω_{mn} are the natural frequencies for the basin and c is the speed of wave propagation defined through the dispersion relationship. If we consider $L>B$, the lowest mode will require $m = 1$ and $n = o$ and therefore $k_{01}^2 = \pi^2/L^2$ which in turn requires the length of the basin, L, to be one-half of the wave length.

However, the concept of a closed basin violates the prerequisite that the harbor be connected in some manner to the open sea. Several theoretical approaches to the problem of harbor resonance have been based on the fact that the solution at the entrance to a basin oscillating at one of its natural frequencies must correspond to this case. That is, a node in velocity occurs at the harbor entrance which leads to the same boundary condition (vanishing normal derivative of f) as the rigid walls used for the rest of the basin. Unfortunately, this has been referred to as a "resonance" motion. This is not the case as may be seen in later theories; i.e., the maximum amplification occurs at other frequencies.

The other extreme—that of a node in displacement ($f = o$)—at the entrance corresponds to the acoustical problem of a quarter wave resonator. Setting f equal to zero at the entrance, $x = o$, and requiring the lowest mode which is again independent of y, we find $p = \pi/2L$, $q = o$, and therefore the length of the basin, L, is one-quarter of the wave length, λ.

The correct boundary condition to be used at the entrance will, of course, lie somewhere between these two cases and will vary with incident wave frequency. All that can be required physically is that the wave heights and particle velocities in the open sea match those in the harbor at this interface.

To illustrate the true problem found in long-period harbor oscillations, consider the relationship between vertical wave height and horizontal particle displacements. As an example, consider a closed rectangular basin oscillating in its lowest mode with a 3-minute period. Depth will be taken as 40 feet, giving $c = 35.6$ feet/second and $\omega = 0.035$ second^{-1}. Then $k = 0.001$ feet^{-1} and the wave length $d = 6280$ feet. The maximum horizontal particle displacement can be found by integrating the horizontal velocity over one-half period. The corresponding ratio of maximum horizontal particle displacement, which occurs at a node in surface height, and the maximum surface height is $\lambda^2/4\pi Lh = 25.0$ where L is the harbor length and in this case is $(1/2)\lambda$. While these maxima do not occur at the same location within the harbor and horizontal water displacement does not mean an equivalent horizontal ship displacement, we can

readily see that a long-period resonance in a harbor can cause serious horizontal motions of ships within the harbor. Corresponding difficulty in navigation within the harbor occurs, as well as difficulty in loading and unloading cargo and preventing collisions between a ship tied to a dock and the dock itself.

Methods of Solution

The means by which harbor oscillation problems can be studied fall naturally into three groups: 1) theoretical studies based on solving the mathematical formulation of water wave theory (which is admittedly an approximation to the physical situation), 2) laboratory wave basin experiments which attempt to model the real basin but which are usually distorted in scale to permit feasible experiments, and 3) analog experiments which use other physical mechanisms such as electrical circuits or acoustical models to represent the water waves.

Theoretical Solutions

The theoretical approaches can be divided into two categories: those which assume boundary conditions at the harbor entrance and those which calculate these conditions from a coupling with the exterior (open sea) region. We shall concentrate on the latter as more representative of the actual physical situation. This second category can be further separated into methods which solve the governing equations "exactly" albeit numerically and those which further approximate the physical situation to allow for simple solutions; examples of the former are the integral equation method of Hwang and Le Méhauté[4], and that of Lee[5], while the method of Carrier, Shaw and Miyata[7] exemplifies the latter approach.

Integral Equation Approaches to Harbor Entrance

A number of integral equation formulations are equivalent to the two-dimensional Helmholtz equation. These may be interpreted in terms of

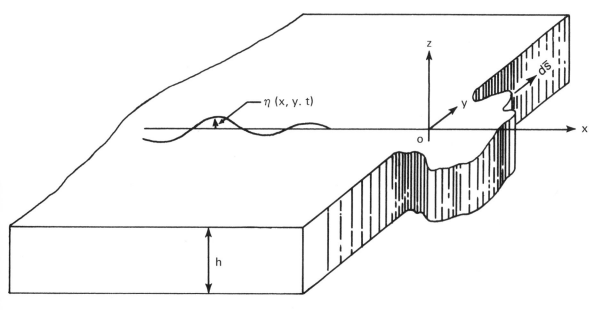

Figure 4-2. Schematic drawing of harbor (from Hwang and Le Méhauté).

sources and/or doublets distributed on the scattering boundary curve.

One such approach is used by Hwang and Le Méhauté in Reference 4. The scattered wave is assumed to be represented by a source distribution of unknown strength along the coastline and harbor boundary, as Figure 4-2. The sources are the Green's functions, i.e., the solutions for a delta function point source, for the two-dimensional Helmholtz equation, $G(x,y; \xi,\eta) = (- i/4) Ho^{(1)} (kR)$ where $R = [(x - \xi)^2 + (y- \eta)^2]^{\frac{1}{2}}$ is the distance between the field point (x,y) and the source point (ξ, η) located on the boundary curve Γ.

The total field f then is composed of the sum of the incident wave, f_0, plus the scattered wave, f_s, as

$$f(x,y) = f_O(x,y) + \int_\Gamma Q(\xi,\eta) \cdot G(x,y;\xi,\eta)\, d\sigma \quad (4\text{-}5)$$

where $\sigma(\xi,\eta)$ describes the boundary curve Γ. Q is now found from the boundary condition that the total normal derivative of f vanishes on Γ.

$$L_{im}_{x,y,\to \xi,\eta} \left[\frac{\partial f_O}{\partial n} + \frac{\partial}{\partial n} \int_\Gamma d\sigma\, Q(\xi,\eta) \cdot G(x,y;\xi,\eta) \right] = 0.$$

Since G is singular on the boundary Γ, this differentiation and limit must be carried out carefully. The resulting contribution from the singularity is $\frac{1}{2} Q(x,y)$ leaving

$$0 = \frac{\partial f_O}{\partial n} + \frac{1}{2} Q(x,y) + \int_\Gamma Q(\xi,\eta) \cdot \frac{\partial G(x,y;\xi,\eta)}{\partial n}\, d\sigma \quad (4\text{-}6)$$

where the integral is now a principal value integral, that is, excludes the point $R = 0$.

For a given f_O and geometry (Γ), this equation may be solved for $Q(x,y)$, which then gives the solution for $f(x,y)$.

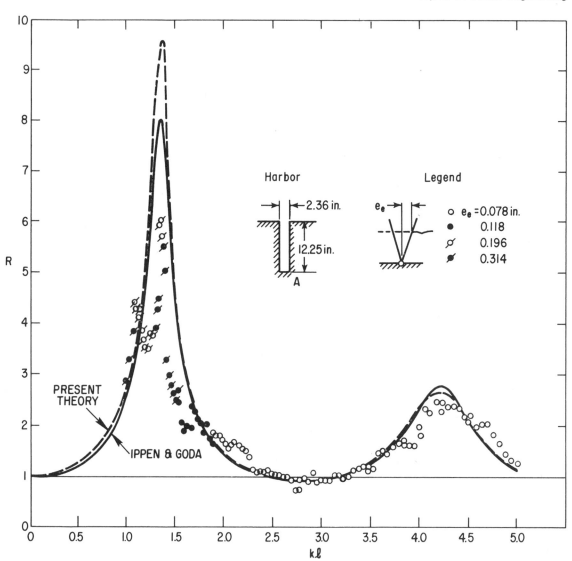

Figure 4-3. A comparison of theoretical and experimental results for response of a fully open harbor (from Hwang and Le Méhauté).

The solution is accomplished numerically, with contributions from the coastline far from the harbor becoming less and less significant as R increases and therefore being terminated at some location to allow a finite number of algebraic equations to represent the integral equation. This in turn is accomplished by assuming that Q remains constant over small but finite segments of the boundary and applying equation (4-6) successively to a field point in each segment. These equations are complex, resulting in a set of $2n$ equations for n segments and therefore n complex unknowns (the sources Q on the boundary). Once the values of Q are known on the boundary, the solution for f either on the boundary or in the field may be found from equation (4-5). Typical results are

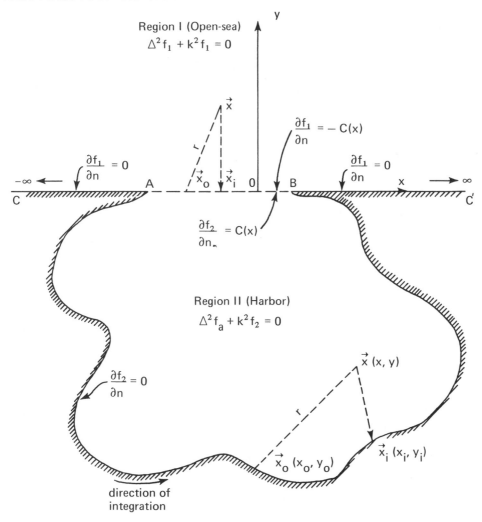

Figure 4-4. Sketch of arbitrary shaped harbor (from Lee).

shown in Figure 4-3 with a comparison to other theoretical solutions and experimental results.

A second integral equation formulation is proposed by Lee in Reference 5. Separate solutions are formulated for the exterior (open sea) and interior (harbor basin) regions; these solutions are then matched at the harbor entrance. The geometry is shown in Figure 4-4.

The solution in the exterior region, f_1, and in the interior region, f_2, both satisfy the two-dimensional Helmholtz equation and therefore the Weber integral form of this equation: for $j = 1, 2$.

$$f_j(x,y) = \frac{-i}{4} \int_{\Gamma_j} \left[f_j(\xi,\eta) \frac{\partial}{\partial n} H_O^{(1)}(kR) \right.$$

$$\left. - H_O^{(1)}(kR) \frac{\partial f_j}{\partial n} \right] d\sigma \quad (4\text{-}7)$$

with matching conditions at the harbor entrance, $\overline{AB}$, which require $f_1 = f_2$ and $\partial f_1/\partial n = -\partial f_2/\partial n$ (minus sign due to sign change in outward normal). For the exterior region, f_1 may be divided into three parts, $f_i + f_r + f_s$, an incident field, a reflected field (which would be the total effect of the coast if the harbor entrance were closed) and a scattered field due to the presence of the harbor entrance, respectively. The incident field, f_i, is specified and f_r is equal to f_i but opposite in direction if the coastline is a rigid straight boundary, i.e., $f_r(x,y) = + f_i(x, -y)$. Then the problem can be written in terms of f_s alone:

1. $\nabla_0^2 f_s + k^2 f_s = 0$
2. $\partial f_s/\partial n = 0$ on the coastline
3. $\partial f_s/\partial n = - \partial f_2/\partial n$ across the entrance $\overline{AB}$
4. $\underset{R\to\infty}{Lim} f_s = 0$, a radiation condition

which requires the effect of the harbor entrance on the exterior field to die out far from the harbor itself. Using Weber's equation for a field point on the boundary composed of the straight coastline plus the harbor entrance, $\overline{AB}$ gives

$$\tfrac{1}{2} f_s(x,o) = \frac{i}{4} \int_{\overline{AB}} \left[H_O^{(1)} (kR) (\partial f_2/\partial n) \right] d\xi$$

$$(4\text{-}8)$$

since the first term in the integral form, equation (4-7), contains a $\partial R/\partial n$ term which is zero on this straight boundary (except for the contribution from the singularity at $R = 0$). The second term has $\partial f_s/\partial n$, which vanishes everywhere except on AB. For the interior region the solution is correspondingly

$$\tfrac{1}{2} f_2 = \frac{i}{4} \int_{\Gamma} \left[f_2 (\partial/\partial n) H_O^{(1)} (kR) \right.$$
$$\left. - H_O^{(1)}(kR) (\partial f_2/\partial n) \right] d\sigma \qquad (4\text{-}9)$$

where $\partial f_2/\partial n$ is different from zero only across the entrance. Finally $f_s = f_2$ across the entrance.

These integral equations may be approximated again by sets of algebraic equations and solved for the unknowns on the boundaries as fol-

lows. Assume the interior boundary to be divided into m segments with p of these segments defining the harbor entrance. Consider one field point in each segment.

Define: $\overline{X} = f_2$ evaluated at m field points
$\overline{P} = \partial f_2/\partial n$ evaluated at m field points
$G_n = \int (\partial/\partial n) H_O^{(1)} (kR) d\sigma$
$G = \int H_O^{(1)} (kR) d\sigma$

For the interior region with m divisions, equation (4-9) is approximated by $\overline{X} = (-i/2)[G_n\overline{X} - G\overline{P}]$ where G_n, G are m x m matrices and $\overline{X}, \overline{P}$ are m vectors.

Then $\overline{X} = (I + i/2 \ G_n)^{-1} (i/2 \ G)\overline{P}$ represents m equations on $m + p$ unknowns. I is the unit matrix.

$\overline{P}$ includes as a subset the unknown normal derivatives across the entrance; all other values are zero.

For the exterior region, $\overline{Y}$ is the p vector representing f_s across $\overline{AB}$ and equation (4-8) is approximated by

$$\overline{Y} = (-\frac{i}{2}) (H) (-\overline{P})$$

where H is a p x p matrix, using p divisions across the entrance, $\overline{AB}$. Then at the entrance $\overline{Y} = \overline{X} - f_i - f_r$, which together with equation (4-10) gives an additional p equation. These equations can be solved for the $m + p$ complex unknowns, m values of f_2 and p values of $\partial f_2/\partial n$.

Results of these calculations are given in Figure 4-5 for the same case as Hwang and Le Méhauté.

Seiche by an Integral Equation Approach

The free oscillations of water in a closed basin of arbitrary shape can also be calculated by these integral equation methods. The fundamental equation is again Weber's form expressed for a field point on a rigid boundary

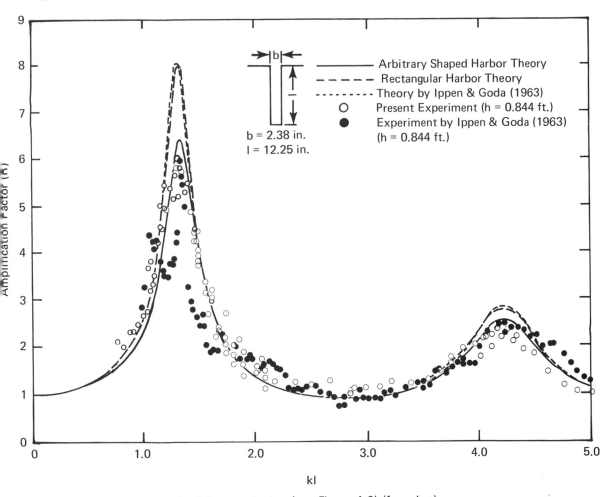

Figure 4-5. Response curve for fully open harbor (see Figure 4-2) (from Lee).

$$\tfrac{1}{2} f(x,y) = -\frac{i}{4} \int_{\Gamma} f(\xi,\eta) \ \frac{\partial}{\partial n}\left(H_O^{(1)}(kR)\right) d\sigma.$$

When approximated in the usual manner with f assumed constant over specified segments of the boundary, this equation becomes a set of homogeneous linear algebraic equations involving the parameter k. The eigenvalues of this system will then be approximations to the natural wave numbers of the closed basin and the corresponding normal modes can be found—first on the boundary through solving the set of algebraic equations for a given eigenvalue (using some point as a reference level) and then in the interior using the original Weber's integral equation formulation.

An Approximate Approach

The two previous integral equation approaches are exact in the sense that the only errors made are in the discretization of the integral equation to a finite set of algebraic equations; these in theory can be made arbitrarily small. Of course, the overall solution is still approximate since the Helmholtz differential equation itself is derived

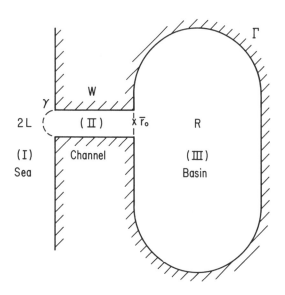

Figure 4-6. Sketch of open sea—entrance channel—harbor basin regions (from Carrier, Shaw and Miyata).

only after several physical approximations such as inviscid irrotational flow, etc. have been made.

Other approaches, however, introduce further approximations into the solution procedure for the advantage of having a simpler problem to solve. When the approximations are valid—which implies knowledge of the physical character of the approximations—this can be a quick and simple way to study harbor problems.

One such approximate technique was introduced by Carrier in the University of Hawaii's 1969 seminar series; several applications of this approach have been carried out (References 6, 7). The method is applicable to constant depth harbors of general shape connected by a straight narrow channel to the straight coastline of an open sea. The physical requirements of the approximations are that the incident wave length be long compared to the width of the entrance channel and that the harbor basin dimensions be large compared to the width of the entrance channel, i.e., long-period oscillations of narrow-mouthed harbors.

The domain of the problem is divided into three parts: the open sea region (I), the entrance channel (II) and the harbor basin (III) (Figure 4-6). Assumptions are made as to the form of the solution in each region. In the open sea the solution is assumed to consist of the incident wave, a wave reflected from the straight coastline and a wave representing the effect of the harbor entrance. Since the entrance dimensions are small compared to the wave lengths involved, the latter wave is approximated by a cylindrical wave of unknown strength; the solution for a point source at the midpoint of the entrance channel/open sea interface. Then $f_I = e^{ikx} + e^{-ikx} + AH_0^{(1)}(kr)$. The solution in the entrance channel is assumed to consist of plane waves of unknown amplitude traveling towards and away from the harbor basin $f_{II} = Be^{ikx} + CE^{-ikx}$. Finally, the solution in the harbor basin is assumed, as an expansion in the normal modes of the completely closed basin, as that for the problem of a point source of unknown strength F_O located just inside the entrance channel/harbor basin interface.

$$f_{III} = \sum_{j=0}^{\infty} b_j \, \varphi_j(\bar{r})$$

where $\varphi_j(\bar{r})$ are the normal modes of the completely closed basin and the b_j are related to the source strength F_O.

The procedure then matches the total fluid fluxes and wave heights at the two interfaces, open sea entrance channel and entrance channel/harbor basin, to obtain four equations on the four unknown quantities A, B, C and F_O. The series solution for f_{III} must be terminated after only a few terms since the point source representation of the entrance channel effect on the harbor basin is valid only for large wave lengths and small entrance channel/harbor dimension ratio, but this is consistent with the original assumptions of long-period oscillations of narrow-mouthed harbors.

Solutions based on this approach are described in Reference 7; some results are shown in Figure 4-7.

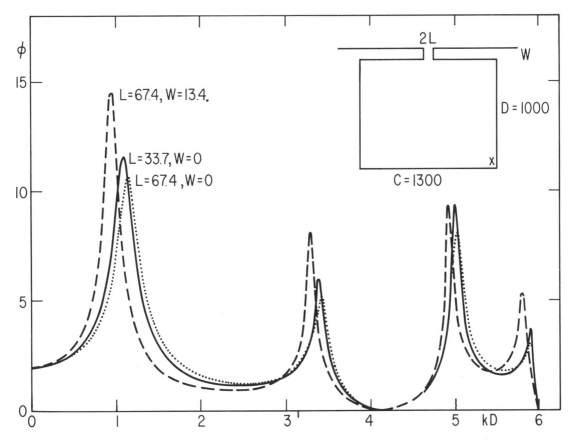

Figure 4-7. Response curve for narrow mouthed harbor (from Carrier, Shaw and Miyata).

Laboratory Models

Many of the proposed theoretical methods have been accompanied by model experiments to check the results. Earlier models used coupled basins to check similar theory; later experiments use absorbers in the main section to simulate an open sea, e.g., (8).

Most laboratory experiments use deep-water models, i.e., models where water depth is large compared to the wave lengths involved. This is done for several practical reasons:

1. Wave lengths of deep-water waves are not affected by wave amplitude or variation in water depth. Shallow-water waves are affected by changes in water depth, e.g., table not per-fectly horizontal, and their wave lengths are affected by their amplitude; e.g., a 5% amplitude/water depth ratio will increase the actual wave length by about 5% over that predicted by small amplitude theory.

2. Wave filters and absorbers are more effective for deep-water waves than for shallow-water waves since deep-water waves can have a reasonably sized steepness.

3. The wave basin can be much smaller for deep-water waves than for shallow-water waves. The basin should be several wave lengths long. For deep-water waves, the wave length can be small, while for shallow-water waves, with depths sufficient to eliminate capillary effects and allow for accurate measurements, the wave lengths must necessarily be several feet, thereby requiring a very large basin.

The two theories differ in wave speeds and frequencies and in vertical velocity distribution but for a given wave length have the same surface elevation and velocity patterns.

Analog Experiments

In addition to water basin model experiments, several other physical systems obey the same governing equations, i.e., two-dimensional Helmholtz equation, boundary conditions, etc., and therefore may be used as analogs to the water wave problem. If these analogs are easier and/or cheaper to construct than the original water wave scale model, they may be quite useful in parametric studies. One handicap of course lies in extending these analogs to include more physically realistic aspects such as turbulence, friction, etc.

One possible analog consists of an acoustic model. Using rigid walls, the acoustic equation and boundary conditions are identical to those for water waves. However, an acoustic model is easier to construct and test.

For this case, acoustic pressure would be analogous to surface wave height. Such a model is described in Reference (9).

Discussion of Results

The effect of energy loss mechanisms on resonant frequencies can clearly be seen in the results found from these solutions. While the solution at the natural frequency of the closed basin can of course be predicted by assuming a velocity mode (potential antinode) at the harbor entrance resulting in a surface height, immediately inside the entrance, of 2 to match the exterior wave doubling solution, this is clearly *not* the resonance frequency. The maximum amplification in the harbor occurs at frequencies significantly different from the closed harbor natural frequency and with amplitudes significantly greater than 2 as seen in Figures

4-3, 4-5 and 4-7. This is borne out both by theory and experiment. It is clear then that energy losses due to radiation from the harbor entrance have an important effect on predicting the harbor response to incident periodic waves and must be taken into account in any harbor design. Although other loss mechanisms may also be significant, this one appears to be the dominant effect.

References

1. Raichlen, F. 1966. Harbor resonance. In *Estuary and coastal hydrodynamics*, A.T. Ippen, ed. New York: McGraw Hill.
2. Wiegel, R.L. 1964. Tsunamis, storm surges and harbor oscillations. In *Oceanographical engineering*. Englewood Cliffs: Prentice Hall.
3. Eagleson, P.S. and R.G. Dean. 1966. Small amplitude wave theory. In *Estuary and coastal hydrodynamics*, A.T. Ippen, ed. New York: McGraw Hill.
4. Hwang, L.S. and B. Le Méhauté. 1968. *On the oscillations of harbors of arbitrary shape*. Tetra Tech Rept. TC-123A. MsoSec *J.F.M.* Vol. 42, No. 3, pp. 447-464 (1970).
5. Lee, J.J. 1969. *Wave induced oscillations in harbors of arbitrary shape*. Rept. KH-R-20. Div. of Engineering and Applied Science, Cal. Inst. of Tech.
6. Carrier, G.F. 1969. Low frequency response of narrow mouth harbors. Delivered at University of Hawaii Dept. of Ocean Engineering Seminar Series.
7. ___, R.P. Shaw and M. Miyata. 1971. The response of narrow mouthed harbors in a straight coastline to periodic incident waves. *J. App. Mech.* Vol 38 No2 pp 335-344, June 1971
8. Ippen, A.T. and Y. Ooda. 1963. *Wave induced oscillations on harbors—the solution for a rectangular harbor connected to the open sea*. MIT Hydrodynamics Laboratory Rept. 59.
9. Shaw, R.P. and A. Parvulescu. An acoustic model for the experimental study of water wave problems. *J. Acoust. Soc. Am.* Vol 50, No 6, Pt 1 pp 1443-1446 Dec. 1971

part 4: taivo laevastu

Oceanographic Processes in Coastal Waters

Predicting Pollution Dispersion with Hansen's Hydrodynamical Numerical Models

Classifying and Forecasting Near-Surface Ocean Thermal Structure

Taivo Laevastu has studied in the universities of Kiel, Gothenburg and Lund (fil. kand. in 1951). He has received his M.S. degree in oceanography from the University of Washington (1954), and his Ph.D. degree from the University of Helsinki (1960).

Besides his earlier research in oceanography and fisheries in Sweden and the U.S.A. (Seattle), he has been a fisheries oceanographer in the Food and Agriculture Organization of the United Nations in Rome from 1955 to 1962. From 1962 to 1964, he was an associate professor of oceanography at the University of Hawaii, where he organized teaching and research in physical oceanography. From 1964 to 1971, he was the Chief of the Oceanographic Research Division of the Fleet Numerical Weather Central, and from 1971 to the present he is Head, Oceanography Department Environmental Prediction Research Facility, Naval Postgraduate School, Monterey.

He has published numerous scientific papers on chemical and fisheries oceanography, sea-air interactions, and oceanographic forecasting. He has published books on fisheries oceanography (with I. Hela) and sea-air interaction. His pioneering works and contributions to science are in fisheries oceanography, sea-air interactions and oceanographic forecasting.

Besides his regular work, Dr. Laevastu has been an expert for the International Atomic

Energy Agency on the disposal of radioactive wastes into the sea, and UNESCO has called him to teach refresher courses in marine sciences in Bombay. He has received the Navy Oceanography Award for 1969, and the Naval Weather Service Special Contribution Award for 1970.

5

oceanographic processes in coastal waters

Introduction

This chapter is a condensed review of processes and conditions in coastal waters which affect oceanographic analyses and forecasts. The coastal problems are of necessity varied and local in nature; thus, only the general characteristics of coastal oceanography will be described. Such a voluminous amount of literature is available on this subject that no attempt is made to give references to all possible sources. It should be emphasized that local descriptions of conditions and processes in coastal waters form an important basis for their analyses and forecasts and should be thoroughly investigated for the area under consideration.

Oceanographic Variables in Coastal Waters

Classification of Coastal Waters

In this chapter we will define coastal waters as the waters between the 200-meter depth contour and the coast. This definition includes in addition a narrow belt of water on the continental slope which is greatly affected by the presence of the continental shelf. The continental shelves of the world are shown on Figure 5-1 and the corre-

sponding names, areas and general characteristics of the major continental shelves are given in Table 5-1.

Several classifications of coastal waters are possible. The continental shelf waters can be classified by the width of the continental shelf and the depth of the water on the shelf, i.e., 1) areas with narrow continental shelf and 2) areas with wide continental shelves, each of which can be subdivided into areas deeper than 50 meters and areas shallower than 50 meters. A further division can be made by latitudinal zones; tropical, temperate and polar coastal water.

Most adjacent and semiclosed seas fall within the category of coastal waters. An oceanographic classification scheme of adjacent seas is shown in Figure 5-2. Again, further classification is possible: 1) semiclosed seas with unrestricted connection with the ocean, 2) semiclosed seas with restricted connection with the ocean, usually brackish, 3) semiclosed bays, and 4) estuarine environments. Large semi-closed seas such at the Mediterranean and some Indonesian seas are often excluded from the classification of coastal waters.

Coastal and oceanic environments are compared in Table 5-2 as to the distribution of elements and the effects of various processes. One of the most characteristic environmental differ-

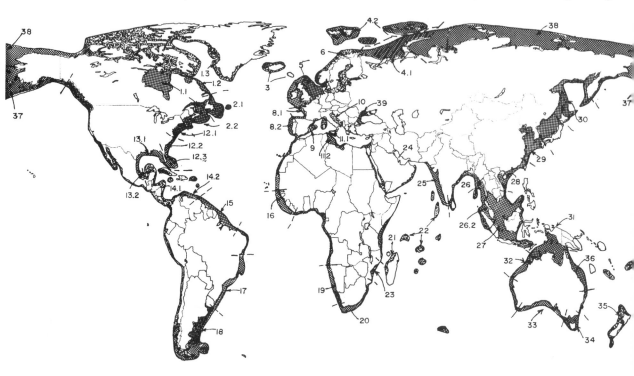

Figure 5-1. The continental shelves of the world.

ences between the coastal and oceanic waters is the difference in salinity. Some schemes have been used in the past for numerically determining the coastal influence by surface salinity. In many areas and conditions the coastal waters are separated from oceanic waters by sharp water mass discontinuities or fronts. The two water masses are differentiated through temperature, salinity, clarity and color of the surface layers.

Dynamical Processes in Coastal Waters

Three dynamic processes have pronounced intensity and influence in coastal waters: 1) tides and tidal currents, 2) windstau effects and 3) bathymetrical and topographical modification of waves and currents. These processes promote turbulent mixing and thus affect the distribution of oceanographic parameters in shallow water.

A number of dynamic processes are peculiar to coastal waters. One of the more common is estuarian circulation in areas where fresh water runoff from the coast enters the ocean. The circulation is characterized by offshore flow of less saline water at the surface and shoreward flow of more saline water along the bottom. There is usually a stronger vertical stratification in coastal water near estuaries than in the open ocean, especially during the summer. During the winter the coastal waters in medium and high latitudes are usually isothermal and isohaline.

Another well-known dynamic process in coastal waters is wave refraction with its associated surf and currents. Longshore and rip currents return water which has been transported to the shore by the mass transport of waves.

The most pronounced dynamic process in shallow water is, however, the tides. Their nature and effects are quite varied from one location to another. Their analyses and prediction are the oldest subjects of environmental predictions in the oceans.

Table 5-1. Continental Shelves of the World				
Shelf	Area in 1000 km^2	Avg. depth or depth range (m)	Avg. benthos biomass g/m^2 (estimate)	Remarks (biological productivity, nature of bottom and coast, etc.)
Hudson Hudson proper	1010	60-150		Low productivity. Rocky coast.
Labrador	160	120		Low to medium productivity. Very rocky coast with numerous offshore islands.
West Greenland	180	75	100	Medium productivity. Very irregular and rocky coast with numerous fjords.
New Foundland– Nova Scotia New Foundland	400	120		High productivity. Rocky coast; many good harbours.
Nova Scotia	370	60-150		High productivity. Bottom very irregular with holes >200 m deep; many good harbours; coast rocky in many places.
Iceland-Faroes	120	150	300	High productivity (except SE part of Iceland shelf). Rocky coast; many bays and fjords.
Spitzbergen– Barents Barents	550	70-150	100	Medium productivity. Low coast.
Spitzbergen	240	(120)		Medium to low productivity. Rocky and irregular coast.
NW European North Sea	570	35-(130)	200	N. part of the shelf is deep. Medium to high productivity. Low coast.
English Channel	90	35-(120)	40	SE part deep. Medium productivity. Many good harbours.
Irish Sea	380	80-150		Medium to low productivity. Rocky, irregular coast with many bays.
Norwegian	120	150		Medium to high productivity. Coast rocky with numerous fjords and offshore islands.
Baltic Sea	390	(100)	35	Low productivity. Fennonscandian coast rocky with numerous small offshore islands; SE coast low.
SW European Gulf of Biscay	80	130		Low productivity.
West Pyrenean	50	(150)		Medium productivity. Narrow shelf with rocky coast.
Balearian and Gulf of Lion	40	50-100	15	Low productivity.
Adriatic	90	30-150 (70)		Low productivity. Eastern coast rocky, but has many good harbors. Central, western and northern parts good for trawling.

Table 5-1 continued

Table 5-1 continued

Shelf	Area in 1000 km^2	Avg. depth or depth range (m)	Avg. benthos biomass g/m^2 (estimate)	Remarks (biological productivity, nature of bottom and coast, etc.)
Sicilian-Sidra Sicilian	30	50-100	10	Low productivity. The shelf is divided into western and eastern parts. Few good harbours.
Sidra	100	60		Low productivity. Southwestern part relatively shallow.
USA-Atlantic New England	100	Coastal part 35 Offshore part 80		Medium to high productivity. Many protected bays with good harbors.
Carolina	120	25		Low to medium productivity. Shallow coast with offshore island and bars; Southern part coralline.
Bahama	130	15		Low productivity. Very shallow and reefy.
Gulf of Mexico Florida-Texas	450	40		Low to medium productivity. Outer part of Florida Shelf $>$120 m deep, shallow coast; many offshore bars.
Campeche	180	30		Low productivity. Shallow shelf, (except the northern corner); rocky and coralline; several offshore bars.
Carribbean Mosquito Venezuela	110 130	25 20-80		Low productivity. Shallow shelf; rocky and coralline; many offshore bars & islands. Eastern part has medium productivity, western low productivity. Rocky and coralline coast; irregular bottom.
Amazon	540	Southern part 40 N. coastal half 15-25 N. offshore half 70		Central and northern part along the coast has medium productivity; southern part and offshore area unproductive. Rocky and coralline, except off the Amazon river; no good harbor for medium or bigger boats.
Guinean	210	25	20	Low productivity. Many good sheltered places for harbors; central part rocky.
S. Brazilian	400	80		Central part has medium productivity; other parts low. Numerous lagoons and offshore bars; coast and N. part shallow (30 m); N. part rocky and coralline.
Patagonian	1000	80-100		Medium to high productivity. Southern coast and SE part rocky; Blanca and La Plata Bays shallow (ca. 20 m); few good harbors.

Table 5-1 continued

Table 5-1 continued				
Shelf	Area in 1000 km²	Avg. depth or depth range (m)	Avg. benthos biomass g/m² (estimate)	Remarks (biological productivity, nature of bottom and coast, etc.)
St. Helena	140	150	200	High productivity. Coast rocky; no good harbors.
Agulhas	110	100		Low productivity. Few good harbors; rocky coast.
Madagascar	180	15		Low productivity. Very coralline and rocky; no good harbors.
Island-shelves in Central Indian Ocean	240	40-80		Medium productivity. Most of the shelves are coralline.
Zambesi	60	30		Low productivity. Generally rocky, except off Zambesi River
Persian Gulf	240	35		NW part has good bottom and medium productivity; SE part rocky and very shallow; N. part rocky and deeper with rocky coast.
Bombay	300	NE - 20 Rest - 80		NE part is shallow, has good trawling grounds and is productive. Central and S. parts are rocky, with medium production (except Wadge Bombs, which has productive trawling grounds)
Bengal-Burma Bengal Burma	170 380	N - 15 Rest 50-100 (80)		Northern part shallow and with medium productivity; remainder has low productivity. Low productivity. Rocky coast with numerous offshore islands; coralline.
South China	2300	50		Low productivity (except Gulf of Siam). Shallow coastal area; bottom rocky & coralline.
Tong-King Hong-Kong East China Yellow Sea Okhotsk	480 1000 580	Coastal half 35 Offshore half 85 W part 30 E part 80 140	175 220	Low productivity. Partly rocky; N part has numerous small offshore islands. Medium productivity. E. and S. coast rocky; shallow and muddy off deltas. Medium productivity. Kamchatka and Sakhalin coast more shallow (60 m); rocky coast; few good harbors.
Arafura	1350	(70)		Low productivity. Rocky and coralline; no good shelters along the coast.
NW Australian	300	80		Low to medium productivity. Irregular bottom; NE coast has many offshore islands; other parts of coast do not have good shelters.

Table 5-1 continued

Shelf	Area in 1000 km²	Avg. depth or depth range (m)	Avg. benthos biomass g/m² (estimate)	Remarks (biological productivity, nature of bottom and coast, etc.)
S. Australian	350	40-130		Medium productivity. Only central part has some good shelters.
Tasmanian	170	60-80		Medium to high productivity. Partly rocky.
New Zealand	250	70		High productivity (except N. part) rocky coast.
Queensland	220	<40		Low productivity. Extremely rocky & coralline.
Bering Sea	1200	NE-35		Medium to high productivity. (NW part low productivity). Many great sheltered bays; coastal area in some places rocky; low bottom water temperatures.
North Siberian	2600	W-25 E-40	25	Low productivity. Low coast, mostly ice covered.
Black Sea			(300 Azov Sea)	Sea of Azov very productive. (Depth 15 m). Relatively large shelf in northern part. Depth of NW shelf *ca.* 50 m.

Table 5-1 continued

The prevailing direction of winds influences the thermal and salinity stratification as schematically shown in Figure 5-3. This figure also demonstrates the need to include within coastal waters the strip of about 100 miles off the continental shelf.

Sea-Air-Land-Bottom Interaction in Coastal Waters

Sea-Air Interactions

The intensity of sea-air interactions with respect to heat exchange is greatly determined by the difference of the air and sea surface properties in coastal waters. Great differences exist between the east and west coasts of the continents in this interaction because of the continental influence. The medium and high latitudes off the eastern coasts of continents are characterized by great heat loss from the water during the fall and winter. This is brought about by cold and relatively dry continental air masses moving over the warmer sea, allowing the sea surface to release large amounts of heat and moisture. At medium and high latitudes on the western coasts of continents the prevailing onshore winds bring maritime air masses over the continents. These air masses are near the surface at equilibrium with the properties of the sea surface; thus little heat and moisture is exchanged with the sea surface.

In lower latitudes along the west coasts, upwelling frequently occurs. Here, relatively dry warm air from the land is often brought over the ocean causing considerable evaporative cooling in the surface layers of the ocean. In addition, the low stratus cover over upwelling areas diminishes the insolation. The cool upwelled water is further cooled by heat exchange. The effects of this heat exchange are extended seaward a considerable distance due to increased stability of the air near the sea surface as it cools.

Heat exchange over coastal waters is also enhanced by the greater turbulence in the lower layers of the atmosphere caused by the large surface roughness of land areas.

The evaporation very near the coast is considerably affected by spray from waves breaking

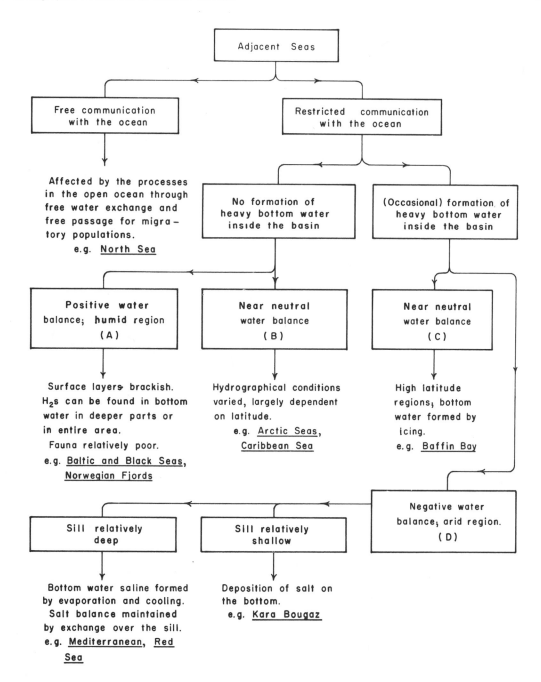

Figure 5-2. Oceanographic classification scheme of the adjacent seas (after Hela and Laevastu[2]).

Table 5-2. Comparison of Coastal and Oceanic Environments		
Parameter and/or Process	Oceanic	Coastal
Temperature	Relatively uniform with moderate to low ambient thermal noise (except near major current boundaries). Vertical and horizontal gradients moderate to low. Changes determined mainly by heat exchange and advection.	Moderate to high ambient thermal noise, greatly influenced by coastal topography, bathymetry and upwelling-anstau. Vertical and horizontal gradients moderate to high. Changes determined by heat exchange, advection; and mixing, especially by tidal current mixing. Horizontal gradients towards coast usually positive in summer, negative in winter.
Salinity	Uniformly high with small horizontal and vertical gradients. Determined mainly by evaporation-precipitation balance.	Relatively low, especially in bays and near estuaries. Horizontal and vertical gradients moderate to high. Determined by runoff and evaporation-precipitation balance.
Permanent Currents	Predominantly thermohaline, speed relatively uniform and low.	Thermohaline component greatly modified by local bathymetry. Considerable accelerations on continental slopes, due to anstau.
Wind Currents	Determined by average prevailing wind fields.	Wind currents greatly modified by coastal topography, anstau and upwelling.
Tidal Currents	Usually weak	Usually dominates other current components
Waves	Determined by local wind fields and by swell propagation.	Modified in shallow water due to bottom influence. Generation affected by sheltering effect of land (limited fetch).
Sea-air exchange process	Determined by surface meteorological systems and processes; patterns correspond to the dimensions of meteorological systems.	Exchange processes can be intense at times due to large differences in temperature and moisture as continental air masses move over the sea surface.

Table 5-2 continued

Table 5-2 continued		
Parameter and/or Process	Oceanic	Coastal
Sea level	Variations usually small, dependent on tides and atmospheric pressure.	Variations usually large, caused by astronomical tides, co-oscillations of bays, atmospheric pressure and wind-induced anstau.
Turbidity (and visibility in water)	Low turbidity; largely determined by standing crop of phytoplankton	High turbidity due to suspended matter, transported by runoff and stirred up from bottom by wave action.
Biological populations	Plankton population relatively uniformly distributed. Productivity medium to low (Except near current boundaries). Standing crop of fish relatively low and dispersed	Plankton population relatively patchy. Productivity generally medium to high. Standing crop of fish high, many species occurring in schools. Diurnal and seasonal migrations pronounced.
Pollution	Low	Can be considerable along populated shores and near estuaries.

on the beach. This spray is orders of magnitude more intense than that formed by waves breaking in offshore water.

Additional differences in heat exchange in coastal waters are caused by coastal fog and stratus which influence insolation. Some coastal areas, especially off mountainous coasts, receive more precipitation than offshore areas, due to orographic lift along mountainous coasts.

Despite the relatively intense exchange in coastal areas, seasonal differences of surface temperature in coastal waters are only moderately higher than in corresponding offshore areas. This is due mainly to the equalizing effect of the water exchange between the shelf and the open ocean. The standard deviation of temperature from the monthly mean in coastal locations is also greatly affected by the width of the continental shelf and the relative sheltered location of the measurement. Some examples of the standard deviation from the monthly mean temperatures in some oceanic and coastal locations are given in Table 5-3.

Sea-Land Interactions

Coastal topography has a pronounced influence on sea-land interactions, especially along mountainous coasts. The coastal topography determines the sheltering and fetch effects which limit the generation of waves. Local coastal winds such as the mistral, tramontana, Santa-Anna, etc., are a function of coastal topography.

An integral part of the sea-land interactions is the sea and land breeze caused by diurnal heating and cooling at the land. It occurs with great regularity on many medium and low latitude coasts but at higher latitudes only during warm seasons.

The most significant oceanographic sea-land interactions manifest themselves in the form of

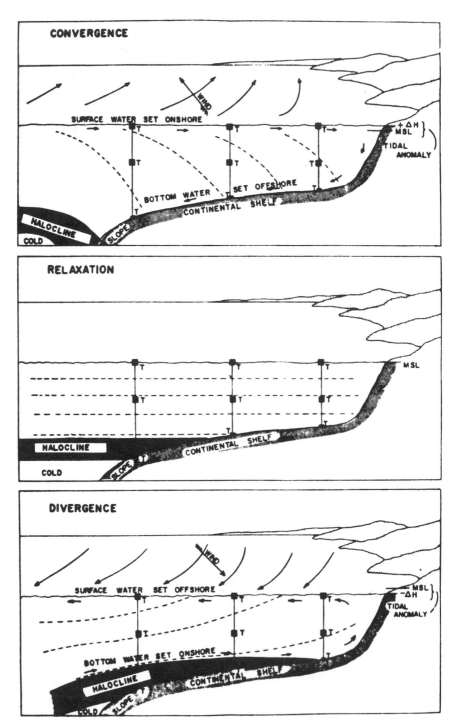

Figure 5-3. Schematic of convergence, relaxation and divergence situations off the west coast of Vancouver Island (after Dodimed and Pickard[1]).

Table 5-3. Examples of Standard Deviations from Monthly Mean Temperature at Some Oceanic and Coastal Locations (After M. Robinson)

Location	Average standard deviation for all months °C
Offshore locations	
49°N, 148°W	1.5
30°N, 140°W	1.5
37°N, 123°W	2.5
38°N, 158°W	1.3
Shore Stations	
La Jolla, California	2.3
Friday Harbor, Washington	1.5
Atlantic City, New Jersey	3.7
Eastport, Maine	1.7

Table 5-4. General Characteristics of Estuarine Conditions

Coastal morphology and bathymetry	Bay usually shallow, intended estuary, sometimes a canyon present off estuary. Presence of sand bars; delta formation off the coast.
Sediments	Usually sand near coast, mud, rich in organic matter off coast. Usually smooth bottom on delta formation.
Salinity	Fresh water mixing in estuary and flowing out on the surface as brackish water, forming river plume offshore. Surface salinity increasing seaward. Deep water salinity usually relatively high. Seasonal variation of salinity depending on the amount of river flow.
Temperature	Surface water temperature usually higher than surrounding in summer, colder in winter. Deep water temperature variations small.
Turbidity	High, especially in surface waters, due to transport of suspended matter with runoff. Turbidity decreases seaward due to coagulation and sedimentation. Usually high phytoplankton content during summer in the plume.

runoff from rivers and streams. This runoff has considerable influence on the oceanographic conditions near the coast, creating special environments in estuaries and in semiclosed bays. The characteristics of estuarine conditions are listed in Table 5-4. The runoff changes the salinity, turbidity and to a lesser degree the temperature of the coastal waters. Thus, the salinity anomalies tend to be more local phenomena than the temperature anomalies in the ocean.

Another major characteristic of coastal waters is the high basic organic production. The proximity of land provides an abundance of nutrients, both by runoff and by wind transport.

Ice formation in coastal waters depends largely on the continentality of the coastal climate, the salinity stratification, water depth and the exchange of water between the offshore and coastal area. The conditions of surface ice formation are given in Figure 5-4.

Sea-Bottom Interactions

The bathymetry (i.e., the water depth) has a direct influence on many processes and affects conditions in coastal waters. It determines the amplitudes of tides and tidal currents and mixing by them. Although an apparent relation between the sea surface temperature and the water depth can be found in many instances, this relation is scarcely direct but is a result of factors affected by topography which in turn affect temperature such as mixing, distance from the coast, wave action, etc.

The effects of bathymetry, especially the effects of the continental slope configuration on the currents and sea surface temperature distribution are apparent from Figures 5-5 and 5-6. The currents over the continental shelf tend also to follow the steeper bottom slopes. Thus, certain

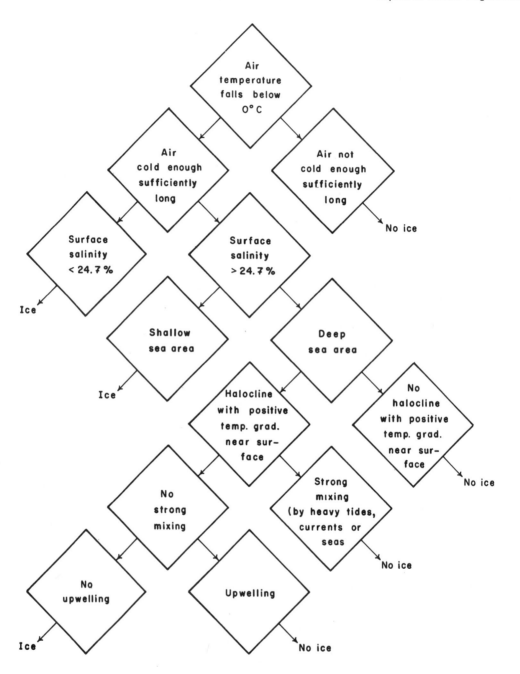

Figure 5-4. Scheme for determination of probability of surface ice formation (after Hela and Laevastu).

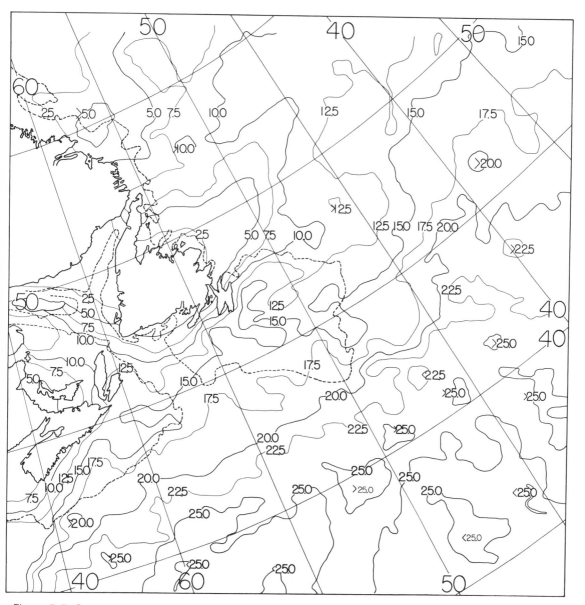

Figure 5-5. Synoptic sea surface temperature analyses of the Grand Banks area. OOZ 5 August 1967.

current characteristics can be obtained from an examination of bathymetric charts.

The type and coarseness of the sediment is greatly affected by the topography. Fine sediments tend to accumulate in deeper holes where bottom currents are weak. Slopes and plateaus swept by swifter currents are usually barren or covered with coarse-grain sediments. Where wave action reaches the bottom, sediments are stirred up causing waters in shallow coastal areas to become especially turbid during storms.

The high turbidity over the continental shelf

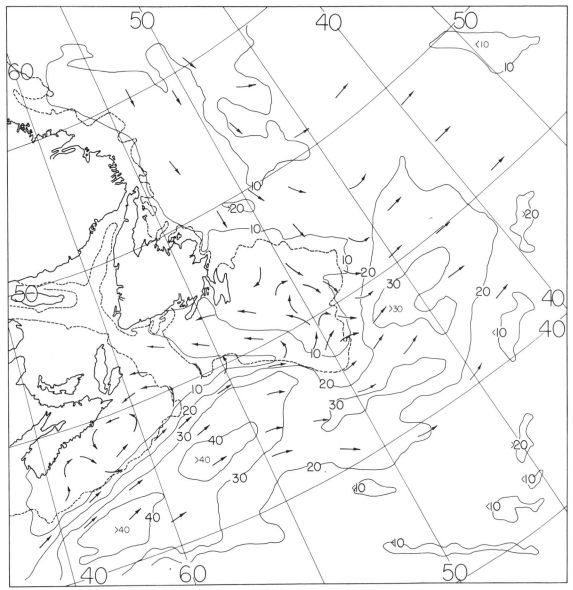

Figure 5-6. Synoptic surface current analyses of the Grand Banks area. OOZ 5 August 1967 (isotachs in nautical miles per 24 hours).

as compared to offshore waters is caused mainly by living and decayed plankton. Maximum turbidities in subsurface layers are usually found at maximum vertical temperature gradients.

The phenomenon of surf is basically a seabottom interaction. The surf parameters at any shallow water location are determined by the water depth and its slope and in conformity with the deep water wave direction and period. Considerable amounts of literature and relatively accurate forecasting methods are available for surf.

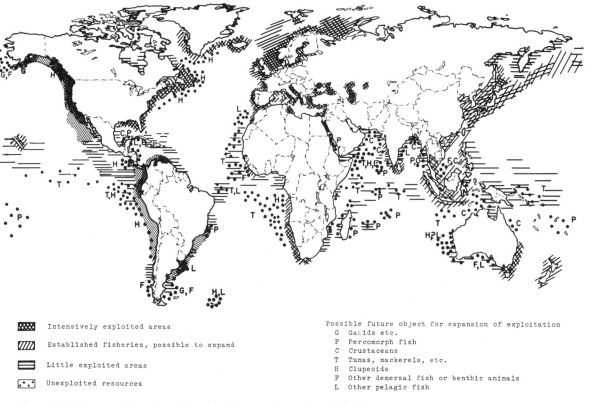

Figure 5-7. Present intensity of world fishery and underexploited resources.

Some Peculiarities of Marine Biology in Coastal Waters

The standing crops of marine ecological groups are considerably higher in coastal waters than in offshore waters due to the higher fertility of the coastal areas. Most of the world's commercial fishing is undertaken on the continental shelf and continental slope. The continental shelf is the major spawning ground for most fishes who might spend part of their lives in offshore waters, but who return for spawning purposes to the shelf, where they are then aggregated into schools over spawning grounds. The fishing grounds of the world and the relative intensity of fishing are shown in Figure 5-7.

The bulk of the biomass on the continental shelf belongs to benthos and/or animals dependent on it such as demersal fish. The quantity and species composition of benthos and demersal fish is quite varied from one location to another and is greatly determined by prevailing environmental conditions such as depth, type of bottom, temperature, salinity, current speed near bottom, etc.

Fisheries and marine biological research have produced a wealth of information on the biology of coastal waters. There is an urgent need to reduce the information into atlases and into numerical form. This reduction has been just barely started, as exemplified by the serial Atlas of the Marine Environment (American Geographical Society).

Sound Propagation Peculiarities in Coastal Waters

Water depth and bottom properties have the most effect on shallow water sound propagation problems, whereas the thermal structure, which has primary importance on sound propagation in deep water, has only a secondary effect in shallow water. It is imperative to know accurately the type of bottom, its roughness and its depths to make accurate propagation loss computations in shallow water.

Coastal waters are also considerably more noisy than offshore waters. This ambient noise is caused by waves breaking in shallow water, by strong currents, high traffic noises and high biological noises due to the higher standing crops of fish and other noise-making marine animals in shallow water. The short-period fluctution of sound in shallow water is also considerably greater than in deep water.

Although theoretical propagation loss computations are possible in shallow water, the conditions are so variable that these computations are not generally applicable. Therefore, empirical approaches derived from theoretical considerations and complemented by parameterization of the affecting environment promise the greatest applicability.

References

Dodimed, A. J. and Pickard, G. L. 1967. Annual changes in the oceanic-coastal waters of the Eastern Subarctic Pacific. *J. Fish, Res. Bd. Canada,* 24(11):2207-2227.

Hela, I. and Laevastu, T. 1962. *Fisheries Hydrography.* London: Fishing News (Books) Ltd.

6

predicting pollution dispersion with hansen's hydrodynamical numerical models

Prospects of Pollutant Dispersion Predictions with HN Models

Problems of diffusion and transport of substances in the sea have received considerable theoretical and experimental attention in recent decades—especially radioactive substances and pollutants (1, 2 and 3). Despite these broad efforts, Okubo (2) concludes that "diffusion is confusion."

Most attempts to find accurate, universal solutions to dispersion formulas have encountered the same difficulties:

1. Numerical solutions require simultaneous treatment of many parameters (diffusion, transport, decay, etc.). All of these vary widely in space and time and are not independent.
2. Velocity field specification has been incomplete. Fluctuations in transport and turbulence vectors in short time steps were not known. Averaging over any extended period introduces unacceptable errors because the processes are nonlinear.

In spite of the simplifications used to surmount these difficulties, some Eulerian approaches to diffusion problems have given satisfactory results (1). In most applied problems treating both diffusion and transport, however, a Lagrangian solution is required.

Two recent developments show promise in handling these complicated problems:

1. Successful application of hydrodynamical numerical models which include diffusion formulas in finite difference form.
2. Use of large computers for synoptic analysis/prediction in the oceans.

The combined effect of these developments permits simultaneous, numerical solution of complex formulations in short time steps. The necessary current vectors and parameter distributions are carried along in a marching process. The solutions are essentially Lagrangian in form although certain Eulerian treatments will be used where possible and necessary.

Hydrodynamical Formulas and Their Solution by Finite Difference Methods

Formulas

Using numerical hydrodynamical method for computation of tides and currents was originally proposed by Hansen (4). With the development and availability of high-speed electronic computers these methods became economically feasible and practically possible.

Derivation of vertically integrated hydrodynamical equations is described by Hansen (5), Sündermann (6), Jensen, Weywadt and Jensen (7) and Bretschneider (8). Additional derivations are available in works by Schmitz (9, 10). The following basic equations are used in the single-layer model:

$$\frac{\partial u}{\partial t} - fv - \gamma \Delta\ u + \frac{r}{H} u \ \sqrt{u^2 + v^2}\ +$$

$$g\,\frac{\partial \zeta}{\partial x}\ = X + \frac{\tau^{(x)}}{H} \tag{6-1}$$

$$\frac{\partial v}{\partial t} + fu - \gamma \Delta v + \frac{r}{H} v \sqrt{u^2 + v^2}\ +$$

$$g\,\frac{\partial \zeta}{\partial y}\ = Y + \frac{\tau^{(y)}}{H} \tag{6-2}$$

$$\frac{\partial \zeta}{\partial t} + \frac{\partial}{\partial x}\,(Hu) + \frac{\partial}{\partial y}\,(Hv) = 0. \tag{6-3}$$

$\tau^{(x)}$ (and $\tau^{(y)}$) are usually expressed as:

$$\tau^{(x)} = \lambda W_x \ \sqrt{W_x^2 + W_y^2}$$

The bottom stress (friction) term in equations 6-1 and 6-2 is

$$\tau^{(b)} = \frac{r}{H}\, u \sqrt{u^2 + v^2}\ \ ; \ \ \frac{r}{H}\, v \sqrt{u^2 + v^2}$$

The following symbols are used in the preceding equations:

x, y	=	space coordinates
t	=	time
u, v	=	components of velocity
H	=	total depth (H = h + ζ)
ζ	=	surface elevation
X, Y	=	components of external forces
$\tau^{(x)}, \tau^{(y)}$ =		components of wind stress (λ = 3.5 x 10^{-6})
g	=	acceleration of gravity
f	=	Coriolis parameter
r	=	friction coefficient (3 x 10^{-3}) (bottom stress)
γ	=	coefficient of horizontal eddy viscosity
Δ	=	Laplace operator
λ	=	coefficient of friction (drag coefficient)
W_x, W_y	=	wind speeds
$\tau^{(b)}$	=	bottom stress

Analytical solutions to equations 6-1 to 6-3 are of little value, as exact solutions are possible only for basins of regular shape, simple depth and simple wind distribution. However, Hansen (5) and his later collaborators (e.g., Sündermann, Ref. 6, and Bretschneider, Ref. 8) have developed an explicit method for achieving time-dependent solutions to these formulas using finite difference approach. The finite approximations are:

$$\zeta^{t+\tau}\ (n,m) = \bar{\zeta}^{t-\tau}\ (n,m) -$$

$$\frac{\tau}{\ell}\Big\{H_u^t\ (n,m)U^t(n,m) - H_u^t(n,m\text{-}1)U^t(n,m\text{-}1) +$$

$$H_v^t\ (n\text{-}1,m)V^t(n\text{-}1,m) - H_v^t\ (n,m)\ V^t\ (n,m)\Big\}$$

$$U^{t+2\tau}(n,m) = \Big\{1 - [2\tau r/H_u^{t\ +2\tau}(n,m)]\ [\bar{U}^t\ (n,m)^2 +$$

$$V^{*t}(n,m)^2\ ]^{0.5}\Big\}\ \bar{U}^t(n,m) + 2\tau\,fV^{*t}\ (n,m) -$$

$$\frac{\tau g}{\ell_{-}}\Big\{\zeta^{t+\tau}\ (n,m+1) - \zeta^{t+\tau}\ (n,m)\Big\} +$$

$$2\tau\,X^{t+2\tau}\ (n,m)$$

$$V^{t+2\tau}(n,m) = \Big\{ 1 -$$
$$[2\tau r / H^{t+2\tau}(n,m)] \; [\bar{V}^t(n,m)^2 +$$
$$U^{*t}(n,m)^2]^{0.5} \Big\} \bar{V}^t(n,m) - 2\tau f U^{*t}(n,m) -$$
$$\frac{\tau g}{\ell} \Big\{ \varsigma^{t+\tau}(n,m) - \varsigma^{t+\tau}(n+1,m) \Big\} +$$
$$2\tau Y^{t+2\tau}(n,m).$$

The "averaged" velocity and water elevation (sea level) components are

$$\vec{U}^t(n,m) = \alpha U^t(n,m) + \frac{1-\alpha}{4} \Big\{ U^t(n-1,m) +$$
$$U^t(n+1,m) + U^t(n,m+1) +$$
$$U^t(n,m-1) \Big\}.$$

$\bar{V}^t(n, m)$ and $\bar{\varsigma}^t(n, m)$ are analogous.

(The factor α can be interpreted as "horizontal viscosity parameter." Its normal value is 0.99.)

$$U^{*t}(n,m) = \frac{1}{4} \Big\{ U^t(n,m-1) + U^t(n+1,m-1) +$$
$$U^t(n,m) + U^t(n+1,m) \Big\}.$$

$V^{*t}(n,m)$ is analogous to $U^{*t}(n,m)$ in the preceding equation.

The time step is 2τ. The total depth (H_u, H_v) is computed as

$$H_u^{t+2\tau}(n,m) = h_u(n,m) + \frac{1}{2} \Big\{ \varsigma^{t+\tau}(n,m) +$$
$$\varsigma^{t+\tau}(n,m+1) \Big\}.$$

The effects of wind (external force) are computed with the following formula:

$$X^t = \frac{\lambda W_x^t \; [(W_x^t)^2 + (W_y^t)^2]^{0.5}}{H} - \frac{1}{\rho r} \frac{\partial P_o}{\partial x}$$

$$Y^t = \frac{\lambda W_y^t \; [(W_x^t)^2 + (W_y^t)^2]^{0.5}}{H} - \frac{1}{\rho r} \frac{\partial P_o}{\partial y}$$

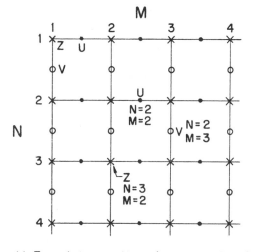

× Z-point, • U-point, ○ V-point.

Figure 6-1. Scheme of the grid net.

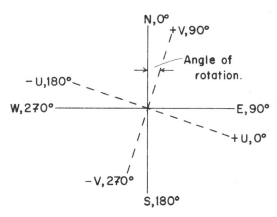

Figure 6-2. Relations between the geographic coordinates and computation coordinates.

In small-area computations the pressure gradient term (acceleration) is usually neglected. A number of slightly modified finite difference schemes are possible for solving the hydro-dynamical equations. Some of these schemes are being tested and the results will be given in another report.

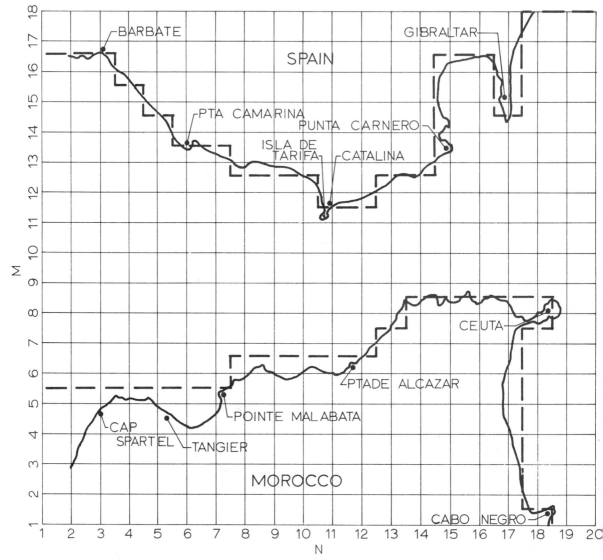

Figure 6-3. Computation grid for Strait of Gibraltar.

The Grid Net and Input of Depths

The grid net is shown in Figure 6-1. It consists of three different sets of grid points: the water elevation points (z) at the intersection of the grid, the point for u-velocity component to the right of z point and point for v-velocity component below the corresponding z point. Each of these three points has the same coordinate designation (n,m).

The geographical orientation of the grid is usually selected so that the x axis (m coordinate) is parallel to the entrance of the bay (or to the initiation of computations with tidal input). The relation between the geographic coordinates and computation coordinates is shown in Figure 6-2. This orientation must be considered in input of wind direction and in obtaining the current direction in geographic coordinates.

The coastline must pass through u and v points and not through z points (see Figure 6-3); it becomes a "step line" along irregular coasts.

Three sets of depth values must be read into the computer. The first set is symbolic depth (sea and land table) at z points. The z points over land are 0 at the boundary on the coast −1 (immediately outside the boundary line), over the water 1 and at the input boundaries −2 (and −3). The second and third set of depth values are read at u and v points, respectively. The u and v depth values over land are 0, at boundary points −1 and over the sea the depth is in centimeters.

Courant-Friedrich-Lewy Criterion, Horizontal Viscosity and Bottom Friction

Grid size selection is usually based on requirements of detail and accuracy, and availability of computer core memory. A 25x25 array can easily be computed using a computer with 32K memory. Large arrays require bigger computers.

The maximum length of time step is determined by grid size and maximum depth in the area according to Courant-Friedrich-Lewy criterion:

$$\Delta t \leqslant \frac{0.5\ell}{(2gH_{max})^{0.5}}$$

where t is time step (second), ℓ is grid size (cm), g is acceleration of gravity and H is maximum depth in the area of computation (cm). Any attempt to increase the time step above the criterion has resulted in "blow up" of the computations.

The smoother ALPHA acts as horizontal viscosity coefficient. The normal value of ALPHA as used by Hansen (11) is 0.99. However, this coefficient can also be used as a "tuning factor." If a lower value (e.g., 0.90) is used, the current speed is decreased, making the water more viscous. The lower value of ALPHA is also useful if depth distribution is irregular over the area.

The bottom stress is assumed to be linearly dependent of u and v. Over deep water the bottom stress becomes small indeed.

The sudden depth change at the continental slope and to some extent the bottom stress in shallower water are responsible for computed stronger currents on steeper slopes. This condition corresponds well with observations in nature, where stronger currents flow along the steeper slopes.

Initial Boundary Values and External Forces

The principal initial boundary values introduced in every time step are tidal heights at open boundaries and wind speed and direction at each grid point (wind stress). In addition, river flow and/or flow through channels entering the area can be added as initial values.

Normally, four tidal components are used for the input of sea level elevation on the open boundary:

$$Z = A_1 cos(\alpha_1 t - \mathcal{H}_1) + A_2 cos(\alpha_2 t - \mathcal{H}_2)$$
$$+ A_3 cos(\alpha_3 t - \mathcal{H}_3) + A_4 cos(\alpha_4 t - \mathcal{H}_4).$$

The speeds of the constituents (α) must be given in radians per second, the epoch of the constituent ($\mathcal{H}$) in radians and the amplitude of the constituent (A) in cm.

All boundary values should be defined if the program is to work correctly. Thus, the boundaries along the coast must be defined as closed boundaries (i.e., no flow into or from the land). This is done by setting either u or v component 0 at the coast (dependent on the direction of the coast).

Finally, it should be pointed out that the model must be run 10 to 60 hours in real time (dependent on size of the area and grid length) before equilibrium is established and correct outputs can be taken. A computational grid for Strait of Gibraltar is shown in Figure 6-3, and Figure 6-4 gives an example of computed currents.

Dispersion Equations and Their Finite Difference Forms Used in HN Models

As HN models provide a current vector at

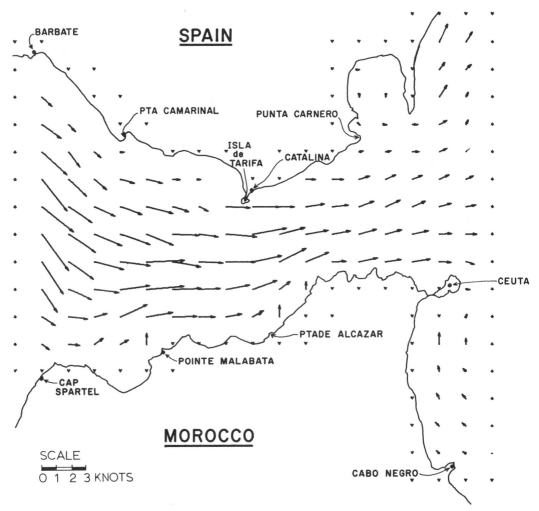

Figure 6-4. Example of synoptic currents computed with HN model (currents in Strait of Gibraltar three hours after high water at Tarifa).

each grid point at small time steps, one is tempted to solve the interacting advection and diffusion problem using time-stepping finite difference method. It should be borne in mind that diffusion (mixing) is irreversible, whereas advection is reversible. In this chapter we treat the dispersion of substances which can be readily mixed in shallow water or in a surface mixed layer. If these problems can be properly solved by the HN model, no difficulties are foreseen in treating specific problems such as oil dispersion, brackish water,

etc. The following were considered in designing the dispersion portion of the model:

1. Vertical diffusion should be neglected (can be added to multilayer models).

2. A basic Lagrangian approach should be used because of its application to computer solutions for grid arrays in finite-difference time stepping.

3. Eulerian modifications should be used in initial state for instantaneous releases where

the size of the diffusing blob is small in comparison with the mesh length.

4. The grid should be fine enough to permit neglect of nonlinear terms.
5. Fickian diffusion with constant diffusivity is used since advection and diffusion equations are solved separately in short, interlocking time steps (similar to the two-step method of Schönfeld and Groen, Ref. 3).
6. The Austausch coefficient for eddy diffusion per se must be related to grid size.

Diffusion in water bodies has been presented by many formulas, a few of which are presented here (neglecting vertical diffusion).

The general diffusion formula is

$$\frac{\partial^2 S}{\partial x^2} + \frac{\partial^2 S}{\partial y^2} - \frac{1}{A}\frac{\partial S}{\partial t} = 0.$$

The basic dispersion formula is

$$\frac{\partial S}{\partial t} = Y - \frac{S}{n} - \frac{\partial}{\partial x} \cdot (U_x + pS_x)$$

$$- \frac{\partial}{\partial y} \cdot (U_y + pSy).$$

The Fickian equation is

$$\frac{\partial S}{\partial t} = Y - \frac{S}{n} - K\nabla^2 S - \frac{\partial}{\partial x}(S_u U)$$

$$- \frac{\partial}{\partial y}(S_v V)$$

where Y is addition (release), n is decay, S is concentration, $p^S_{x,y}$ is concentration velocity component, K is $\beta a V_r$; a is depth, $V_r = (u^2 + v^2)^{0.5}$, β is 0.003 and A is diffusion coefficient.

The "statistical" diffusion formula is

$$\frac{\partial S}{\partial t} = \frac{1}{r}\frac{\partial}{\partial r}\left(\cdot Pr^2 \frac{\partial S}{\partial t}\right).$$

The Lagrangian approach we have adopted is given by

$$S^{t+\tau}_{n,m} = S^t_{n,m}\left(1 - \frac{4\tau A}{\ell^2}\right) + \frac{A\tau}{\ell^2}\ (S_{n-1,m}$$

$$+ S_{n+1,m} + S_{n,m-1} + S_{n,m+1}$$

$$- 4S_{n,m}) + \frac{A\tau}{\ell^2}\ (S_{n-1,\ m-1}$$

$$+ S_{n+1,\ m-1} + S_{n-1,\ m+1} + S_{n+1,\ m+1})$$

The preceding finite equation deviates from the usual finite difference diffusion formula in the addition of the term $(-4S)$, which makes the solution similar to solution of the Laplacian $(k\ \nabla^2 S)$, and the addition of the last term, which conserves the substance.

The advection is computed linearly in finite difference form:

$$S^{t+\tau}_{n,m} = S^t_{n,m} - \tau |U_{n,m}|\ \frac{(S_{n,m} - S_{n,m\pm1})}{\ell}$$

$$- \tau |V_{n,m}|\ \frac{(S_{n,m} - S_{n\pm1,\ m})}{\ell}$$

where $Sn, m-1$ or $n, m+1$ (respectively $n-1$, $n+1$) are used, depending on the direction (sign) of U and V.

Due to the short space and time steps in the HN model, secondary terms such as the following are neglected:

$$\left[\frac{1}{2}\left(U_{n,m} - U_{n,m-1}\right)\left(S_{n,m} - S_{n,m-1}\right) + \frac{1}{2}\left(U_{n,m+1} - U_{n,m}\right)\left(S_{n,m+1} - S_{n,m}\right)\right]$$

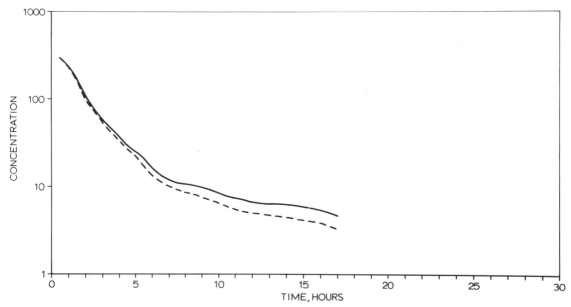

Figure 6-5. Example of dispersion from an instantaneous release of 300 units in Strait of Gibraltar. (Change of maximum concentration with time; full line: A = 3.0 X 10⁴; dashed line: A = 1.4 X 10⁵.)

If the grid size is large in relation to the dispersing cloud (or the time step is long), linear interpolation of gradients introduces errors. In this case, an additional term (empirical) has been added to the transport components:

$$\frac{C\left(S_{n,m} - S_{n,m\pm1}\right)}{|u,v|\tau\ell}$$

Numerical experiment shows, however, that the last term is relatively unimportant.

If the diffusion formula alone is used, A can be determined from the relations given by Joseph and Sendner (6-1)

$$A = \frac{P\ell}{2}$$

where P is approximately equal to 1.5. This formula requires a diffusion coefficient of 1.5 x 10^5 for our numerical runs for the Strait of

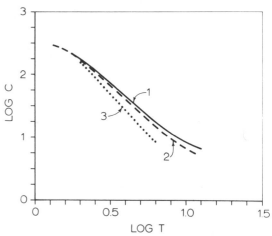

Figure 6-6. Change of concentration comparison with time from experiments in Figure 6-5 and field observations by Hela and Voipio (1960). (Full line: A = 3.0 X 10⁴, m = 1.5; dashed line: A = 1.4 X 10⁵, m = 1.65; dotted line, Hela and Voipio (1960), m = 2.4.)

Gibraltar. If diffusion and advection are treated separately, a much smaller A is needed ($A = 3.0$ x 10^3).

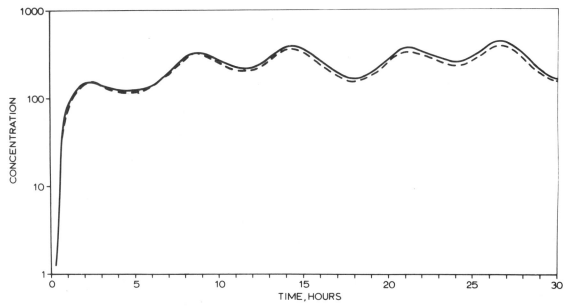

Figure 6-7. Example of dispersion from a continuous release of 5 units per minute in Strait of Gibraltar. (Change of maximum concentration with time; full line: A = 3.0 X 10⁴; dashed line: A = 1.4 X 10⁵.)

Results of Numerical Computation of Dispersion

Figures 6-5 to 6-8 present some preliminary dispersion computation results in the Strait of Gibraltar. Figure 6-5 shows the dispersion from an instantaneous release of 300 arbitrary units. The concentration values are maximum values. If we compare the results with an experiment conducted by Hela and Voipio (12) and express the maximum concentration change with time with their formula (see Figure 6-6) $\log C = \log C_1 - m \log t$, we find that the coefficient m in our computation is 1.5 with $A = 3.0 \times 10^4$ and 1.65 with $A = 1.4 \times 10^5$ as compared to 2.4 in Hela and Voipio's experiment. The difference is apparently due to the two causes: 1) our computation did not include decay and 2) there was some eddy diffusion in the vertical in the Hela-Voipio experiment.

Figure 6-7 shows the change of maximum concentration with time from a continuous release of 5 units per minute. It is interesting to note that the maximum concentration is reached asymptotically and that it fluctuates in a tidal rhythm.

With this model it is thus possible to compute the maximum concentration from a given continuous release (or to determine the release level if a maximum permissible concentration is prescribed by regulations). The areal distribution of the concentration after 7 hours from instantaneous release of 10,000 units in another computation in Lower Bay of New York is shown in Figure 6-8. The original release points are marked with X.

If the dimensions of a dispersing blob are small in relation to grid size (in case of an instantaneous release), it is preferable to use an Eulerian approach in the initial time stage. In this treatment the center of the blob is advected by the currents, and its position is recorded in every time step. The concentrations at neighboring grid points as well as the change of central concentration are computed from the Joseph-Sendner (1) formula. This approach is repeated until a number of surrounding grid points have some concentrations above a predetermined minimum level, whereafter the Lagrangian treatment is continued.

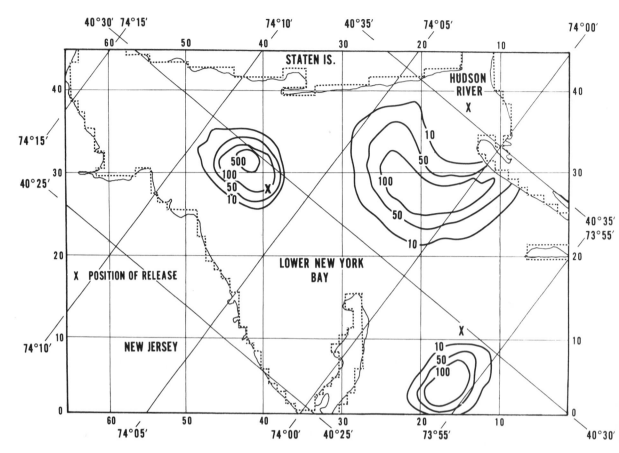

Figure 6-8. Example of computed distribution of pollutants in Lower Bay of New York 7 hours after instantaneous release of 10,000 units at the points marked with X.

Decay (uptake by biota, remineralization, etc.) was not included in the tests reported here; however, inclusion would not cause any difficulty or specially difficult programming. Furthermore, "thermal pollution" can be treated by including the heat exchange terms into the model, as done synoptically (Laevastu and Hubert, Ref. 13). Obviously, a considerable amount of testing and tuning must be done on this model to make it universally applicable.

Summary

1. The hydrodynamical numerical (HN) models provide the velocity field in small time steps and are suited for inclusion of dispersion formulas in finite difference form.

2. The HN models reproduce the changes of sea level and currents very well. As the treatment of open boundaries as well as multiple layers has been successfully solved recently, these

models have wide application.

3. The dispersion formulas (including diffusion, transport and decay) have been programmed into HN models in Lagrangian form and solved in small time steps. These dispersion formulas can also be used in Eulerian form when a blob of pollutant is small initially in relation to grid size.

4. Some preliminary test results show that this model can be used for computing pollutant distribution at any given time and for computing maximum concentration from a given continuous release (or for determining the level of release if a maximum permissible concentration is given).

References

1. Joseph, J. and H. Sendner. 1958. Über die horizontale Diffusion im Meere. *Dtsch. Hydrogr. Zeitschr.* 11(2): 49-77.

2. Okubo, A. 1970. *Oceanic mixing.* Chesapeake Bay Inst. Rept. 62, Ref. 70-1:119.

3. Schönfeld, J.C. and P. Groen. 1961. Mixing and exchange processes. Radioactive waste disposal into the sea. *IAEA Safety Series* 5:100-132.

4. Hansen, W. 1938. Amplitudenverhältnis und Phasenunterschied der harmonischen Konstanten in der Nordsee. *Ann. Hydr. Marit. Met.* 66(9): 429-443.

5. _____. 1956. Theorie zur Errechnung des Wasserstandes und der Strömungen in Randmeeren nebst Anwendungen. *Tellus* 8:287-300.

6. Sündermann, J. 1966. *Ein Vergleich zwischen der analytischen und der numerischen Berechnung winderzeugter Strömungen und Wasserstande in einem Modellmeer mit Anwendungen auf die Nordsee.* Hamburg: *Mitteil. Inst. Meeresk.,* 4:73 pp

7. Jensen, H.E., S. Weywadt and A. Jensen. 1966. *Forecasting of storm surges in the North Sea.* Part 1. NATO Subcom. Oceanogr. Res. Tech. Rept. 28 (Mimeo).

8. Bretschneider, C. 1967. *Anwendung des hydrodynamisch-numerischen Verfahrens zur Ermittlung der M_2—Mitschwingungsgezeit der Nordsee.* Hamburg: *Mitteil. Inst. Meeresk.,* f:65 pp

9. Schmitz, H.P. 1964. Modellrechnungen zu winderzengten Bewegungen in einem Meer mit Sprungschicht. *Dtsch. Hydrogr. Zeitschr.* 17(5):201-232.

10. _____. 1965. Ein Differenzengleichungssystem zur Ermittlung instationärer Bewegungen in einem Meer mit geringer Turbulenzreibung. *Dtsch. Hydrogr. Zeitschr.* 18(3):97-113.

11. Hansen, W. 1966. *The reproduction of the motion in the sea by means of hydrodynamical-numerical methods.* Hamburg: *Mitteil. Inst. Meeresk.,* 5:57 pp.

12. Hela, I. and A. Voipio. 1960. Tracer dyes as means of studying turbulent diffusion in the sea. *Ann. Acad. Scien. Fennicae,* Ser. A, VI Physica, 69:9.

13. Laevastu, T. and W.E. Hubert. 1965. Analysis and prediction of the depth of the thermocline and near-surface thermal structure. FNWC, Monterey, Tech. Note 10:90.

7

classifying and forecasting near-surface ocean thermal structure

Ocean Thermal Structure Forecasting— An Engineering Endeavor

A thorough knowledge of subsurface thermal structure is becoming more and more important for naval and fishery problems. The recent demonstration of relatively large, short-term fluctuations in near-surface temperature structure has led to synoptic analysis and forecasting of this structure on hemispheric as well as on smaller scale.

At times the temperature structure itself might not be the direct affecting factor, but it might be used to indicate other changes and conditions in the sea. Examples of indirect uses are estimating upwelling intensities, computing current boundaries, estimating the depth of turbulent mixing and a number of uses in sound propagation problems.

In many applied and scientific problems it is necessary to have a shortcut description of the relatively variable and complex thermal structure of the ocean surface layers. Such condensed presentations are possible only if a systematization of the often complex thermal structure is meaningfully accomplished. A system of classification of the ocean thermal structure from the surface to 400 meters is presented in this chapter.

This chapter also describes in condensed form a procedure for analyzing and predicting temperature structure in the sea, based mainly on consideration of physical cause-effect principles and using parameters which are either contained in routine meteorological and oceanographic observations or can be derived easily from them.

The oceanographic analysis/forecasting is thus essentially an engineering endeavor. It takes the available knowledge and data, screens them for usefulness and application and attempts to model the state and behavior of nature.

The analysis system described in this chapter is suitable for both manual and computerized analysis/forecasting and is, with minor differences, in use at FNWC. A more sophisticated numerical computerized analysis method based on hydrodynamical numerical models is in use for special tasks in Environmental Prediction Research Facility (EPRF) in Monterey.

Bathythermograph Profile (BTP) Types

An example of a generalized thermal structure is given in Figure 7-1. The main features of this structure are:

70

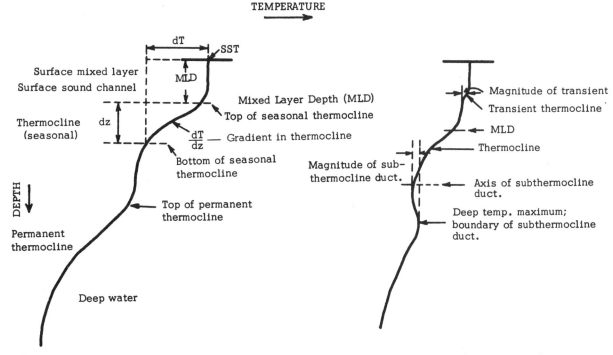

Figure 7-1. Some general definitions of near-surface thermal structure and examples of the parameterization of BT trace.

1. *Surface Mixed Layer*, which reaches from the surface to the top of the seasonal thermocline.
2. *Thermocline Region*, which is the layer of more rapid decline of temperature with depth.
3. *Intermediate and Deep Waters*, layers below the permanent thermocline.

Note that the surface mixed layer is not always turbulent down to the seasonal thermocline and that the surface layers can contain several transient thermoclines. Furthermore, in some areas a permanent thermocline might be found below the seasonal thermocline.

Another classification of the near-surface layers of the sea is also possible. We can call the surface layers above 100 meters *synoptic layers*, because the conditions in these layers change relatively rapidly, in the rhythm of changes of the driving meteorological force at the surface. The layers between 100 and 400 meters can be called *seasonal layers*, because the changes in these layers

are relatively small and follow often, but not necessarily always, a seasonal rhythm. The layers below 400 meters can be called the *deep water layers*.

The different thermal structures of the near-surface layers, "BTP types," found in the oceans are shown in Figure 7-2. Listed below are brief characteristics of the more frequently found types.

1. *Normal (or Standard) BTP Type* consists of a mixed layer above a thermocline and deep water below it. The mixed layer can contain transient thermoclines, especially during the spring and early summer. During the fall, the mixed layer is usually isothermal due to convective stirring caused by heat loss (or a higher evaporation) at the surface.
2. *The Seasonal-Permanent Type* is in the upper part similar to the normal type but contains another, permanent thermocline in the deeper layers. During the spring a seasonal thermo-

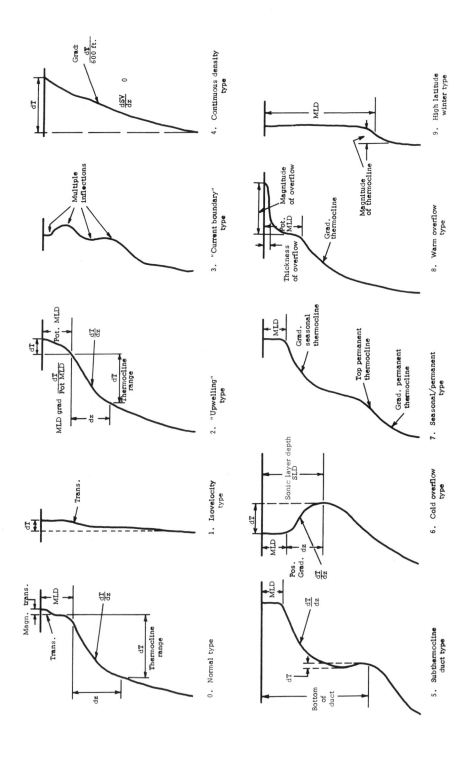

Figure 7-2. Near-surface thermal structure types (BTP types) and their criteria.

cline starts to form as a transient thermocline in the medium and high latitudes caused by warming at the surface. The depth of the transient is usually variable and determined by prevailing wave conditions. The mixed layer depth slowly deepens during the course of the summer and starts to descend more rapidly in October and November, when cooling occurs at the surface. It usually reaches the permanent thermocline between January and March. Thereafter this BTP type might be considered as a normal during the winter. Thus, in some areas there may be a change from one BTP type to another during the year. This seasonal cycle does not occur in some low-latitude areas and in areas of strong currents as well as in upwelling areas.

3. *Upwelling Type* features, in the areas of upwelling, a mixed layer depth that is usually shallow. There can be a considerable gradient between the top of the thermocline, which is at times ill defined, and the temperature at the surface. The thermocline range and gradient can be at times sharp indeed, especially at the continental slopes, where the main upwelling occurs.

4. *Continuous Density Type* occurs often in upwelling areas and also during spring and summer in high-latitude areas.

5. *Sub-Thermocline Duct Type* occurs in relatively large areas in the North Pacific and in smaller areas in the North Atlantic. Some areas in the Sargasso Sea and northeast of the Japanese Islands might also contain this thermocline type in a less pronounced form. It is characterized by a slight temperature inversion below the seasonal thermocline. The temperature inversion usually contains water of a slightly lower salinity.

6. *High-Latitude Winter Type* in medium and high latitudes in the North Atlantic is characterized by a very deep mixed layer depth (MLD) and a small thermocline in the deeper layers. In still higher latitudes this small thermocline might be absent, often resulting in the *isothermal type*.

7. *Current Boundary Type* is found at the boundaries of major currents such as the Kuro Shio-Oya Shio and the Gulf Stream-Labrador Current boundaries. It can contain a very thin mixed layer and a multiple of inflections and temperature inversions. These inversions and inflections are rapidly variable in space and time.

8. *Cold Overflow Type* is created by colder, less saline water on top of warmer, more saline water. This type occurs during the winter in the Gulf of Maine and in the St. Lawrence Gulf.

9. *Warm Overflow Type* occurs when relatively warm water is advected over cold water. It occurs near current boundaries and is usually of short duration.

It is often necessary to determine the bottom of the thermocline. This determination is rather ambiguous, but can be accomplished by graphically extending the thermocline as well as the temperature trend line from the deep waters and projecting the intersection of these lines horizontally to the temperature trace.

Mixed Layer Depth and Its Changes

The mixed layer depth (MLD) is defined as the depth of the turbulent surface layer. However, when this simple definition is applied to the complex thermal structure of the oceans many ambiguous cases are found (see Figure 7-2). There are three main classes of BT traces in which this definition is ambiguous: 1) where the temperature is nearly isothermal to great depths; 2) during the early stages of the seasonal thermocline formation in areas where a permanent thermocline occurs; and 3) where strong currents exist.

A nearly isothermal temperature from the surface to a relatively great depth occurs in some higher latitude areas and in the Mediterranean. At times, when the MLD can be located by small discontinuities in the temperature trace, the MLD is usually called the potential MLD (1).

In most ocean areas a seasonal thermocline starts to form in the near-surface layers during the spring. When this occurs in an area with a permanent thermocline, either the top of the permanent thermocline or the depth of the seasonal thermocline or any transient thermocline can be considered as the MLD. However, as soon as the seasonal thermocline has a magnitude of approximately 1°C, its depth is considered as the MLD.

In areas of strong currents, especially at sharp current boundaries, the thermal structure becomes quite complicated and the simple description of a mixed layer above a smooth thermocline is no longer accurate.

The surface layer is usually isothermal during the fall and winter seasons (when the sea is losing heat to the atmosphere) except in areas near current boundaries. During the spring and summer seasons (when the sea is gaining heat) the surface layers often are not isothermal but contain small transient thermoclines.

In some cases other definitions of the MLD might be used, depending on the purpose for which the information is applied. For climatological purposes the mixed layer depth is often defined as the depth at which the temperature is 1°C different, usually cooler, from that at the surface. For sonar purposes, the mixed layer depth is often defined as the surface sound channel above the sonic layer depth (SLD), which is the depth of maximum sound velocity. In the absence of SLD the surface sound channel is defined as the region with isovelocity profile (or a sound speed profile showing an increase with depth). In still other cases the surface layer is defined to the top of the seasonal thermocline or to any other significant temperature or salinity discontinuity near the surface.

Figure 7-3 summarizes the major causes for the mixed layer depth changes, which also cause its fluctuations. The first major cause of a well-mixed surface layer is the mixing by wave action, both by particle movement by waves as well as by turbulence caused by breaking waves, especially by winds strong enough to generate wave heights in excess of 3 meters. The depth of the wave mixing

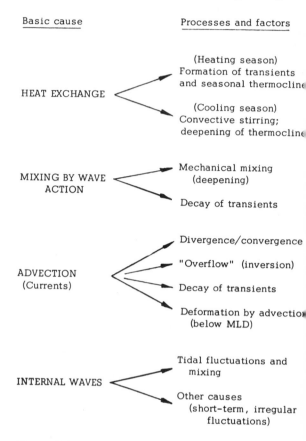

Figure 7-3. Factors determining the mixed layer depth (MLD) and its changes.

is influenced by the stability in the thermocline (i.e., the sharpness of the vertical temperature and salinity gradients).

The second major cause for a change in the MLD is heat exchange. A positive heat exchange means the gain of heat by the sea, creating the seasonal thermocline during the spring and summer. On the other hand, a loss of heat during the fall and winter causes a convective turnover, thus creating a well-mixed surface layer and deepening the MLD.

In areas of strong currents the thermal structure can be complex indeed and is greatly influenced by different currents with different directions at different depths. Short-term fluctuations

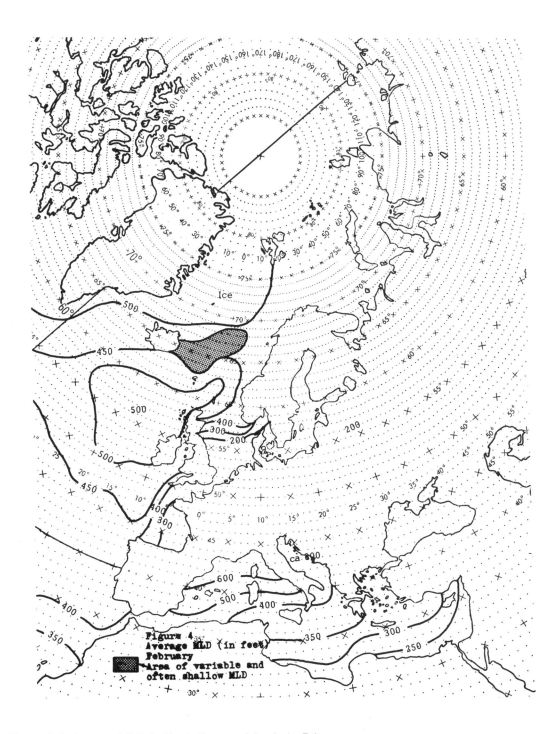

Figure 7-4. Average MLD in North Eastern Atlantic in February.

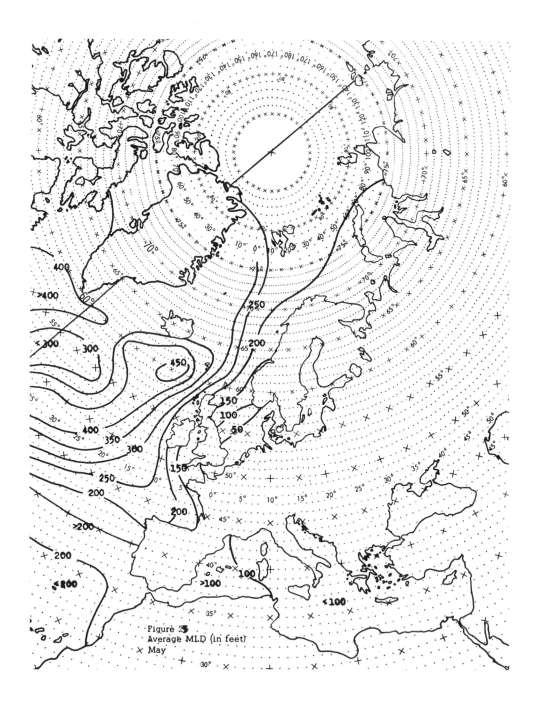

Figure 7-5. Average MLD in North Eastern Atlantic in May.

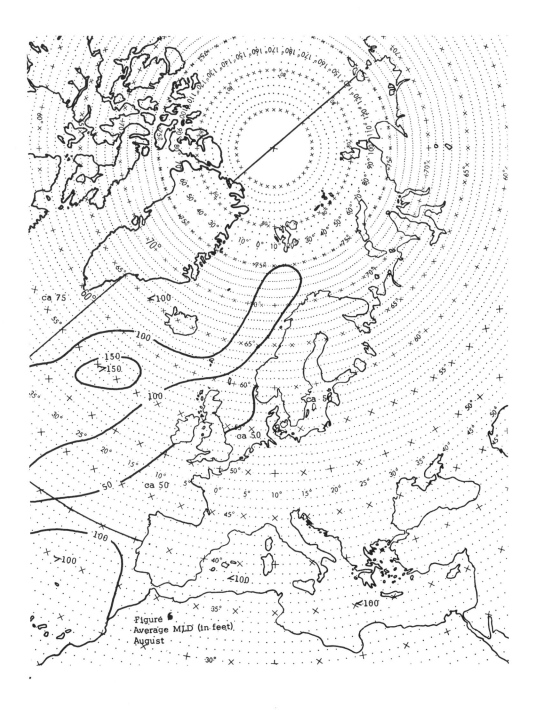

Figure 7-6. Average MLD in North Eastern Atlantic in August.

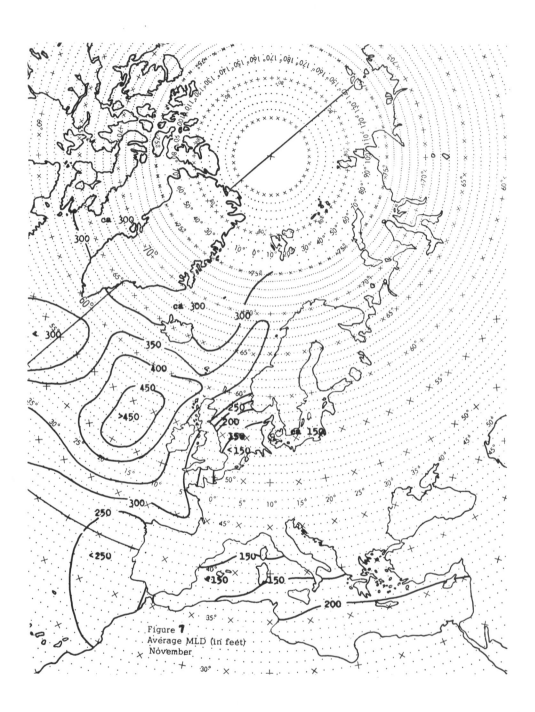

Figure 7-7. Average MLD in North Eastern Atlantic in November.

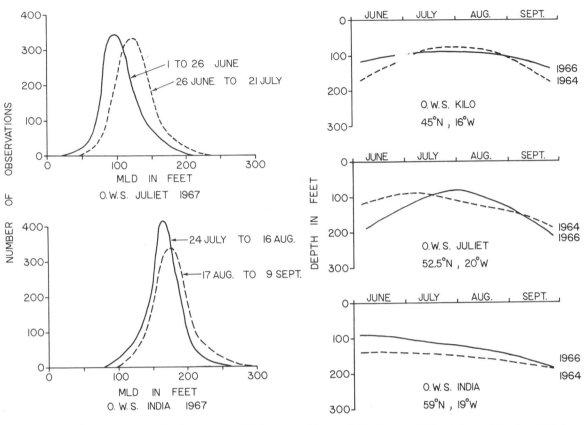

Figure 7-8. Frequency distribution of MLD at O.W.S. Juliet and India, based on BT casts every 15 minutes (after Wood).

Figure 7-9. Mean MLD at three Atlantic O.W.S. during the summer months of 1964 and 1966 (after Wood).

of thermocline depth are affected by so-called "internal tides," divergence/convergence and various other causes.

It should be noted that the whole thermal structure moves up and down with divergence and convergence with the gradients unchanged but changing the mixed layer depth (2).

Four seasonal examples of the monthly mean MLD in the eastern Atlantic and Mediterranean are given in Figures 7-4 to 7-7. This climatology is based only on available BT records and should not be considered final for a number of reasons such as the unequal distribution of BT casts in time and space, ambiguous definitions of mixed layer depth in certain areas and seasons and the large short-term and year-to-year variations of mixed layer depth.

Figures 7-4 to 7-7 illustrate that the mixed layer depth is greatly affected by factors other than the seasonal cycle. The major contributing factor is large-scale advection and its divergence and convergence (3). The mixed layer depth is usually shallower in the centers of divergent cyclonic gyrals. There is a slight "downwelling" and a slight lowering of the MLD in the convergent anticyclonic gyral centers.

The frequency distribution of the mixed layer depth is usually normal in a given location. Exceptions to this rule are the areas near the water type and current boundaries and near the continental shelves. An example of the frequency distribution of the mixed layer depth is shown in Figure 7-8. If a larger area is used for statistical purposes the

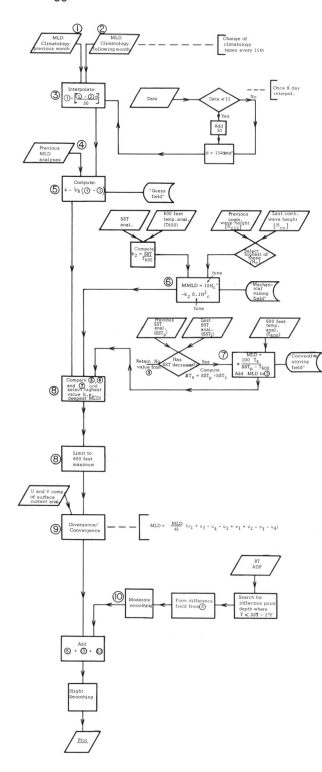

frequency distribution is "flattened" by the existing mixed layer depth gradients in the statistical area. An example of year-to-year differences in the MLD at some weather ships is shown in Figure 7-9.

Synoptic Analyses
Of Subsurface Thermal Structure

Various methods for analyzing subsurface thermal structure and their scientific background are described by Laevastu and Hubert (4). However, the numerical method presently in use is not described in sufficient detail. Therefore, a general description of these methods with their corresponding flow diagrams is given below.

The steps in a manual and computerized analysis of the subsurface thermal structure are almost identical. However, due to the labor involved, manual methods do not allow the detailed consideration of all factors as do those of the computer; consequently, the manual methods are simplified versions of the numerical ones.

There are several essential "anchor" points in a thermal structure analysis. Determination of these can be carried out in separate programs: 1) sea-surface temperature, 2) occurrence and magnitude of transients, 3) mixed layer depth and temperature, and 4) subsurface thermal structure proper.

The deep end of the thermal profile is at present the 1200-feet (about 400-meters) temperature field. This field can be accepted as a constant due to the small temperature variation during the course of a year. The sea-surface temperature (SST) analysis is described by Wolff (5) and will not be repeated in this chapter.

The MLD analysis flow diagram is shown in Figure 7-10. The forecast scheme is the same as that of the analysis with the exception that forecast, rather than analyzed meteorological and wave parameters are used; BTs are not entered directly into the forecast, except those included in the

Figure 7-10. Flow diagram of MLD analyses and forecasting.

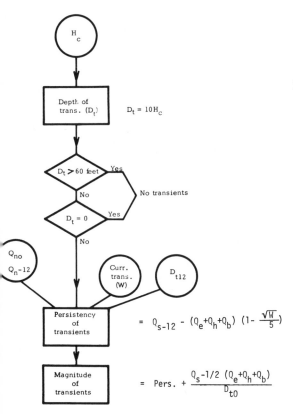

Figure 7-11. Flow diagram for the computation of transient thermoclines.

analysis, which form the first-guess field for the forecast.

The three principal steps in the MLD analysis are

1. Three different MLD fields are computed and compared to each other. The deepest MLD value at each grid point is selected as the true MLD before the BT input and before computing divergence/convergence. These three fields are a) the first-guess field, which is the previous analysis but moved a small fraction toward the interpolated climatology; b) mechanical mixing by wave action, influenced by stability; and c) convective stirring, caused by cooling at the surface.
2. In the second step of analysis, the conver-

gence/divergence is computed from the surface current analysis with equation for continuity and the MLD is moved up or down by a corresponding amount.

3. Finally, the synoptic BT messages which were received during the last 60 hours are placed in proper position, the MLD ascertained and the difference between the observed and computed values formed. The difference field so obtained is then smoothed and added to the field obtained in Step 2.

The magnitude of transients is computed in another program, using heat exchange and wave mixing as the essential parameters (see Figure 7-11). The magnitude of the transient is subtracted from the SST analysis and the result is taken as the temperature at the MLD.

The temperature analysis flow diagram (for the standard levels) is given in Figure 7-12. The principal steps in this analysis are

1. The previous analysis is decayed toward the interpolated climatology at 1/8 of the difference each day.
2. Temperatures below the MLD are modified by divergence/convergence and synoptic BT observations.
3. Temperatures at the standard levels are controlled by tolerance fields, which are computed for each grid point and each analysis period.
4. The standard levels of analysis are 0, 100, 200, 300, 400, 600, 800 and 1200 feet, MLD, D_t.
5. The thermocline gradient and the bottom of the thermocline are computed from analyzed fields (see Figure 7-13). The latter is to some extent a subject of definition.
6. The short-term fluctuations are at present determined from an empirical relation between wave height, MLD and thermocline gradient (stability).
7. The depth of the transient thermocline is computed from actual wave mixing, the magnitude of total heat exchange and mixing

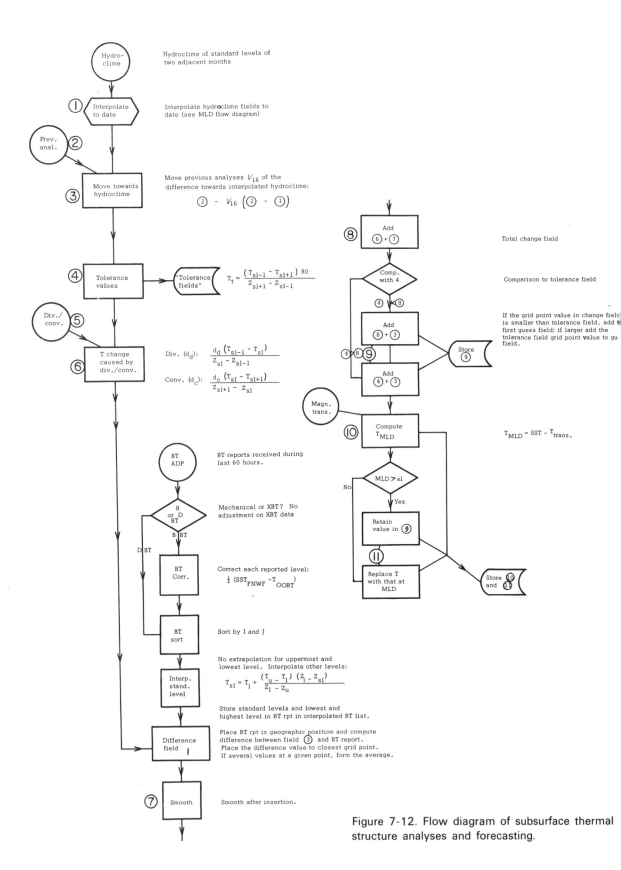

Hydroclime of standard levels of two adjacent months

① Interpolate to date — Interpolate hydroclime fields to date (see MLD flow diagram)

② Prev. anal.

③ Move towards hydroclime — Move previous analyses V_{16} of the difference towards interpolated hydroclime:

$$② - V_{16} \left(② - ① \right)$$

④ Tolerance values → "Tolerance fields"

$$T_t = \frac{(T_{sl-1} - T_{sl+1})\,80}{Z_{sl+1} - Z_{sl-1}}$$

⑤ Div./conv.

⑥ T change caused by div./conv.

Div. (d_d): $\dfrac{d_d\,(T_{sl-1} - T_{sl})}{Z_{sl} - Z_{sl-1}}$

Conv. (d_c): $\dfrac{d_c\,(T_{sl} - T_{sl+1})}{Z_{sl+1} - Z_{sl}}$

⑧ Add ⑥ + ⑦ — Total change field

Comp. with 4 — Comparison to tolerance field

④ > ⑧ — Add ⑧ + ③ → Store ⑨

If the grid point value in change field is smaller than tolerance field, add the first guess field; if larger add the tolerance field grid point value to guess field.

④ < ⑧ ⑨ — Add ④ + ③ → Store ⑨

Magn. trans.

⑩ Compute T_{MLD} — $T_{MLD} = SST - T_{trans.}$

MLD > sl — No / Yes

Retain value in ⑨

⑪ Replace T with that at MLD → Store ⑩ and ⑪

BT ADP — BT reports received during last 60 hours.

B or D BT — Mechanical or XBT? No adjustment on XBT data

B BT

D BT

BT Corr. — Correct each reported level: $\frac{1}{2}\,(SST_{FNWF} - T_{OOBT})$

BT sort — Sort by I and J

Interp. stand. level — No extrapolation for uppermost and lowest level. Interpolate other levels:

$$T_{sl} = T_1 + \frac{(T_u - T_1)\,(Z_1 - Z_{sl})}{Z_1 - Z_u}$$

Store standard levels and lowest and highest level in BT rpt in interpolated BT list.

Difference field — Place BT rpt in geographic position and compute difference between field ③ and BT report. Place the difference value to closest grid point. If several values at a given point, form the average.

⑦ Smooth — Smooth after insertion.

Figure 7-12. Flow diagram of subsurface thermal structure analyses and forecasting.

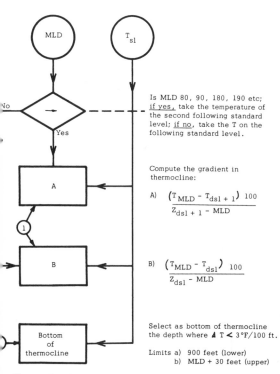

Is MLD 80, 90, 180, 190 etc; if yes, take the temperature of the second following standard level; if no, take the T on the following standard level.

Compute the gradient in thermocline:

A) $\dfrac{\left(T_{MLD} - T_{dsl+1}\right) 100}{Z_{dsl+1} - MLD}$

B) $\dfrac{\left(T_{MLD} - T_{dsl}\right) 100}{Z_{dsl} - MLD}$

Select as bottom of thermocline the depth where $\Delta T < 3°F/100$ ft.

Limits a) 900 feet (lower)
b) MLD + 30 feet (upper)

Figure 7-13. Flow diagram for the computation of thermocline gradient and bottom of the thermocline.

depth (see Figure 7-11). The decay of a transient is determined by the change of wave mixing and currents.

It is, at present, not possible to analyze the currents below the MLD; thus any advective change below the MLD is determined by synoptic BT observations alone.

For small scale (zoomed) analyses a number of additional factors can be considered. Most are local in nature and thus are not of general interest for this brief summary.

A thermal structure profile can be extracted at any oceanic position in the Northern Hemisphere from these analyses. Examples of such extractions are given in Figure 7-14.

The analysis method described above is suitable for medium-size computers. However, advanced methods and programs (multilayer hydrodynamical numerical models) have been programmed and tested by the author and are ready for use on large computers.

Symbols and Abbreviations

BT	—	bathythermograph
BTP	—	bathythermal profile
d_c	—	amount of convergence
d_d	—	amount of divergence
dsl	—	first deeper standard level after MLD (Example: MLD = 235 feet; dsl = 300 feet)
dsl+1	—	second deeper standard level after MLD (Example: MDL = 235 feet; dsl+1 = 400 feet)
D_t	—	depth of the top of transient thermocline
D_{to}	—	current depth of transient analysis
H_c	—	combined wave height
GT	—	gradient of transients
$K_1 K_2$	—	constants
1	—	next depth below a given standard level where temperature is reported in BT message
1	—	length of grid mesh
MLD	—	mixed layer depth
MLD_{an}	—	MLD analysis
MLD_{fcst}	—	MLD forecast
sl	—	standard levels in feet (0,100,200,300,400,600, 800,1000,1200,1400)
SLD	—	sonic layer depth
SST	—	sea surface temperature
SST_p	—	previous SST analysis (12 hours)
SST_1	—	current SST analysis
ST	—	standard
T	—	temperature
T_{BT}	—	temperature from BT report
T_{OOBT}	—	surface temperature from BT
T_1	—	temperature at a given level reported in BT message
T_{MLD}	—	temperature at MLD
T_s	—	surface temperature forecast
T_t	—	tolerance field

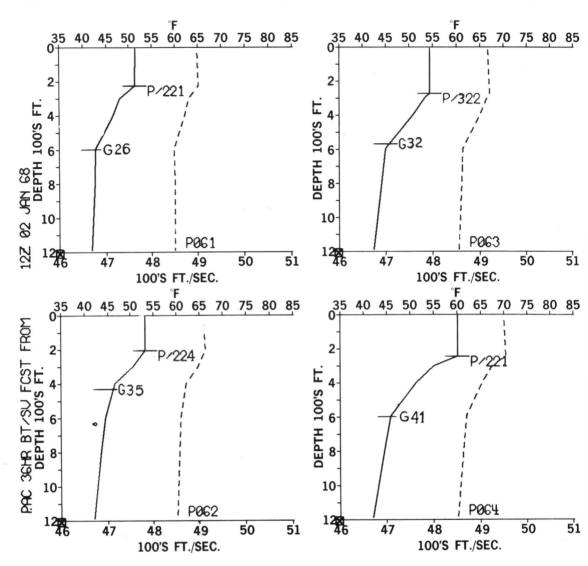

Figure 7-14. Example of BT/SV profile forecast.

T_{trans}	—	temperature magnitude of transient thermocline	u_1	—	next depth above a given standard level where the temperature is reported in BT message
T_{tt}	—	temperature at the top of the thermocline			
T_u	—	temperature of a level above T_1 reported in BT message	u	—	u component of current
			v	—	v component of current
T_{grad}	—	temperature gradient below thermocline	W	—	current transport (nautical miles per day)
			XBT	—	expendable bathythermograph
T_{600}	—	temperature at 600 feet depth	Z_{sl}	—	depth of standard level

Z_{sl+1} — depth of lower standard level

Z_c — level in a BT message below that under consideration

Z_u — level in a BT message above that under consideration

ΔT_s — surface temperature change

ΔMLD — change in mixed layer depth

References

1. Tully, J.P. and L.F. Giovando. 1963. *Seasonal temperature structure in the Eastern Pacific Ocean.* Royal Soc. Canada, Spec. Publ. 5:10-32.

2. Patullo, J.G. and J.D. Cochrane. 1951. *Monthly thermal condition charts for the North Pacific Ocean.* MS Rept. Scripps Inst. Oceanogr.

3. Robinson, M. 1966. Summary of computer-analyzed temperature data for the Pacific and Atlantic Oceans. *Proc. Third U.S. Navy Symp. Milit. Oceanogr.* 243-262.

4. Laevastu, T. and W.E. Hubert. 1965. *Analyses and prediction of the depth of the thermocline and near-surface thermal structure.* FNWC Tech. Note 10.

5. Wolff, P.M. 1967. Numerical synoptic analyses of sea surface temperature. *Int. J. Oceanol. and Limnol.* 1(4):277-290.

part 5: koji hidaka

Dr. Koji Hidaka is a member of the Ocean Research Institute, University of Tokyo, Nakano, Tokyo, Japan. He received his Bachelor of Science degree in 1926 from the Tokyo Imperial University and his Doctor of Science degree in 1933, also from the Tokyo Imperial University. Dr. Hidaka is a member of the Limnological Society of Japan, Meteorological Society of Japan, Seismological Society of Japan, American Meteorological Society, Marine Biological Association of the United Kingdom, Pacific Science Association, and an elected member of The National Geographic Society. He served as president of the Oceanographical Society of Japan from 1948 to 1967 and, since 1972, has been president of the Association of Marine Industry and Engineering. Dr. Hidaka's writings number more than 150 and have been printed in five languages. He is the recipient of numerous honors and awards including the Oceanographical Society of Japan Medal for 1970.

8

numerical computations on poincare waves

Introduction

In 1910, Henri Poincaré (1) discovered long waves propagating in a rotating straight canal of rectangular section, subject to the boundary condition that the cross-canal component of velocity vanishes *only at the walls* on both banks of the canal. There exists a difference between Poincaré waves and the Kelvin wave which was explained by Sir William Thomson (2) as the long waves derived by assuming that no cross-canal component exists *at all points* in the canal, however broad the canal may be. Under this assumption we can show that the velocity of propagation of Kelvin waves is exactly $\sqrt{(gh)}$, where h is depth of the canal supposed to be uniform, and g the acceleration of gravity. On the contrary, we can by no means have the velocity of propagation equal to $\sqrt{(gh)}$ in Poincaré waves.

Although it is almost 60 years since Poincaré published his result in *"Mécanique Céleste"* it does not appear to the author if any satisfactory manifestation of the behavior of Poincaré waves is ever given, except that Albert Defant gave some accounts on this type of waves (3).

Theory

Suppose we have the x-axis coinciding with the axis of the canal, positive in the direction of the wave propagation, y-axis *contra solem* perpendicular to it, while z-axis is taken positive vertically downwards. Let the components of velocity in x- and y-axes be u and v respectively, while ζ is the elevation of the water surface above the undisturbed average level. For small oscillations in which squares of, and products between u and v can be neglected as small, the equations of motion will be

$$\left.\begin{aligned} \frac{\partial u}{\partial t} - 2\omega v &= -g\frac{\partial \zeta}{\partial x}\,, \\ \frac{\partial v}{\partial t} + 2\omega u &= -g\frac{\partial \zeta}{\partial y}\,, \end{aligned}\right\} \tag{8-1}$$

while the equation of continuity is

$$\frac{\partial \zeta}{\partial t} + \frac{\partial}{\partial x}(hu) + \frac{\partial}{\partial y}(hv) = 0\,. \tag{8-2}$$

Let the breadth of the canal be $2a$. The boundary conditons to be satisfied on both sides of the canal are then

$$y = \pm a : v = 0\,, \tag{8-3}$$

since the walls are vertical on both sides. In these equations, t is the time, ω the angular velocity of the canal and g the acceleration of gravity. Assume a progressive wave in the positive direction of x and put

$$u = u_0(y) \cos\left(\sigma t - \frac{\pi x}{2\lambda}\right), \qquad (8\text{-}4)$$

$$v = v_0(y) \sin\left(\sigma t - \frac{\pi x}{2\lambda}\right), \qquad (8\text{-}5)$$

$$\zeta = \zeta_0(y) \cos\left(\sigma t - \frac{\pi x}{2\lambda}\right), \qquad (8\text{-}6)$$

where σ is the frequency of wave, λ the wavelength, and $u_0(y)$, $v_0(y)$ and $\zeta_0(y)$ are functions of y only.

Substitution of (8-3), (8-4) and (8-5) in (8-1) and (8-2) gives

$$\left. \begin{aligned} u_0(y) &= \frac{g/2\omega a}{f^2-1}\left\{\left(\frac{\pi a}{2\lambda}\right)f \cdot \zeta_0 + \frac{d\zeta_0}{d\eta}\right\}, \\ v_0(y) &= -\frac{g/2\omega a}{f^2-1}\left\{\left(\frac{\pi a}{2\lambda}\right)\zeta_0 + f\frac{d\zeta_0}{d\eta}\right\}, \end{aligned} \right\} \;\; (8\text{-}7)$$

and ζ_0 is the solution of

$$\frac{d^2\zeta_0}{d\eta^2} + \left\{\beta(f^2-1) - \left(\frac{\pi a}{2\lambda}\right)^2\right\}\zeta_0 = 0, \qquad (8\text{-}8)$$

where

$$\eta = \frac{y}{a}$$

and

$$f = \frac{\sigma}{2\omega},$$

$$\beta = \frac{4\omega^2 a^2}{gh},$$

subject to the boundary conditions that the cross-canal velocity component vanishes at the walls on both banks, or

$$\eta = \pm 1: \; v = 0$$

or, from (8-7)

$$\eta = \pm 1: \left(\frac{\pi a}{2\lambda}\right)\zeta_0 + f\frac{d\zeta_0}{d\eta} = 0. \qquad (8\text{-}9)$$

General solution of the equation (8-8) is

$$\begin{aligned} \zeta_0 &= A \cos\left\{\sqrt{\beta(f^2-1) - \left(\frac{\pi a}{2\lambda}\right)^2} \cdot \eta\right\} \\ &\quad + B \sin\left\{\sqrt{\beta(f^2-1) - \left(\frac{\pi a}{2\lambda}\right)^2} \cdot \eta\right\} \end{aligned} \qquad (8\text{-}10)$$

where A and B are constants. Now that it follows (see bottom of page) the conditions in (8-9) gives

$$\begin{aligned} &\left(\frac{\pi a}{2\lambda}\right)\Bigg[A \cos\left\{\sqrt{\beta(f^2-1) - \left(\frac{\pi a}{2\lambda}\right)^2} \cdot \eta\right\} \\ &\quad + B \sin\left\{\sqrt{\beta(f^2-1) - \left(\frac{\pi a}{2\lambda}\right)^2} \cdot \eta\right\} \Bigg] \\ &\quad + f\sqrt{\beta(f^2-1) - \left(\frac{\pi a}{2\lambda}\right)^2}\Bigg[\\ &\qquad - A \sin\left\{\sqrt{\beta(f^2-1) - \left(\frac{\pi a}{2\lambda}\right)^2} \cdot \eta\right\} \\ &\qquad + B \cos\left\{\sqrt{\beta(f^2-1) - \left(\frac{\pi a}{2\lambda}\right)^2} \cdot \eta\right\} \Bigg]\Bigg]\Bigg|_{\eta = \pm 1} \\ &= 0, \end{aligned}$$

$$\begin{aligned} \frac{d\zeta_0}{d\eta} &= -A\sqrt{\beta(f^2-1) - \left(\frac{\pi a}{2\lambda}\right)^2} \sin\left\{\sqrt{\beta(f^2-1) - \left(\frac{\pi a}{2\lambda}\right)^2} \cdot \eta\right\} \\ &\quad + B\sqrt{\beta(f^2-1) - \left(\frac{\pi a}{2\lambda}\right)^2} \cos\left\{\sqrt{\beta(f^2-1) - \left(\frac{\pi a}{2\lambda}\right)^2} \cdot \eta\right\} \end{aligned}$$

or (see Box A)

and (see Box B)

Eliminating A and B between these two equations, we have (see Box C)

A

$$\left\{\left(\frac{\pi a}{2\lambda}\right)\cos\sqrt{\beta(f^2-1)-\left(\frac{\pi a}{2\lambda}\right)^2}\ -f\sqrt{\beta(f^2-1)-\left(\frac{\pi a}{2\lambda}\right)^2}\sin\sqrt{\beta(f^2-1)-\left(\frac{\pi a}{2\lambda}\right)^2}\right\}A$$

$$+\left\{\left(\frac{\pi a}{2\lambda}\right)\sin\sqrt{\beta(f^2-1)-\left(\frac{\pi a}{2\lambda}\right)^2}\right.$$

$$\left.+f\sqrt{\beta(f^2-1)-\left(\frac{\pi a}{2\lambda}\right)^2}\cos\sqrt{\beta(f^2-1)-\left(\frac{\pi a}{2\lambda}\right)^2}\right\}B=0,$$

(8-11)

B

$$\left\{\left(\frac{\pi a}{2\lambda}\right)\cos\sqrt{\beta(f^2-1)-\left(\frac{\pi a}{2\lambda}\right)^2}\ +f\sqrt{\beta(f^2-1)-\left(\frac{\pi a}{2\lambda}\right)^2}\sin\sqrt{\beta(f^2-1)-\left(\frac{\pi a}{2\lambda}\right)^2}\right\}A$$

$$-\left\{\left(\frac{\pi a}{2\lambda}\right)\sin\sqrt{\beta(f^2-1)-\left(\frac{\pi a}{2\lambda}\right)^2}\right.$$

$$\left.-f\sqrt{\beta(f^2-1)-\left(\frac{\pi a}{2\lambda}\right)^2}\cos\sqrt{\beta(f^2-1)-\left(\frac{\pi a}{2\lambda}\right)^2}\right\}B=0.$$

C

$$\left\{\left(\frac{\pi a}{2\lambda}\right)\cos\sqrt{\beta(f^2-1)-\left(\frac{\pi a}{2\lambda}\right)^2}\ -f\sqrt{\beta(f^2-1)-\left(\frac{\pi a}{2\lambda}\right)^2}\sin\sqrt{\beta(f^2-1)-\left(\frac{\pi a}{2\lambda}\right)^2}\right\}$$

$$\times\left\{\left(\frac{\pi a}{2\lambda}\right)\sin\sqrt{\beta(f^2-1)-\left(\frac{\pi a}{2\lambda}\right)^2}\right.$$

$$\left.+f\sqrt{\beta(f^2-1)-\left(\frac{\pi a}{2\lambda}\right)^2}\cos\sqrt{\beta(f^2-1)-\left(\frac{\pi a}{2\lambda}\right)^2}\right\}$$

$$-\left\{\left(\frac{\pi a}{2\lambda}\right)\sin\sqrt{\beta(f^2-1)-\left(\frac{\pi a}{2\lambda}\right)^2}\right.$$

$$\left.+f\sqrt{\beta(f^2-1)-\left(\frac{\pi a}{2\lambda}\right)^2}\cos\sqrt{\beta(f^2-1)-\left(\frac{\pi a}{2\lambda}\right)^2}\right\}$$

$$\times\left\{\left(\frac{\pi a}{2\lambda}\right)\cos\sqrt{\beta(f^2-1)-\left(\frac{\pi a}{2\lambda}\right)^2}\right.$$

$$\left.+f\sqrt{\beta(f^2-1)-\left(\frac{\pi a}{2\lambda}\right)^2}\sin\sqrt{\beta(f^2-1)-\left(\frac{\pi a}{2\lambda}\right)^2}\right\}=0,$$

Table 8-1. Computation of Frequencies f					
$\left(\dfrac{\pi a}{2\lambda}\right)$	β	$f=\dfrac{\sigma}{2\omega}$			
		$n=1$	$n=2$	$n=3$	$n=4$
1	100	1.0171893	1.0529463	1.1099847	1.1852359
	10	1.1604913	1.4446316	1.8222681	2.2467403
	1	2.1136227	3.4452292	4.9200213	6.4403740
	1/10	5.9727725	10.4735879	15.2665025	20.1440854
	1/100	18.6477910	32.9842453	48.1836173	63.6305096
$\dfrac{1}{10}$	100	1.0123112	1.0482347	1.1055162	1.1810522
	10	1.1170229	1.4099505	1.7948986	2.2245992
	1	1.8647791	3.2984245	4.8183617	6.3630510
	1/10	5.0768111	9.9897970	14.9387449	19.8968383
	1/100	15.7714967	31.4477414	47.1451056	62.8477665
$\dfrac{1}{100}$	100	1.0122623	1.0481875	1.1054714	1.1810102
	10	1.1165796	1.4095994	1.7946228	2.2243767
	1	1.8621227	3.2969235	4.8173343	6.3622730
	1/10	5.0670515	9.9848407	14.9354310	19.8943503
	1/100	15.7400797	31.4319971	47.1346049	62.8398898

or (see the box at the bottom of the page).

From this equation, we have

$$f^2\left\{\beta(f^2-1)-\left(\frac{\pi a}{2\lambda}\right)^2\right\}+\left(\frac{\pi a}{2\lambda}\right)^2=0 ,$$

or

$$\beta f^4-\left\{\left(\frac{\pi a}{2\lambda}\right)^2+\beta\right\}f^2+\left(\frac{\pi a}{2\lambda}\right)^2=0 , \quad (8\text{-}13)$$

and

$$\sin\left\{2\sqrt{\beta(f^2-1)-\left(\frac{\pi a}{2\lambda}\right)^2}\right\}=0 . \qquad (8\text{-}14)$$

Equation (8-13) gives

$$f^2=\frac{1}{\beta}\left(\frac{\pi a}{2\lambda}\right)^2 \quad \text{or} \quad f=\pm\frac{1}{\sqrt{\beta}}\left(\frac{\pi a}{2\lambda}\right) ,$$
$$(8\text{-}15)$$

and

$$f^2=1 \quad \text{or} \quad f=\pm1 . \qquad (8\text{-}16)$$

Further, equation (8-14) gives

$$\beta(f^2-1)-\left(\frac{\pi a}{2\lambda}\right)^2=\left(\frac{n\pi}{2}\right)^2 , \quad (n=1,2,3,\cdots)$$

from which an infinite number of frequencies is given by

$$f=\pm\left\{1+\frac{\left(\frac{\pi a}{2\lambda}\right)^2+\left(\frac{n\pi}{2}\right)^2}{\beta}\right\}^{\frac{1}{2}} . \qquad (8\text{-}17)$$

No wave exists for $n = 0$.

The values of f were computed after the expression (8-17) for 15 combinations between $(\pi a/2\lambda) = 1$, 1/10 and 1/100, and $\beta = 100$, 10, 1,

$$\frac{1}{2}\left[\left(\frac{\pi a}{2\lambda}\right)^2-f^2\left\{\beta(f^2-1)-\left(\frac{\pi a}{2\lambda}\right)^2\right\}\right] \cdot \sin\left\{2\sqrt{\beta(f^2-1)-\left(\frac{\pi a}{2\lambda}\right)^2}\right\}=0 . \qquad (8\text{-}12)$$

1/10 and 1/100. Fundamental wave, the second, third and fourth harmonics ($n = 1, 2, 3$ and 4) were worked out as compiled in Table 8-1.

As has been anticipated from the expression (8-17), the values of f are approximately proportional to $\beta^{-\frac{1}{2}}$, but do not depend much upon $(\pi a/2\lambda)$ as long as $(\pi a/2\lambda)$ remains less than 1.

Poincaré Wave

From the first equation of (8-11), we have for the frequencies given by (8-17),

$$B = -\frac{\left(\frac{\pi a}{2\lambda}\right)\cos\frac{n\pi}{2} - \frac{n\pi}{2}f\sin\frac{n\pi}{2}}{\left(\frac{\pi a}{2\lambda}\right)\sin\frac{n\pi}{2} + \frac{n\pi}{2}f\cos\frac{n\pi}{2}} \cdot A,$$

$$(n = 1, 2, 3, \cdots)$$

and therefore (see Box A)

Substitution of (8-18) in (8-10) gives (see Box B)

where A is an arbitrary constant. Substitution of (8-19) in (8-7) gives (see Box C)

These expressions are further simplified as (see the box at the bottom of page 94) The velocity of propagation C is

$$C = \left\{1 + \frac{\beta + \left(\frac{n\pi}{2}\right)^2}{\left(\frac{\pi a}{2\lambda}\right)^2}\right\}^{\frac{1}{2}} \sqrt{gh}.$$

$$(n = 1, 2, 3, \cdots)$$

A

$$\zeta = A \cdot \frac{\left(\frac{\pi a}{2\lambda}\right)\sin\left\{\frac{n\pi}{2}\left(1 - \frac{y}{a}\right)\right\} + \frac{n\pi}{2}f\cos\left\{\frac{n\pi}{2}\left(1 - \frac{y}{a}\right)\right\}}{\left(\frac{\pi a}{2\lambda}\right)\sin\frac{n\pi}{2} + \frac{n\pi}{2}f\cos\frac{n\pi}{2}} \cdot \qquad (8\text{-}18)$$

$$\cos\left(\sigma t - \frac{\pi x}{2\lambda}\right), \qquad (n = 1, 2, 3, \cdots)$$

B

$$\zeta_0 = \frac{\left(\frac{\pi a}{2\lambda}\right)\sin\left\{\frac{n\pi}{2}\left(1 - \frac{y}{a}\right)\right\} + \frac{n\pi}{2}f\cos\left\{\frac{n\pi}{2}\left(1 - \frac{y}{a}\right)\right\}}{\left(\frac{\pi a}{2\lambda}\right)\sin\frac{n\pi}{2} + \frac{n\pi}{2}f\cos\frac{n\pi}{2}} \cdot A \qquad (8\text{-}19)$$

$$(n = 1, 2, 3, \cdots).$$

C

$$u = \frac{gA}{a} \cdot \frac{(1/2\omega)}{f^2 - 1}$$

$$\frac{\left\{\left(\frac{\pi a}{2\lambda}\right)^2 + \left(\frac{n\pi}{2}\right)^2\right\}f\sin\left\{\frac{n\pi}{2}\left(1 - \frac{y}{a}\right)\right\} + \left(\frac{\pi a}{2\lambda}\right)\cdot\frac{n\pi}{2}(f^2 - 1)\cos\left\{\frac{n\pi}{2}\left(1 - \frac{y}{a}\right)\right\}}{\left(\frac{\pi a}{2\lambda}\right)\sin\frac{n\pi}{2} + \frac{n\pi}{2}f\cos\frac{n\pi}{2}}$$

$$\cdot \left(2\omega f t - \frac{\pi x}{2\lambda}\right),$$

$$v = \frac{gA}{a} \cdot \frac{(1/2\omega)}{f^2 - 1} \cdot \frac{\left(\frac{\pi a}{2\lambda}\right)^2 + \left(\frac{n\pi}{2}\right)^2 f^2}{\left(\frac{\pi a}{2\lambda}\right)\sin\frac{n\pi}{2} + \frac{n\pi}{2}f\cos\frac{n\pi}{2}}$$

$$\cdot \sin\left\{\frac{n\pi}{2}\left(1 - \frac{y}{a}\right)\right\}\sin\left(2\omega f t - \frac{\pi x}{2\lambda}\right). \qquad (n = 1, 2, 3, \cdots)$$

This expression suggests that the velocity of waves approaches $\sqrt{(gh)}$ when $(\pi a/2\lambda)$ is indefinitely large, or the canal is idefinitely wide ($a \to \infty$), provided n is not large. Wave speed is approximately

$$C=\left(gh+\frac{16}{\pi^2}\cdot\omega^2\lambda^2\right)^{\frac{1}{2}}+\left(\frac{n\lambda}{a}\right)^2. \qquad (8\text{-}20)$$

Kelvin Wave

Now adopt the frequency given by (8-15), or

$$f=\frac{1}{\sqrt{\beta}}\left(\frac{\pi a}{2\lambda}\right);$$

then it follows

$$\beta(f^2-1)-\left(\frac{\pi a}{2\lambda}\right)^2=-\beta\,,$$

and (8-8) becomes

$$\frac{d^2\zeta_0}{d\eta^2}-\beta\zeta_0=0\,,$$

giving

$$\zeta_0(\eta)=A\cosh\left(\sqrt{\beta}\cdot\eta\right)+B\sinh\left(\sqrt{\beta}\cdot\eta\right).$$

Condition (8-9) gives

$$B=-A$$

and

$$\zeta_0=A\exp\left(-\frac{2\omega a}{\sqrt{gh}}\cdot\frac{y}{a}\right). \qquad (8\text{-}21)$$

The velocity of propagation is from (8-20)

$$C=\frac{2\omega f}{(\pi/2\lambda)}=\frac{2\omega a}{(\pi a/2\lambda)}\cdot\frac{(\pi a/2\lambda)}{\sqrt{\beta}}\,,$$

or

$$C=\sqrt{gh} \qquad (8\text{-}22)$$

exactly.

Now it can be shown from (8-7) that cross-canal velocity vanishes at all points in the canal, or

$$\begin{aligned}
v_0&=\frac{g/2\omega a}{f^2-1}\left\{-\left(\frac{\pi a}{2\lambda}\right)\zeta_0-f\frac{d\zeta_0}{d\eta}\right\}\\
&=\frac{Ag}{a}\cdot\frac{1/2\omega}{f^2-1}\cdot\left\{\left(\frac{\pi a}{2\lambda}\right)\cdot\exp\left(-\frac{2\omega a}{\sqrt{gh}}\cdot\frac{y}{a}\right)\right.\\
&\quad\left.-\frac{1}{\sqrt{\beta}}\left(\frac{\pi a}{2\lambda}\right)^2\left(-\frac{2\omega a}{\sqrt{gh}}\right)\exp\left(-\frac{2\omega a}{\sqrt{gh}}\cdot\frac{y}{a}\right)\right\}\\
&=-\frac{Ag}{a}\cdot\frac{1/2\omega}{f^2-1}\left\{\left(\frac{\pi a}{2\lambda}\right)\exp\left(-\frac{2\omega a}{\sqrt{gh}}\cdot\frac{y}{a}\right)\right.\\
&\quad\left.-\left(\frac{\pi a}{2\lambda}\right)\exp\left(-\frac{2\omega a}{\sqrt{gh}}\cdot\frac{y}{a}\right)\right\},
\end{aligned}$$

$$u=\frac{gA}{a}\cdot\frac{1}{2\omega}\cdot\frac{\beta f\sin\left\{\frac{n\pi}{2}\left(1-\frac{y}{a}\right)\right\}+\left(\frac{\pi a}{2\lambda}\right)\cdot\frac{n\pi}{2}\cos\left\{\frac{n\pi}{2}\left(1-\frac{y}{a}\right)\right\}}{\left(\frac{\pi a}{2\lambda}\right)\sin\frac{n\pi}{2}+\frac{n\pi}{2}f\cos\frac{n\pi}{2}}$$

$$\cdot\cos\left(2\omega ft-\frac{\pi x}{2\lambda}\right),$$

$$v=-\frac{gA}{a}\cdot\frac{1}{2\omega}\cdot\frac{\left\{\beta+\left(\frac{n\pi}{2}\right)^2\right\}\sin\left\{\frac{n\pi}{2}\left(1-\frac{y}{a}\right)\right\}}{\left(\frac{\pi a}{2\lambda}\right)\sin\frac{n\pi}{2}+\frac{n\pi}{2}f\cos\frac{n\pi}{2}}$$

$$\cdot\sin\left(2\omega ft-\frac{\pi x}{2\lambda}\right). \qquad (n=1,2,3,\cdots)$$

or

$$v=0. \qquad (8\text{-}23)$$

The expressions (8-21), (8-22) and (8-23) show *this is nothing but a Kelvin wave*.

The case $f^2 = 1$ suggested by (8-16) does not give wave motions because the factor f^2-1 in the denominators of (8-6) makes both $u_0(y)$ and $v_0(y)$ infinite.

Case $f^2 < 1$

If $f^2 < 1$, the coefficient $\beta(f^2-1)-(\pi a/2\lambda)^2$ is a negative quantity, so that equation (8-8) can be transformed as

$$\frac{d^2\zeta_0}{d\eta^2} - \left\{\left(\frac{\pi a}{2\lambda}\right)^2 + \beta(1-f^2)\right\}\zeta_0 = 0,$$

where $\left\{(\pi a/2\lambda)^2 - \beta(1-f^2)\right\} > 0$. In this case, general solution of the equation is

$$\zeta_0 = A\,\cosh\left[\left\{\left(\frac{\pi a}{2\lambda}\right)^2 + \beta(1-f^2)\right\}^{\frac{1}{2}}\cdot\eta\right]$$

$$+ B\,\sinh\left[\left\{\left(\frac{\pi a}{2\lambda}\right)^2 + \beta(1-f^2)\right\}^{\frac{1}{2}}\cdot\eta\right].$$

By the same process as represented by (8-11) and (8-12) we have

$$f^2 = 1 + \frac{\left(\frac{\pi a}{2\lambda}\right)^2}{\beta},\quad 1\quad\text{and}\quad \frac{1}{\beta}\left(\frac{\pi a}{2\lambda}\right)^2.$$

Out of these three sets of roots,

$$f^2 = 1 + \frac{\left(\frac{\pi a}{2\lambda}\right)^2}{\beta}\quad\text{and}\quad 1$$

do not satisfy the condition $f^2 < 1$. Only root fitting this condition is

$$f^2 = \frac{1}{\beta}\left(\frac{\pi a}{2\lambda}\right)^2.$$

However, this root is the same as that given by (8-20). The corresponding motion is nothing but the Kelvin wave.

Numerical Computation on Poincaré Waves

In order to visualize Poincaré waves of regular type specified by the frequencies:

$$f = \pm\left\{1 + \frac{\left(\frac{\pi a}{2\lambda}\right)^2 + \left(\frac{n\pi}{2}\right)^2}{\beta}\right\}^{\frac{1}{2}}\quad (n=1,2,3,\cdots)$$

$$\zeta(y) = A\cdot\frac{\left(\frac{\pi a}{2\lambda}\right)\sin\left\{\frac{n\pi}{2}\left(1-\frac{y}{a}\right)\right\} + \frac{n\pi}{2}f\cdot\cos\left\{\frac{n\pi}{2}\left(1-\frac{y}{a}\right)\right\}}{\frac{1}{\sqrt{2}}\left\{\left(\frac{\pi a}{2\lambda}\right)^2 + \left(\frac{n\pi}{2}\right)^2 f^2\right\}^{\frac{1}{2}}}\cdot\cos\left(2\omega ft - \frac{\pi x}{2\lambda}\right), \tag{8-24}$$

$$u(y) = -\frac{g}{2\omega a}A\cdot\frac{\beta f\sin\left\{\frac{n\pi}{2}\left(1-\frac{y}{a}\right)\right\} + \left(\frac{\pi a}{2\lambda}\right)\cdot\frac{n\pi}{2}\cos\left\{\frac{n\pi}{2}\left(1-\frac{y}{a}\right)\right\}}{\frac{1}{\sqrt{2}}\left\{\left(\frac{\pi a}{2\lambda}\right)^2 + \left(\frac{n\pi}{2}\right)^2 f^2\right\}^{\frac{1}{2}}}\;.$$

$$\cos\left(2\omega ft - \frac{\pi x}{2\lambda}\right), \tag{8-25}$$

$$v(y) = -\frac{g}{2\omega a}A\cdot\frac{\left\{\beta + \left(\frac{n\pi}{2}\right)^2\right\}\sin\left\{\frac{n\pi}{2}\left(1-\frac{y}{a}\right)\right\}}{\frac{1}{\sqrt{2}}\left\{\left(\frac{\pi a}{2\lambda}\right)^2 + \left(\frac{n\pi}{2}\right)^2 f^2\right\}^{\frac{1}{2}}}\sin\left(2\omega ft - \frac{\pi x}{2\lambda}\right), \tag{8-26}$$

numerical values of the frequencies f and wave velocity C together with the cross-canal distribution of $\zeta_0(y)$, $u_0(y)$ and $v_0(y)$ for $n = 1, 2, 3$ and 4 were calculated in terms of an arbitrary constant A, which is equal to the amplitude of free waves along the median line ($y = 0$) of the canal. No wave motion can be expected for $n = 0$.

The functions $\zeta_0(y)$ may be most preferably normalized by dividing it by

$$\left[\frac{1}{2a} \int_{-a}^{a} \{\zeta_0(y)\}^2 dy \right]^{\frac{1}{2}}.$$

In this case, $u_0(y)$ and $v_0(y)$ should be also divided by this quantity. After normalization, we have (see the box at the bottom of page 97).

where A is an arbitrary constant common to the three expressions (8-24), (8-25) and (8-26), being equal to the amplitude of free waves along the median line ($y = 0$). In Table 8-2, the coefficients of sines and cosines in the functions are compiled.

The equations shown at the bottom of the page and

$$v_0(y) = -\frac{g}{2\omega a} A \cdot \frac{\left\{ \beta + \left(\frac{n\pi}{2} \right)^2 \right\} \sin\left\{ \frac{n\pi}{2} \left(1 - \frac{y}{a} \right) \right\}}{\frac{1}{\sqrt{2}} \left\{ \left(\frac{\pi a}{2\lambda} \right)^2 + \left(\frac{n\pi}{2} \right)^2 f^2 \right\}^{\frac{1}{2}}}$$

where the angular velocity of the canal is

$$\omega = 2 \times 0.00007292 \sin \varphi$$

and φ is the average geographic latitude of the canal.

Figures 8-1, 8-2 and 8-3 give the cross-canal distributions of wave amplitude $\zeta_0(y)$ for $(\pi a/2\lambda) = 1, 1/10$ and $1/100$ ($n = 1$ only). In each of these diagrams, curves for $\beta = 100, 10, 1, 1/10$ and $1/100$ are shown. It is a general tendency that these sine curves are displaced to the right-hand side as both $(\pi a/2\lambda)$ and β decrease, although all of them converge to a sine curve passing through $y = 0$ or the median line and cannot be traced in the space to the right of it.

Acknowledgments

The author is indebted to Miss Mariko Karasawa who worked out most of these tedious computations.

$$\zeta_0(y) = A \cdot \frac{\left(\frac{\pi a}{2\lambda} \right) \sin\left\{ \frac{n\pi}{2} \left(1 - \frac{y}{a} \right) \right\} + \frac{n\pi}{2} f \cdot \cos\left\{ \frac{n\pi}{2} \left(1 - \frac{y}{a} \right) \right\}}{\frac{1}{\sqrt{2}} \left\{ \left(\frac{\pi a}{2\lambda} \right)^2 + \left(\frac{n\pi}{2} \right)^2 f^2 \right\}^{\frac{1}{2}}},$$

$$u_0(y) = -\frac{g}{2\omega a} A \cdot \frac{\beta f \sin\left\{ \frac{n\pi}{2} \left(1 - \frac{y}{a} \right) \right\} + \left(\frac{\pi a}{2\lambda} \right) \cdot \frac{n\pi}{2} \cos\left\{ \frac{n\pi}{2} \left(1 - \frac{y}{a} \right) \right\}}{\frac{1}{\sqrt{2}} \left\{ \left(\frac{\pi a}{2\lambda} \right)^2 + \left(\frac{n\pi}{2} \right)^2 f^2 \right\}^{\frac{1}{2}}}$$

			Table 8-2. Coefficients of $\sin\dfrac{n\pi}{2}\left(1-\dfrac{y}{a}\right)$ and $\cos\dfrac{n\pi}{2}\left(1-\dfrac{y}{a}\right)$ in the Functions $\zeta_o(y),\, u_o(y)$ and $v_o(y)$	
$\left(\dfrac{\pi a}{2\lambda}\right)$	β	n	$\dfrac{\left(\dfrac{\pi a}{2\lambda}\right)}{\dfrac{1}{\sqrt{2}}\left\{\left(\dfrac{\pi a}{2\lambda}\right)^2+\left(\dfrac{n\pi}{2}\right)^2 f^2\right\}^{\frac{1}{2}}}$ $\text{(unit}:A)$	$\dfrac{\dfrac{n\pi}{2}\cdot f}{\dfrac{1}{\sqrt{2}}\left\{\left(\dfrac{\pi a}{2\lambda}\right)^2+\left(\dfrac{n\pi}{2}\right)^2 f^2\right\}^{\frac{1}{2}}}$ $\text{(unit}:A)$
1	100	1	0.7502743	1.1987862
		2	0.4092318	1.3537094
		3	0.2655595	1.3890564
		4	0.1882130	1.4016333
	10	1	0.6801818	1.2399003
		2	0.3043081	1.3810854
		3	0.1635824	1.4047209
		4	0.0999305	1.4106786
	1	1	0.4078598	1.3541234
		2	0.1301072	1.4082159
		3	0.0609401	1.4129000
		4	0.0349375	1.4137819
	1/10	1	0.1498877	1.4062481
		2	0.0429605	1.4135609
		3	0.0196559	1.4140770
		4	0.0111731	1.4141694
	1/100	1	0.0482519	1.4133902
		2	0.0136470	1.4141477
		3	0.0062283	1.4141998
		4	0.0035373	1.4142091
1/10	100	1	0.0887614	1.4114252
		2	0.0429246	1.4135619
		3	0.0271412	1.4139531
		4	0.0190558	1.4140852
	10	1	0.0804691	1.4119224
		2	0.0319191	1.4138533
		3	0.0167187	1.4141147
		4	0.0101175	1.4141773
	1	1	0.0482519	1.4133902
		2	0.0136470	1.4141477
		3	0.0062283	1.4141998
		4	0.0035373	1.4142092
	1/10	1	0.0177325	1.4141024
		2	0.0045062	1.4142064
		3	0.0020089	1.4142121
		4	0.0011312	1.4142131
	1/100	1	0.0057085	1.4142015
		2	0.0014314	1.4142128
		3	0.0006366	1.4142134
		4	0.0003581	1.4142135

Table 8-2 continued

Table 8-2 continued

$\dfrac{\beta f}{\dfrac{1}{\sqrt{2}}\left\{\left(\dfrac{\pi a}{2\lambda}\right)^2+\left(\dfrac{n\pi}{2}\right)^2 f^2\right\}^{\frac{1}{2}}}$	$\dfrac{\left(\dfrac{\pi a}{2\lambda}\right)\cdot\dfrac{n\pi}{2}}{\dfrac{1}{\sqrt{2}}\left\{\left(\dfrac{\pi a}{2\lambda}\right)^2+\left(\dfrac{n\pi}{2}\right)^2 f^2\right\}^{\frac{1}{2}}}$	$\dfrac{\beta+\left(\dfrac{n\pi}{2}\right)^2}{\dfrac{1}{\sqrt{2}}\left\{\left(\dfrac{\pi a}{2\lambda}\right)^2+\left(\dfrac{n\pi}{2}\right)^2 f^2\right\}^{\frac{1}{2}}}$
unit: $(g/2\omega a)\cdot A$	unit: $(g/2\omega a)\cdot A$	unit: $(g/2\omega a)\cdot A$
76.317103	1.178528	76.878662
43.089909	1.285640	44.962134
29.476694	1.251419	32.453121
22.307686	1.182577	26.251658
7.893450	1.068427	8.480099
4.396131	0.956012	6.046482
2.980910	0.770864	5.268435
2.245164	0.627878	4.944372
0.862062	0.640665	1.414214
0.448249	0.408744	1.414214
0.299827	0.287174	1.414214
0.225010	0.219519	1.414214
0.089525	0.235443	0.384822
0.044995	0.134969	0.428299
0.030008	0.092626	0.438456
0.022507	0.070203	0.442214
0.008998	0.075794	0.119539
0.004501	0.042873	0.134827
0.003001	0.028461	0.138372
0.002250	0.022225	0.139681
89.854127	0.139426	90.951467
44.995074	0.134852	47.161108
30.005016	0.127900	33.168318
22.505865	0.119731	26.578693
8.988577	0.126400	10.032399
4.500435	0.100277	6.342198
3.000845	0.078785	5.384540
2.250733	0.063570	5.005967
0.899792	0.075794	1.673088
0.450137	0.042873	1.483379
0.300103	0.029350	1.445380
0.225078	0.022225	1.431831
0.090024	0.027854	0.455264
0.045016	0.014157	0.449246
0.030011	0.009467	0.448119
0.022508	0.007108	0.447723
0.009003	0.008967	0.141421
0.004502	0.004497	0.141421
0.003001	0.003000	0.141421
0.002251	0.002250	0.141421

Table 8-2 continued

Table 8-2 continued

$\left(\dfrac{\pi a}{2\lambda}\right)$	β	n	$\dfrac{\left(\dfrac{\pi a}{2\lambda}\right)}{\dfrac{1}{\sqrt{2}}\left\{\left(\dfrac{\pi a}{2\lambda}\right)^2+\left(\dfrac{n\pi}{2}\right)^2f^2\right\}^{\frac{1}{2}}}$ (unit: A)	$\dfrac{\dfrac{n\pi}{2}\cdot f}{\dfrac{1}{\sqrt{2}}\left\{\left(\dfrac{\pi a}{2\lambda}\right)^2+\left(\dfrac{n\pi}{2}\right)^2f^2\right\}^{\frac{1}{2}}}$ (unit: A)
1/100	100	1	0.0088939	1.4141855
		2	0.0042946	1.4142071
		3	0.0027147	1.4142109
		4	0.0019058	1.4142122
	10	1	0.0080630	1.4141902
		2	0.0031935	1.4142100
		3	0.0016722	1.4142126
		4	0.0010119	1.4142132
	1	1	0.0048349	1.4142052
		2	0.0013654	1.4142129
		2	0.0006230	1.4142135
		4	0.0003538	1.4142135
	1/10	1	0.0017768	1.4142124
		2	0.0004508	1.4142135
		3	0.0002009	1.4142136
		4	0.0001131	1.4142136
	1/100	1	0.0005720	1.4142135
		2	0.0001432	1.4142136
		3	0.0000637	1.4142136
		4	0.0000358	1.4142136

$\dfrac{\beta f}{\dfrac{1}{\sqrt{2}}\left\{\left(\dfrac{\pi a}{2\lambda}\right)^2+\left(\dfrac{n\pi}{2}\right)^2f^2\right\}^{\frac{1}{2}}}$ unit: $(g/2\omega a)\cdot A$	$\dfrac{\left(\dfrac{\pi a}{2\lambda}\right)\cdot\dfrac{n\pi}{2}}{\dfrac{1}{\sqrt{2}}\left\{\left(\dfrac{\pi a}{2\lambda}\right)^2+\left(\dfrac{n\pi}{2}\right)^2f^2\right\}^{\frac{1}{2}}}$ unit: $(g/2\omega a)\cdot A$	$\dfrac{\beta+\left(\dfrac{n\pi}{2}\right)^2}{\dfrac{1}{\sqrt{2}}\left\{\left(\dfrac{\pi a}{2\lambda}\right)^2+\left(\dfrac{n\pi}{2}\right)^2f^2\right\}^{\frac{1}{2}}}$ unit: $(g/2\omega a)\cdot A$
90.029852	0.013971	91.133740
45.015608	0.013492	47.184755
30.010487	0.012793	33.175711
22.507886	0.011975	26.582026
9.003015	0.012665	10.052503
4.501570	0.010033	6.345379
3.001052	0.007880	5.385740
2.250790	0.006358	5.006595
0.900311	0.007595	1.676441
0.450158	0.004289	1.484123
0.300105	0.002936	1.445702
0.225079	0.002223	1.432011
0.090032	0.002791	0.456177
0.045016	0.001416	0.449471
0.030011	0.000947	0.448218
0.022508	0.000711	0.447779
0.0090032	0.0008985	0.141705
0.0045016	0.0004499	0.141492
0.0030011	0.0003000	0.141453
0.0022508	0.0002251	0.141439

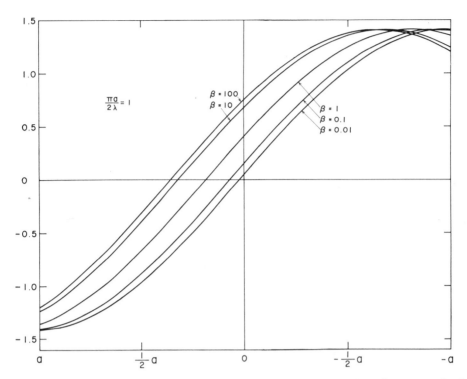

Figure 8-1. Cross-canal variation of wave amplitude, o(*y*).

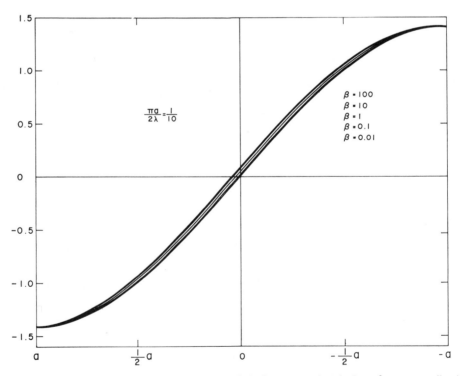

Figure 8-2. Cross-canal variation of wave amplitude, o(*y*).

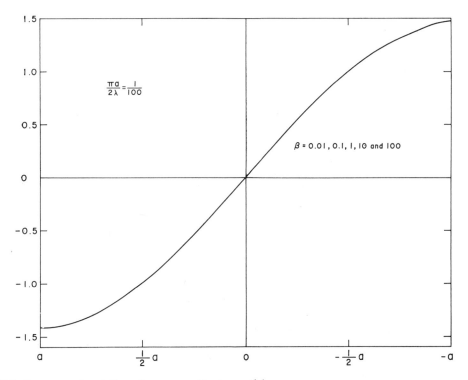

Figure 8-3. Cross-canal variation of wave amplitude,   0(y).

References

1. Poincaré, Henri (1910) Lecons de Mécanique Céléste 3, Théorie des Marées, pp. 122-123.
2. Thomson, Sir William (1879, 1880) On gravitational oscillations of rotating water. Proc. Roy. Soc. Edinburgh, March 17, 1879. (Reprinted in *Phil. Mag.* s.4 (10): 97-104; Papers 4, pp. 141-148.)
3. Defant, Albert 1961. *Physical oceanography* 2: pp. 208-210.

9

poincare and kelvin waves in a rotating straight canal

Introduction

This chapter will discuss the problem of tidal waves propagating along a rotating straight canal of parabolic section. In 1879, Sir William Thomson (1) discussed a similar problem for a straight canal of uniform depth and this motion is now popular as "Kelvin wave." In Sir William Thomson's problem, it is assumed in advance that there is no motion of water across the canal however broad the canal may be. In Kelvin wave, the velocity of propagation is exactly equal to $\sqrt{(gh)}$, where g is the acceleration of gravity and h the depth of canal. Afterward, Henri Poincaré (2) discovered another type of wave motion in the same type of canal, assuming that the transverse or cross-canal velocity component vanishes only at the boundaries. In Poincaré wave, the velocity of propagation is not always equal to $\sqrt{(gh)}$.

In the present discussion, these two kinds of waves are treated when the section of the canal is given by a parabola, or when the canal is deepest along the median line and gradually shelves toward the boundaries according to a parabolic law. Solution for a nonrotating canal of the same type was given by J. Proudman (3).

Theory

If x-axis is taken positive in the direction of wave propagation, y-axis *contra solem* perpendic-

ular to it, while z-axis is chosen positive vertically downward respectively, the law of depth h will be expressed as

$$h = h_0\left(1 - \frac{y^2}{a^2}\right),$$

where h is the depth at a distance y from the median line on which it is largest $h = h_0$, $2a$ is the breadth of the canal, so that the depth is *nil* along the coasts on both sides $y = \pm a$. This feature is represented in the diagram in Figure 9-1

If ζ is the elevation of water surface above an average level $z = z_0$, while u and v are components of velocity in x- and y-directions, the equations of motion and equation of continuity between ζ and u, v are given by

$$\frac{\partial u}{\partial t} - 2\omega v = -g\frac{\partial \zeta}{\partial x}, \qquad \frac{\partial v}{\partial t} + 2\omega u = -g\frac{\partial \zeta}{\partial y}$$

$$(9\text{-}1)$$

and

$$\frac{\partial \zeta}{\partial t} + \frac{\partial}{\partial x}(hu) + \frac{\partial}{\partial y}(hv) = 0, \qquad (9\text{-}2)$$

where t is the time, h the depth, g the gravity and ω the angular velocity of the canal.

The boundary conditions to be satisfied at both banks $y = \pm a$ are

$$hv = 0 \qquad \text{at} \qquad y = \pm a.$$

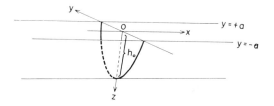

Figure 9-1. Straight canal of parabolic section.

and

$$f = \frac{\sigma}{2\omega} .$$

In order to solve equation (9-5), assume

$$\zeta_0(\eta) = \sum_{n=0}^{\infty} A_n P_n(\eta) , \qquad (9\text{-}7)$$

where $Pn(\eta)$ is a zonal harmonic of order n, and A_n's are constants to be determined.

Substitution of (9-7) in (9-5) gives

$$\sum_{n=0}^{\infty} A_n \Big[\frac{d}{d\eta} \Big\{ (1-\eta^2) \frac{dP_n}{d\eta} \Big\}$$

$$+ \beta(f^2-1) P_n(\eta) - \Big(\frac{\pi a}{2\lambda} \Big) \frac{\eta}{f} P_n(\eta)$$

$$- \Big(\frac{\pi a}{2\lambda} \Big) (1-\eta^2) P_n(\eta) \Big] = 0 . \qquad (9\text{-}8)$$

From the theory of spherical harmonics, the following expressions are valid,

$$\frac{d}{d\eta} \Big\{ (1-\eta^2) \frac{dP_n}{d\eta} \Big\} = -n(n+1) P_n(\eta) , \qquad (9\text{-}9)$$

However, because $h = 0$ at $y = \pm a$, we shall have $v \neq \infty$ at $y = \pm a$, or that transverse or cross-canal component of velocity remains finite at $y = \pm a$.

Let us consider that the wave travels in the positive direction of x, and because v cannot be *nil* even at the boundaries, assume

$$u = u_0(y) \cos\Big(\sigma t - \frac{\pi x}{2\lambda} \Big) ,$$

$$v = v_0(y) \sin\Big(\sigma t - \frac{\pi x}{2\lambda} \Big) , \qquad (9\text{-}3)$$

$$\zeta = \zeta_0(y) \cos\Big(\sigma t - \frac{\pi x}{2\lambda} \Big)$$

where λ is the wavelength, and $u_0(y)$, $v_0(y)$ and $\zeta_0(y)$ are functions of y only, finite at $y = \pm a$.

Substitution of equations (9-3) in (9-1) gives

$$\left. \begin{array}{l} u = -\dfrac{g/2\omega a}{f^2-1} \Big\{ \Big(\dfrac{\pi a}{2\lambda} \Big) f \cdot \zeta_0 + \dfrac{d\zeta_0}{d\eta} \Big\} \cdot \cos\Big(2\omega ft - \dfrac{\pi x}{2\lambda} \Big) , \\[3mm] v = -\dfrac{g/2\omega a}{f^2-1} \Big\{ +\Big(\dfrac{\pi a}{2\lambda} \Big) \zeta_0 + f \dfrac{d\zeta_0}{d\eta} \Big\} \cdot \sin\Big(2\omega ft - \dfrac{\pi x}{2\lambda} \Big) . \end{array} \right\} \qquad (9\text{-}4)$$

Substituting (9-4) in (9-2) and taking into account the law of depth $h = h_0[1 - y^2/a^2)]$ give the following differential equation:

$$\frac{d}{d\eta} \Big\{ (1-\eta^2) \frac{d\zeta_0}{d\eta} \Big\} + \Big\{ \beta(f^2-1) - \Big(\frac{\pi a}{2\lambda} \Big) \frac{\eta}{f} - \Big(\frac{\pi a}{2\lambda} \Big)^2 (1-\eta^2)^2 \Big\} \zeta_0 = 0 , \qquad (9\text{-}5)$$

where

$$\eta = \frac{y}{a} ,$$

$$\beta = \frac{4\omega^2 a^2}{gh_0} , \qquad (9\text{-}6)$$

$$(1-\eta^2)P_n(\eta) = -\frac{(n-1)\cdot n}{(2n-1)(2n+1)}\,P_{n-2}(\eta)$$

$$+\left\{1-\frac{n^2}{(2n-1)(2n+1)}-\frac{(n+1)^2}{(2n+1)(2n+3)}\right\}P_n(\eta)-\frac{(n+1)(n+2)}{(2n+1)(2n+3)}\,P_{n-2}\,,\qquad(9\text{-}10)$$

$$\eta P_n(\eta) = \frac{n+1}{2n+1}\,P_{n+1}(\eta)-\frac{n}{2n+1}\,P_{n-1}(\eta)\,.\qquad(9\text{-}11)$$

Substitution of (9-9), (9-10) and (9-11) in (9-8) gives

$$\sum_{n=0}^{\infty}\Bigg[-\frac{(n-1)\cdot n}{(2n-3)(2n-1)}\Big(\frac{\pi a}{2\lambda}\Big)\cdot f\cdot A_{n-2}-\frac{n}{2n-1}\,A_{n-1}$$

$$+\left\{\frac{\beta f(f^2-1)}{(\pi a/2\lambda)}-\frac{n(n+1)}{(\pi a/2\lambda)}\cdot f-\Big(\frac{\pi a}{2\lambda}\Big)f\Big(1-\frac{n^2}{(2n-1)(2n+1)}-\frac{(n+1)^2}{(2n+1)(2n+3)}\Big)\right\}A_n$$

$$-\frac{n+1}{2n+3}\,A_{n+1}-\frac{(n+1)(n+2)}{(2n+1)(2n+3)}\Big(\frac{\pi a}{2\lambda}\Big)\cdot f\cdot A_{n+2}\Bigg]P_n(\eta)=0\,,$$

from which the following sequence equation can be derived.

$$\left.\begin{array}{l}-\dfrac{(n-1)\cdot n}{(2n-3)(2n-1)}\Big(\dfrac{\pi a}{2\lambda}\Big)\cdot f\cdot A_{n-2}-\dfrac{n}{2n-1}\,A_{n-1}\\[2mm]
+\left\{\dfrac{\beta f(f^2-1)}{(\pi a/2\lambda)}-\dfrac{n(n+1)}{(\pi a/2\lambda)}\cdot f-\Big(\dfrac{\pi a}{2\lambda}\Big)f\Big(1-\dfrac{n^2}{(2n-1)(2n+1)}-\dfrac{(n+1)^2}{(2n+1)(2n+3)}\Big)\right\}A_n\\[2mm]
-\dfrac{n+1}{2n+2}\,A_{n+1}-\dfrac{(n+1)(n+2)}{(2n+3)(2n+5)}\Big(\dfrac{\pi a}{2\lambda}\Big)\cdot f\cdot A_{n+2}=0\,.\qquad(n=0,1,2,\cdots)\end{array}\right\}\qquad(9\text{-}12)$$

These equations give the relations between $A_0,\,A_1,\,A_2,\,\ldots,A_n\ldots$. For the negative suffices we have exclusively $A_n=0$ by theory. Substituting $n = 0, 1, 2, \ldots$ in order, we have from (9-12)

$$\left.\begin{array}{l}\left\{\dfrac{\beta f(f^2-1)}{(\pi a/2\lambda)}-\dfrac{2}{3}\Big(\dfrac{\pi a}{2\lambda}\Big)f\right\}A_0-\dfrac{1}{3}\,A_1-\dfrac{2}{15}\cdot\Big(\dfrac{\pi a}{2\lambda}\Big)f\cdot A_2=0\,,\\[3mm]
-A_0+\left\{\dfrac{\beta f(f^2-1)}{(\pi a/2\lambda)}-2\dfrac{f}{(\pi a/2\lambda)}-\dfrac{2}{5}\Big(\dfrac{\pi a}{2\lambda}\Big)f\right\}A_1+\dfrac{2}{5}\,A_2-\dfrac{6}{35}\Big(\dfrac{\pi a}{2\lambda}\Big)f\cdot A_3=0\,,\\[3mm]
\dfrac{2}{3}\Big(\dfrac{\pi a}{2\lambda}\Big)f\cdot A_0-\dfrac{2}{3}\,A_1+\left\{\dfrac{\beta f(f^2-1)}{(\pi a/2\lambda)}-6\dfrac{f}{(\pi a/2\lambda)}-\dfrac{10}{21}\Big(\dfrac{\pi a}{2\lambda}\Big)f\right\}A_2-\dfrac{3}{7}\,A_3=0\,,\\[3mm]
\dfrac{2}{5}\Big(\dfrac{\pi a}{2\lambda}\Big)f\cdot A_1-\dfrac{3}{5}\,A_2+\left\{\dfrac{\beta f(f^2-1)}{(\pi a/2\lambda)}-12\dfrac{f}{(\pi a/2\lambda)}-\dfrac{22}{45}\Big(\dfrac{\pi a}{2\lambda}\Big)f\right\}A_3+\dfrac{4}{9}\,A_4=0\,,\end{array}\right\}$$

Numerical Computation

Retaining A_0 and A_1 only and assuming $A_2 = A_3 = ... = 0$, we have the sequence equations:

$$\left\{ \frac{\beta f(f^2-1)}{(\pi a/2\lambda)} - \frac{2}{3}\left(\frac{\pi a}{2\lambda}\right)f \right\} A_0 - \frac{1}{3} A_1 = 0 ,$$

$$-A_0 + \left\{ \frac{\beta f(f^2-1)}{(\pi a/2\lambda)} - \frac{2}{(\pi a/2\lambda)}f - \frac{2}{5}\left(\frac{\pi a}{2\lambda}\right)f \right\} A_1 = 0 .$$

Eliminating A_0 and A_1 out of these two equations, the following frequency equation will be derived:

$$\left\{ \frac{\beta f(f^2-1)}{(\pi a/2\lambda)} - \frac{2}{3}\left(\frac{\pi a}{2\lambda}\right)f \right\} \cdot \left\{ \frac{\beta f(f^2-1)}{(\pi a/2\lambda)} - \frac{2}{(\pi a/2\lambda)}f - \frac{2}{5}\left(\frac{\pi a}{2\lambda}\right)f \right\} - \frac{1}{3} = 0 ,$$

or

$$\beta^2 f^2(f^2-1)^2 - \left\{ 2\beta + \frac{16}{15}\left(\frac{\pi a}{2\lambda}\right)^2 \right\} \beta f^2(f^2-1)$$

$$+ \left\{ \frac{4}{3}\left(\frac{\pi a}{2\lambda}\right)^2 + \frac{4}{15}\left(\frac{\pi a}{2\lambda}\right)^4 \right\} f^2 - \frac{1}{3}\left(\frac{\pi a}{2\lambda}\right) = 0 . \tag{9-13}$$

This equation is cubic in f^2. For positive values of $(\pi a/2\lambda)$ and β, the values of f were calculated as compiled in Table 9-1.

In this case the approximate expression for ζ_0 is

$$\zeta_0 = A_0 P_0(\eta) + A_1 P_1(\eta) .$$

Since the series for ζ_0 must converge, the coefficient A_1 must be small compared with A_0 for the gravest oscillations, and the values of f which do not make $|A_1/A_0|$ considerably smaller than unity must be replaced by the corresponding values of f obtained by adopting one more term, or

$$\zeta_0(y) = A_0 P_0(\eta) + A_1 P_1(\eta) + A_2 P_2(\eta) . \tag{9-14}$$

In this case the frequency equation will be given in a determinant form by

$$\begin{vmatrix} \dfrac{\beta f(f^2-1)}{(\pi a/2\lambda)} - \dfrac{2}{3}(\pi a/2\lambda)f, & -\dfrac{1}{3}, & \dfrac{2}{15}(\pi a/2\lambda)f \\[2mm] -1, & \dfrac{\beta f(f^2-1)}{(\pi a/2\lambda)} - \dfrac{2}{(\pi a/2\lambda)}f - \dfrac{2}{5}(\pi a/2\lambda)f, & -\dfrac{2}{5} \\[2mm] \dfrac{2}{5}(\pi a/2\lambda)f, & -\dfrac{2}{3}, & \dfrac{\beta f(f^2-1)}{(\pi a/2\lambda)} - \dfrac{6}{(\pi a/2\lambda)}f - \dfrac{10}{21}(\pi a/2\lambda)f \end{vmatrix} = 0. \qquad (9\text{-}15)$$

The frequency determinant in this case is

$$\zeta_0(y) = A_0 P_0(\eta) + A_1 P_1(\eta) + A_2 P_2(\eta) + A_3 P_3(\eta). \qquad (9\text{-}16)$$

Further accuracy will be obtained by assuming

$$\begin{vmatrix} \dfrac{\beta f(f^2-1)}{(\pi a/2\lambda)} - \dfrac{2}{3}(\pi a/2\lambda)f, & -\dfrac{1}{3}, & \dfrac{2}{15}(\pi a/2\lambda)f, & 0 \\[2mm] -1, & \dfrac{\beta f(f^2-1)}{(\pi a/2\lambda)} - 2\dfrac{f}{(\pi a/2\lambda)} - \dfrac{2}{5}(\pi a/2\lambda)f, & -\dfrac{2}{5}, & \dfrac{2\cdot 3}{5\cdot 7}(\pi a/2\lambda)f \\[2mm] \dfrac{2}{3}(\pi a/2\lambda)f, & -\dfrac{2}{3}, & \dfrac{\beta f(f^2-1)}{(\pi a/2\lambda)} - 6\dfrac{f}{(\pi a/2\lambda)} - \dfrac{10}{21}(\pi a/2\lambda)f, & -\dfrac{3}{7} \\[2mm] 0, & \dfrac{2}{5}(\pi a/2\lambda)f, & -\dfrac{3}{5} & \dfrac{\beta f(f^2-1)}{(\pi a/2\lambda)} - 12\dfrac{f}{(\pi a/2\lambda)} - \dfrac{22}{45}(\pi a/2\lambda)f \end{vmatrix} = 0. \qquad (9\text{-}17)$$

It is advisable to insert particular values for $(\pi a/2\lambda)$ and β in these determinants beforehand and reduce them into equations of 9th and 12th order respectively before solving them for f^2.

Numerical Result for Poincaré Waves

Equation (9-13) gives three pairs of distinct positive and negative roots f for given values of $(\pi a/2\lambda)$ and $\beta = 4\omega^2 a^2/gh$, so that it gives three different frequencies together with the corresponding ratios A_1/A_0. Computations were made for several pairs of $(\pi a/2\lambda)$ and β, or the combination of one of $(\pi a/2\lambda) = 1, 1/10$ and $1/100$ into one of $\beta = 1$, $1/10$ and $1/100$. The result is compiled in Table 9-1.

The result shows that for a given set of parameters $(\pi a/2\lambda)$ and β, there are always three pairs of definite frequencies f. In other words, if the wavelength is given in a canal of definite breadth, there are at least three waves of different periods. Frequencies of all these three waves appear with positive and negative signs and of the same absolute magnitudes, giving progressive and regressive waves. So actually it is sufficient to discuss positive frequencies only.

These triads of oscillations consist of (1) waves of frequencies smaller than 1, approximately proportional to $(\pi a/2\lambda)$ and inversely proportional to β, (2) ordinary waves which tend to $f^2 = 1$ as both $(\pi a/2\lambda)$ and β decrease indefinitely, and (3) waves of larger frequencies which increase as β decreases.

Out of these triads, the last ones may be regarded as higher harmonics, for the ratio A_1/A_0 is large for them, meaning that further terms A_2,

Table 9-1

Approximate Frequencies f and Coefficients, A_0 and A_1 Obtained by Solving Equation (9-13)

$(\pi a/2\lambda)$	β		f	A_0	:	A_1
1	1	f_1	$\pm$ 0.24949	1	:	-1.20086
		f_2	$\pm$ 1.24516	1	:	-0.43421
		f_3	$\pm$ 1.85849	1	:	9.96521
1	1/10	f_1	$\pm$ 0.42352	1	:	-0.95130
		f_2	$\pm$ 2.72231	1	:	-0.20884
		f_3	$\pm$ 5.00763	1	:	26.15428
1	1/100	f_1	$\pm$ 0.45299	1	:	-0.91678
		f_2	$\pm$ 8.20864	1	:	-0.07017
		f_3	$\pm$15.52674	1	:	80.77746
1/10	1	f_1	$\pm$ 0.03323	1	:	-1.00229
		f_2	$\pm$ 1.00250	1	:	-0.04990
		f_3	$\pm$ 1.73337	1	:	103.89196
1/10	1/10	f_1	$\pm$ 0.12279	1	:	-0.38736
		f_2	$\pm$ 1.02508	1	:	-0.04880
		f_3	$\pm$ 4.58702	1	:	274.83743
1/10	1/100	f_1	$\pm$ 0.32575	1	:	-0.15251
		f_2	$\pm$ 1.24889	1	:	-0.04007
		f_3	$\pm$14.19158	1	:	848.12129
1/100	1	f_1	$\pm$ 0.00333	1	:	-1.00003
		f_2	$\pm$ 1.00003	1	:	-0.00500
		f_3	$\pm$ 1.73206	1	:	1035.07770
1/100	1/10	f_1	$\pm$ 0.01259	1	:	-0.37806
		f_2	$\pm$ 1.00025	1	:	-0.00500
		f_3	$\pm$ 4.58262	1	:	2738.65262
1/100	1/100	f_1	$\pm$ 0.04062	1	:	-0.12247
		f_2	$\pm$ 1.00250	1	:	-0.00499
		f_3	$\pm$14.17759	1	:	8501.31254

Table 9-2

Frequencies. f, Coefficients A_0, A_1, A_2 and A_3 in the Expansion of $0(y)$ in Series of Zonal Harmonics, and Velocity of Propagation. C

$(\pi a/2\lambda)$	β		f^2	f	A_0	:	A_1	:	A_2	:	A_3	Wave velocity C
1	1	f_1	0.0805294	0.2837770	1	:	-1.2907443	:	0.5249191	:	-0.1212790	$0.28378\sqrt{gh_0} = 0.34755\sqrt{gh}$
		f_2	1.5216096	1.2335354	1	:	-0.4603184	:	0.1550036	:	-0.0216860	$1.23354\sqrt{gh_0} = 1.51077\sqrt{gh}$
1	1/10	f_1	0.2176108	0.4664877	1	:	-0.9818556	:	0.3248354	:	-0.0644974	$0.14752\sqrt{gh_0} = 0.18067\sqrt{gh}$
		f_2	7.2285284	2.6885923	1	:	-0.2205718	:	0.1235048	:	-0.0097582	$0.85021\sqrt{gh_0} = 1.04129\sqrt{gh}$
1	1/100	f_1	0.2464782	0.4964657	1	:	-0.9433555	:	0.3061596	:	-0.0598051	$0.04965\sqrt{gh_0} = 0.06080\sqrt{gh}$
		f_2	65.8234395	8.1131646	1	:	-0.0739256	:	0.1154626	:	-0.0032185	$0.81132\sqrt{gh_0} = 0.99366\sqrt{gh}$
1/10	1	f_1	0.0012347	0.0351383	1	:	-1.0594968	:	0.2947685	:	-0.0390326	$0.35138\sqrt{gh_0} = 0.43035\sqrt{gh}$
		f_2	1.0050041	1.0024989	1	:	-0.0499343	:	0.0016648	:	-0.0000249	$10.02499\sqrt{gh_0} = 12.27806\sqrt{gh}$
1/10	1/10	f_1	0.0153016	0.1236996	1	:	-0.3899842	:	0.0356178	:	-0.0015563	$0.39117\sqrt{gh_0} = 0.47909\sqrt{gh}$
		f_2	1.0507644	1.0250680	1	:	-0.0488357	:	0.0516407	:	-0.0000243	$3.24155\sqrt{gh_0} = 3.97007\sqrt{gh}$
1/10	1/100	f_1	0.1064616	0.3262845	1	:	-0.1526392	:	0.0062947	:	———	$0.32628\sqrt{gh_0} = 0.39962\sqrt{gh}$
		f_2	1.5594466	1.2487780	1	:	-0.0400947	:	0.0014681	:	———	$1.24878\sqrt{gh_0} = 1.52943\sqrt{gh}$
1/100	1	f_1	0.000012380	0.0035185	1	:	-1.0556250	:	0.2857412	:	-0.0374850	$0.35185\sqrt{gh_0} = 0.43093\sqrt{gh}$
		f_2	1.0000500	1.0000250	1	:	-0.0050000	:	0.0000167	:	0	$100.00250\sqrt{gh_0} = 122.47756\sqrt{gh}$
1/100	1/10	f_1	0.000160736	0.012678170	1	:	-0.3805355	:	0.0328853	:	-0.0012875	$0.40092\sqrt{gh_0} = 0.49102\sqrt{gh}$
		f_2	1.0005001	1.0002500	1	:	-0.0049985	:	0.0000167	:	0	$31.63068\sqrt{gh_0} = 38.73952\sqrt{gh}$
1/100	1/100	f_1	0.0016523	0.0406485	1	:	-0.1225565	:	0.0033010	:	———	$0.40649\sqrt{gh_0} = 0.49784\sqrt{gh}$
		f_2	1.0050083	1.0025015	1	:	-0.0049876	:	0.0000166	:	0	$10.02501\sqrt{gh_0} = 12.278.8\sqrt{gh}$

A_3,... have to be computed in order to have satisfactory convergence of the series. For this reason, those waves with large frequencies were left for future computation.

As second approximations, the equation (9-15) and (9-17) were solved for the same combination of $(\pi a/2\lambda)$ and β. The result is given in Table 9-2.

For smaller values of $(\pi a/2\lambda)$, the ratio A_2/A_0 is mostly small, meaning that the solutions are represented by the series (9-14) very closely. Yet for large values of $(\pi a/2\lambda)$ and β or in case the ratio is still not small enough, it is recommended to compute as far as A_3, so that $\zeta_0(y)$ is represented by (9-16) with the frequency equation (9-17). The result is also given in Table 9-2.

Now let us try to explain why these slow oscillations do not appear in Poincaré waves in a rectangular channel of uniform depth. In Poincaré waves, we have

$$f=\pm\left\{1+\frac{\left(\frac{\pi a}{2\lambda}\right)^2+\left(\frac{n\pi}{2}\right)^2}{\beta}\right\}^{\frac{1}{2}}$$

(Chapter 8.) If we put $\beta = 0$ in this expression, we shall have $f = \pm\infty$, whereas, in the present case, by putting $\beta = 0$ in (9-13), we have

$$f=\pm\left\{\frac{1}{4+\frac{6}{5}\left(\frac{\pi a}{2\lambda}\right)^2}\right\}^{\frac{1}{2}}$$

although to a rough approximation. From this expression, we learn that the frequency always lies between $-1/2$ and $+1/2$, the absolute value decreasing indefinitely as $\beta\to0$. This will explain why such slow oscillations often called the "geostrophic modes" do not appear in Poincaré waves in a rectangular canal of uniform depth. At the same time, it can be mentioned that these small frequencies are by no means similar to Hough's oscillations of second class (Hough, 1879) or waves of Rossby type (Rossby, 1938), often called the "planetary waves," because no higher frequencies exist for the wave under consideration, and the motion remains as a wave even if the rotation of

the canal comes to a stop. In this study the frequency smaller than 1 will be denoted as f_1, that larger than 1 as f_2.

Velocity of Propagation

The velocity of wave propagation c is given by $c = \sigma/(\pi a/2\lambda)$ from (9-3) and (9-4).

Now that $\sigma = 2\omega f$, it follows

$$c=\frac{2\omega a}{(\pi a/2\lambda)}\cdot f. \tag{9-18}$$

Further, the expression (9-8), or

$$\beta=\frac{4\omega^2 a^2}{gh_0}$$

gives

$$2\omega a=\sqrt{\beta}\cdot\sqrt{gh_0}\ .$$

Substitution of this expression in (9-18) gives

$$c=\frac{\sqrt{\beta}\cdot f}{(\pi a/2\lambda)}\cdot\sqrt{gh_0}\ .$$

Since the average depth $\bar{h}$ of the canal is given by $h = (1/2a)\int[1-(y^2/a^2)]\,dy = (2/3)ho$, the velocity of propagation can be also given by

$$c=\frac{\sqrt{\frac{3}{2}\beta}\cdot f}{(\pi a/2\lambda)}\cdot\sqrt{g\bar{h}_0}\ .$$

For long waves traveling along an open ocean, the velocity of propagation is supposed approximately equal to $\sqrt{(gh)}$, so that the factor:

$$\frac{\sqrt{\frac{3}{2}\beta}\cdot f}{(\pi a/2\lambda)}$$

is supposed to be very close to unity. The velocity of propagation in terms of $\sqrt{(gh_0)}$ and $\sqrt{(g\bar{h})}$ are

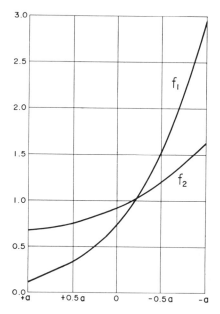

Figure 9-2. Distribution of waveheight across the channel.

$$(\pi a/2\lambda) = 1, \quad \beta = 1 \quad \begin{cases} f_1 = 0.2837770 \\ f_2 = 1.2335354 \end{cases}$$

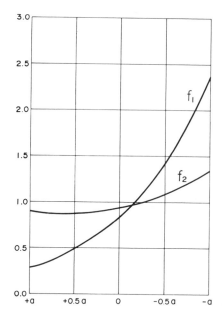

Figure 9-3. Distribution of waveheight across the channel.

$$(\pi a/2\lambda) = 1, \quad \beta = \frac{1}{10} \quad \begin{cases} f_1 = 0.4664877 \\ f_2 = 2.6885923 \end{cases}$$

computed for the several combinations of $(\pi a/2\lambda)$ and β, and compiled in Table 9-2. The result shows that the factor above sometimes differs greatly from unity, so that the speed of wave propagation is seldom close to $\sqrt{(g\bar{h})}$ where $\bar{h}$ is the average depth of the canal.

Discussion of Poincaré Waves

It is a striking feature that A_1 has always opposite signs to A_0 in these tables except for f_3, which is excluded from discussions in this chapter. This means that the wave amplitude $\zeta_0(y)$ increases from one side to the other, being smaller on the side which is forward in respect to the rotation. This feature is the same as in the cases of Kelvin and Poincaré waves in a canal of rectangular section.

The wave amplitude $\zeta_0(y)$ was computed for the above result.

The wave amplitude, $\zeta_0(y)$ is larger for larger values of $(\pi a/2\lambda)$ and β. If $(\pi a/2\lambda)$ is given, the amplitude becomes smaller as β decreases. This feature is illustrated by diagrams of Figure 9-2. If β is given, the wave amplitude decreases as $(\pi a/2\lambda)$ decreases and tends to zero. However, the ratio of wave amplitude at the side of the canal which is backward in respect to the rotation, over that at the other side is always far greater for f_2 than for f_1 (Figures 9-2, 9-3, 9-4, 9-5 and 9-6).

Moreover the wave amplitude for f_2-waves for the case $(\pi a/2\lambda) = 1$, $\beta = 1/10$ and $\beta = 1/100$ have a minimum at about $y = +0.6a$ and $y = 0.25a$ respectively and increases as we approach both banks (Figures 9-3 and 9-4).

For $\beta = 1$, the ratio of the amplitude of f_1-waves at the right bank over that at the left is large and almost independent of $(\pi a/2\lambda)$ (Figures 9-2, 9-5 and 9-8). For smaller values of $(\pi a/2\lambda)$, this ratio for f_2-waves is nearly constant across the canal (Figures 9-5, 9-6, 9-7, 9-8, 9-9, and 9-10).

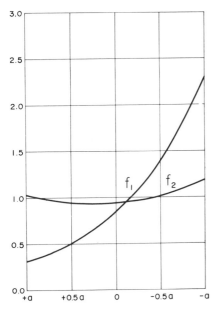

Figure 9-4. Distribution of waveheight across the channel.

$$(\pi a/2\lambda)=1, \quad \beta=\frac{1}{100} \quad \begin{cases} f_1=0.4964657 \\ f_2=8.1131646 \end{cases}$$

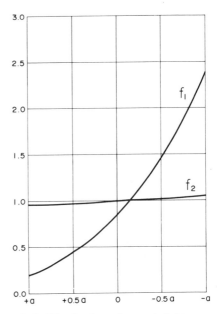

Figure 9-5. Distribution of waveheight across the channel.

$$(\pi a/2\lambda)=\frac{1}{10}, \quad \beta=1 \quad \begin{cases} f_1=0.0351383 \\ f_2=1.0024989 \end{cases}$$

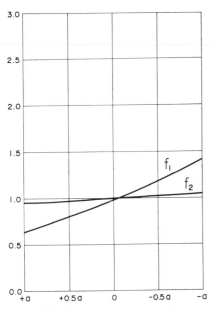

Figure 9-6. Distribution of waveheight across the channel.

$$(\pi a/2\lambda)=\frac{1}{10}, \quad \beta=\frac{1}{10} \quad \begin{cases} f_1=0.1236996 \\ f_2=1.0250680 \end{cases}$$

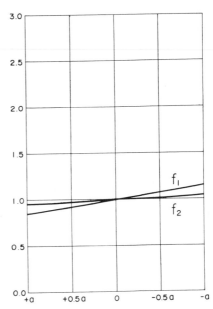

Figure 9-7. Distribution of waveheight across the channel.

$$(\pi a/2\lambda)=\frac{1}{10}, \quad \beta=\frac{1}{100} \quad \begin{cases} f_1=0.3262845 \\ f_2=1.2487780 \end{cases}$$

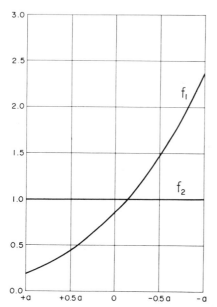

Figure 9-8. Distribution of waveheight across the channel.

$$(\pi a/2\lambda)=\frac{1}{100}, \quad \beta=1 \quad \begin{cases} f_1=0.0035185 \\ f_2=1.0000250 \end{cases}$$

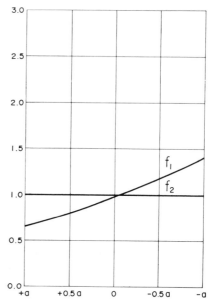

Figure 9-9. Distribution of waveheight across the channel.

$$(\pi a/2\lambda)=\frac{1}{100}, \quad \beta=\frac{1}{10} \quad \begin{cases} f_1=0.0126782 \\ f_2=1.0002500 \end{cases}$$

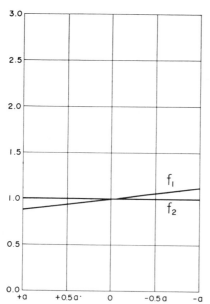

Figure 9-10. Distribution of waveheight across the channel.

$$(\pi a/2\lambda)=\frac{1}{100}, \quad \beta=\frac{1}{100} \quad \begin{cases} f_1=0.0406485 \\ f_2=1.0025010 \end{cases}$$

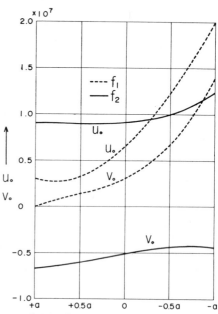

Figure 9-11. Distribution of amplitudes of long-channel and cross-channel velocity components u_0 and v_0 across the channel.

$$(\pi a/2\lambda)=1, \quad \beta=1$$

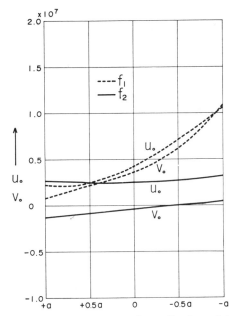

Figure 9-12. Distribution of amplitudes of long-channel and cross-channel velocity components u_0 and v_0 across the channel.

$$(\pi a/2\lambda)=1, \quad \beta=\frac{1}{10}$$

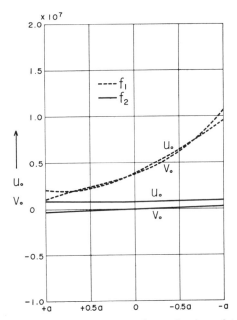

Figure 9-13. Distribution of amplitudes of long-channel and cross-channel velocity components u_0 and v_0 across the channel.

$$(\pi a/2\lambda)=1, \quad \beta=\frac{1}{100}$$

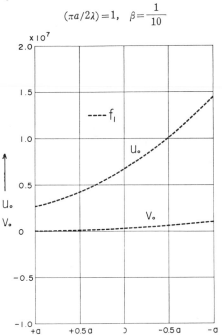

Figure 9-14. Distribution of amplitudes of long-channel and cross-channel velocity components u_0 and v_0 across the channel.

$$(\pi a/2\lambda)=\frac{1}{10}, \quad \beta=1$$

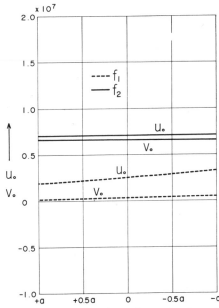

Figure 9-15. Distribution of amplitudes of long-channel and cross-channel velocity components u_0 and v_0 across the channel.

$$(\pi a/2\lambda)=\frac{1}{10}, \quad \beta=\frac{1}{10}$$

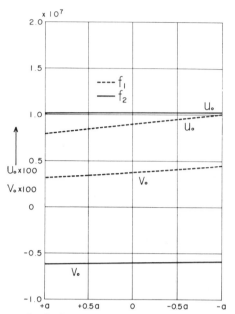

Figure 9-16. Distribution of amplitudes of long-channel and cross-channel velocity components u_0 and v_0 across the channel.

$$(\pi a/2\lambda) = \frac{1}{10}, \quad \beta = \frac{1}{100}$$

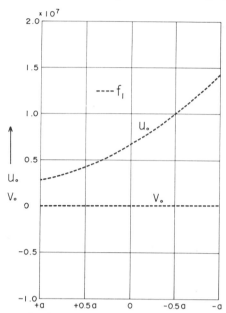

Figure 9-17. Distribution of amplitudes of long-channel and cross-channel velocity components u_0 and v_0 across the channel.

$$(\pi a/2\lambda) = \frac{1}{100}, \quad \beta = 1$$

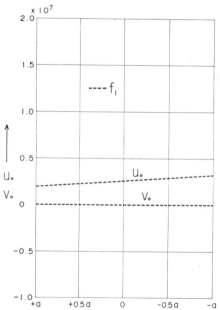

Figure 9-18. Distribution of amplitudes of long-channel and cross-channel velocity components u_0 and v_0 across the channel.

$$(\pi a/2\lambda) = \frac{1}{100}, \quad \beta = \frac{1}{10}$$

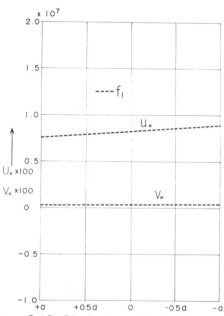

Figure 9-19. Distribution of amplitudes of long-channel and cross-channel velocity components u_0 and v_0 across the channel.

$$(\pi a/2\lambda) = \frac{1}{100}, \quad \beta = \frac{1}{100}$$

Figures 9-11 through 9-19 represent the particle velocity or current along and across the canal. In these diagrams, dotted lines give $u_0(y)$ and $v_0(y)$ for the f_1-waves of geostrophic modes having frequencies less than unity, while full lines mean the same for f_2-waves having frequencies larger than unity. The cross-canal component does not vanish at both banks because the depth vanishes there.

For the waves $(\pi a/2\lambda) = 1/10$, $\beta = 1$ $[u_0 = 6.76 \times 10^7 \, (A_0/a), v_0 = -6.71 \times 10^7 \, (A_0/a)]$, $(\pi a/2\lambda) = 1/100$, $\beta = 1$ $[u_0 = 67.2 \times 10^7 \, (A_0/a)]$, $v_0 =)67.2 \times 10^7 \, (A_0/a)]$ $(\pi a/2\lambda) = 1/100$, $\beta = 1/10$ $[u_0 = 6.724 \times 10^7 \, (A_0/a), v_0 = -6.709 \times 10^7 \, (A_0/a)]$, both the components $u_0(y)$ and $v_0(y)$ for f_2 are very nearly constant all the way across the canal so that no curve will be given for them.

Rotating Canal of Parabolic Section

Now let us show that no Kelvin wave exists in a rotating canal of parabolic section. If we assume $v = 0$ in (9-1) and (9-2), we have

$$\frac{\partial u}{\partial t}=-g\frac{\partial \zeta}{\partial x}, \quad 2\omega u=-g\frac{\partial \zeta}{\partial y}, \qquad (9\text{-}19)$$

and

$$\frac{\partial \zeta}{\partial t}+h_0\left(1-\frac{y^2}{a^2}\right)\frac{\partial u}{\partial x}=0 \qquad (9\text{-}20)$$

since $h = h_0[1 - (y^2/a^2)]$. Assume a progressive wave and put

$$\zeta=\zeta_0(y)\cos\left(\sigma t-\frac{\pi x}{2\lambda}\right), \qquad (9\text{-}21)$$

$$u=u_0(y)\cos\left(\sigma t-\frac{\pi x}{2\lambda}\right), \qquad (9\text{-}22)$$

where σ is assumed to be a constant, and the wave propagation satisfies Kelland's condition (6) that wave speed is independent of y. Substitution of (9-21) and (9-22) in (9-19) and (9-20) gives

$$\sigma u_0=\frac{\pi}{2}\cdot\frac{g}{\lambda}\zeta_0, \qquad (9\text{-}23)$$

$$2\omega u_0=-g\frac{d\zeta_0}{dy}, \qquad (9\text{-}24)$$

and

$$-\sigma\zeta_0+\frac{\pi}{2}\cdot\frac{h_0}{\lambda}\left(1-\frac{y^2}{a^2}\right)u_0=0. \qquad (9\text{-}25)$$

Eliminating u_0 between (9-24) and (9-25), we have

$$\frac{1}{\zeta_0}\frac{d\zeta_0}{dy}=-\frac{2\omega\cdot2\lambda\cdot\sigma}{\frac{\pi}{2}gh_0}\cdot\frac{1}{1-\frac{y^2}{a^2}}.$$

Integrating with respect to y, assuming that σ is independent of y,

$$\log\zeta_0=$$

$$-\frac{2\omega a\cdot\lambda\cdot\sigma}{\pi gh_0}\int_0^{\frac{y}{a}}\left(\frac{1}{1+\frac{y}{a}}-\frac{1}{1-\frac{y}{a}}\right)d\left(\frac{y}{a}\right)$$

$$=\log\left(\frac{1-\frac{y}{a}}{1+\frac{y}{a}}\right)^{-\frac{2\omega a\cdot\lambda\cdot\sigma}{\pi gh_0}}$$

Thus we have

$$\zeta_0=\left(\frac{a-y}{a+y}\right)^{\frac{\lambda\cdot\sigma\sqrt{\beta}}{\pi\sqrt{gh_0}}}\cdot A_0 \qquad (9\text{-}26)$$

where A_0 is a constant, being equal to the wave-height along the median line.

Substitution of (9-26) in (9-23) gives

$$u_0=A_0\cdot\frac{\pi}{2}\frac{g}{\lambda\cdot\sigma}\left(\frac{a-y}{a+y}\right)^{\frac{\lambda\cdot\sigma\sqrt{\beta}}{\pi\sqrt{gh_0}}}.$$

Thus Kelvin wave in this canal is given by

$$\zeta = A_0 \cdot \left(\frac{a-y}{a+y} \right)^{\frac{\lambda \cdot \sigma \sqrt{\bar{\beta}}}{\pi \sqrt{gh_0}}} \cdot \cos\left(\sigma t - \frac{\pi x}{2\lambda} \right),$$

$$u = A \cdot \frac{\pi}{2} \cdot \frac{g}{\lambda \cdot \sigma} \left(\frac{a-y}{a+y} \right)^{\frac{\lambda \cdot \sigma \sqrt{\bar{\beta}}}{\pi \sqrt{gh_0}}} \cdot \cos\left(\sigma t - \frac{\pi x}{2\lambda} \right), \quad v = 0.$$

From (9-23) and (9-25) the velocity of propagation can be obtained as

$$c = \frac{\sigma}{(\pi/2\lambda)} = \sqrt{gh_0 \left(1 - \frac{y^2}{a^2} \right)},$$

which is a result contrary to the assumption that σ or c is independent of y. This result shows that Kelvin wave does not exist in a rotating canal of parabolic section.

Acknowledgments

The author is much indebted to Miss Mariko Karasawa who carried out most of these tedious calculations.

References

1. Thomson, Sir William 1879, 1880. *On gravitational oscillations of rotating water.* Proc. Roy. Soc. Edinburgh, March 17, 1879. (Reprinted in *Phil. Mag.,* 10, 1880, pp. 97–104: Papers 4, pp. 141–148.)
2. Poincaré, Henri. 1910. Lecons de Mécanique Céleste 3, Théorie des Marées. pp. 122–123.
3. Proudman, J. 1925. Tides in a channel. *Phil. Mag.* s. 6 (49), pp. 465–475.
4. Kelland, P. 1839. On the theory of waves. *Trans. Roy. Soc. Edinburgh,* 14: 497–546.

10

a contribution to the theory of upwelling and coastal currents

Introduction

There have been several descriptions of upwelling observed off Californian, Peruvian, West African, and other coasts by Thorade (1), McEwen (2, 3), Gunther (4), Defant (5, 6), Sverdrup (7, 8), and Sverdrup and Fleming (9). The explanation of this process given by Sverdrup is worth attention. According to Sverdrup et al. (10), it is known from analysis of the water masses that the water taking part in the process of upwelling off the coast of Southern California originates primarily in the layers from 200 to 300 meters below the surface. These are, however, all qualitative discussions and it has not yet been possible to explain this phenomenon theoretically and predict the velocity and width of coastal currents produced by the prevailing winds. Recently Defant (6) made a theoretical explanation assuming a sea consisting of two layers of water with different densities. The present research is an attempt to solve this problem mathematically and to draw some quantitative conclusions concerning this process which is very important in all fields of oceanography.

The only satisfactory explanation of upwelling seems to come from treating the problem thermodynamically as well as hydrodynamically. The following discussion will, however, be made purely from the hydrodynamical standpoint, on the assumption that sea water is of uniform density. Even though this mathematical simplification has been employed, the author feels that the results obtained are consistent with observed phenomena.

It has been noticed that upwelling occurs most favorably when, in the northern (southern) hemisphere, a wind blows in such a manner that the coast is on the left-hand (right-hand) side of an observer who looks in the direction of the wind. Thus, upwelling off the coast of California is most noticeable in early summer when northwesterly winds prevail for several weeks nearly parallel to the coast.

In this investigation the effect of the earth's rotation and the frictional forces due to both vertical and lateral mixing are taken into account. It should be stressed that horizontal mixing seems to play the most important role in the process of upwelling.

Theory of Upwelling Produced by a Wind Parallel to the Coast

Consider a straight coast coincident with the axis of y, and take the x-axis perpendicular to it in the offshore direction (Figure 10-1). Suppose a wind of constant force and direction is blowing steadily in a belt of limited width L parallel to the coast from negative to positive direction of y. This is a disposition favorable for the upwelling to

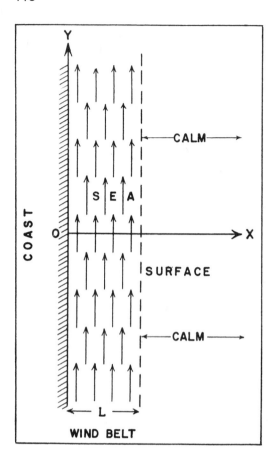

Figure 10-1. Schematic diagram of wind-to-coast relationship for the development of upwelling in the northern hemisphere.

occur actually. Take the z-axis vertically downwards.

If a constant wind blows for a sufficiently long period, a steady condition is attained in which the motion of water is independent of time. We assume that the wind stress cannot vary in the y-direction, but may be a function of x. This means that all the vertical and horizontal components of currents can be determined as functions of x and z only. Moreover, the surface of the sea will not be a plane but will have a slope in the x-direction. In such a case, the hydrodynamical equations of motion are, after several reasonable simplifications:

$$(A_h/\rho)\, \delta^2 u/\delta x^2 + (A_v/\rho)\, \delta^2 u/\delta z^2$$
$$+ 2\,\omega \sin\phi\; v - g\,\delta\,\xi/\delta x = 0$$
$$(A_h/\rho)\, \delta^2 v/\delta x^2 + (A_v/\rho)\, \delta^2 v/\delta z^2$$
$$- 2\,\omega \sin\phi\, u = 0 \qquad (10\text{-}1)$$

where u and v are the horizontal components of the current velocity in the x- and y-directions ζ the surface elevation depending on x only, ρ the density, A_v and A_h the coefficients of vertical and horizontal mixing of sea water, ω the angular velocity of the earth and ϕ the geographic latitude. In addition to these, we have the equation of continuity in the form

$$\partial u/\partial x + \partial w/\partial z = 0 \qquad (10\text{-}2)$$

since $\partial v/\partial y = 0$. Here w is the vertical component of currents and represents the intensity of the upwelling.

Suppose the wind blows in the positive direction of y in a belt between the coast $x = 0$ and $x = L$. The conditions to be satisfied on the surface of the sea are therefore

$$z = 0 \quad \begin{cases} -A_v \partial u/\partial z = 0 \\ -A_v \partial v/\partial z = \tau \quad \text{for } 0 \leqq x \leqq L \\ \qquad\qquad = 0 \quad \text{for } L \leqq x < \infty \end{cases}$$

where τ is the stress of wind and may be either a constant or a function of x. On the bottom $z = h$ we must have

$$z = h \quad u = v = 0$$

because of vertical friction. Along the coast which is considered to consist of vertical cliffs

$$x = 0 \quad u = v = 0$$

because of horizontal friction. In the region very far from both the coast and wind region, we have

$$x = \infty \quad u = v = \partial\zeta/\partial x = 0$$

Let us define D_v and D_h by

$$D_v = \pi \sqrt{A_v/\rho \; \omega \sin \phi}$$

$$D_h = \pi \sqrt{A_h/\rho \; \omega \sin \phi}$$

D_v is the depth of frictional influence defined by Ekman (11) in his theory of ocean currents, and D_h is a quantity having a dimension of a length and which may be called "frictional distance." This is a measure of the horizontal turbulence. Thus (10-1) now becomes

$$\left.\begin{array}{c} \partial^2 u/\partial\xi^2 + D_v{}^2 \; \partial^2 u/\partial z^2 + 2\pi^2 v - (\pi^2 g/\omega \sin \phi) \; \partial\xi/\partial\xi = 0 \\[2mm] \partial^2 v/\partial\xi^2 + D_v{}^2 \partial^2 v/\partial z^2 - 2\pi^2 u = 0 \end{array}\right\} \qquad (10\text{-}3)$$

where

$$\xi = x/D_h$$

In order to solve (10-3), suppose (Takegami, 12)

$$u = (2/\pi) \int_0^\infty u_1(\lambda) \sin \lambda\xi d\lambda \qquad\qquad u_1(\lambda) = \int_0^\infty u(\alpha,z) \sin \lambda\alpha d\alpha \qquad (10\text{-}4)$$

$$v = (2/\pi) \int_0^\infty v_1(\lambda) \sin \lambda\xi d\lambda \qquad\qquad v_1(\lambda) = \int_0^\infty v(\alpha,z) \sin \lambda\alpha d\alpha \qquad (10\text{-}5)$$

$$\partial\zeta/\partial\xi = (2/\pi) \int_0^\infty \gamma(\lambda) \sin \lambda\xi d\lambda \qquad\qquad \gamma(\lambda) = \int_0^\infty (\partial\zeta/\partial\alpha) \sin \lambda\alpha d\alpha \qquad (10\text{-}6)$$

Next assume for the wind stress

$$-A_v \; \partial v/\partial z = (2/\pi) \int_0^\infty T(\lambda) \sin \lambda\xi d\lambda$$

and

$$T(\lambda) = \int_{0}^{\infty} (-A_v \, \partial v/\partial \alpha) \sin \lambda \alpha d\alpha = \int_{0}^{L/D_h} \tau \; \sin \lambda \alpha \lambda \alpha = [1 - \cos(\lambda L/D_h)] \; \tau/\lambda$$

If τ is independent of x, substituting (10-4), (10-5) and (10-6) into (10-3) and writing

$$u_1 + iv_1 = \overline{W}$$

the two equations of (10-3) are combined into

$$D_v^2 \; d^2 \overline{W}/dz^2 - (\lambda^2 + 2\pi^2 i) \; \overline{W} - (\pi^2 g/\omega \sin \phi \, D_h) \, \gamma(\lambda) = 0 \qquad (10\text{-}7)$$

and the conditions to be satisfied along the boundaries now become

$$-A_v \, d\overline{W}/dz \; \Big|_{z=0} = [1 - \cos(\lambda L/D_h)] \; i \, \tau/\lambda \qquad (10\text{-}8)$$

and

$$\overline{W} \; \Big|_{z=h} = 0 \qquad (10\text{-}9)$$

The solution of (10-7) subject to conditions (10-8) and (10-9) is

$$\overline{W} = -\left[\pi^2 g\gamma(\lambda)/(\lambda^2 + 2\pi^2 i) \; \omega \sin \phi \, D_h\right] \left\{ 1 - [\cos h \,(\sqrt{\lambda^2 + 2\pi^2 i} \; z/D_v)]/[\cos h \,(\sqrt{\lambda^2 + 2\pi^2 i} \; h/D_v)]\right\}$$

$$+ \left[(i\tau D_v/A_v)/\sqrt{\lambda^2 + 2\pi^2 i}\,\right] \left\{ \sin h \,[\sqrt{\lambda^2 + 2\pi^2 i} \,(h-z)/D_v]\right\} \left\{[1 - \cos(\lambda L/D_h)]/\lambda\right\}/[\cos h \,(\sqrt{\lambda^2 + 2\pi^2 i} \; h/D_v)]$$

$$(10\text{-}10)$$

If we separate the real part P of $\sqrt{(\lambda_2 + 2\pi^2 i)}$ from the imaginary part Q we have

$$P = +\sqrt{(\sqrt{\lambda^4 + 4\pi^4} + \lambda^2)/2} \qquad\qquad Q = +\sqrt{(\sqrt{\lambda^4 + 4\pi^4} - \lambda^2)/2} \qquad (10\text{-}11)$$

Thus the real part of $\sqrt{(\lambda^2 + 2\pi^2 i)}$ is always greater than π. Hence, if the depth of the sea is sufficiently large $(h/D_v > 2)$, (10-10) can be given very accurately by

$$\overline{W} = -[\pi^2 g \, \tau \, (\lambda)/(\lambda^2 + 2\pi^2 i) \, \omega \, \sin \phi \, D_h] \; [1 - e^{-\sqrt{\lambda^2 + 2\pi^2 i} \; (h-z)/D_v}]$$

$$+ (i \, \tau \, D_v/A_v) \, (1/\sqrt{\lambda^2 + 2\pi^2 i}) \, e^{-\sqrt{\lambda^2 + 2\pi^2 i} \; z/D_v} \; [1 - \cos \, (\lambda \, L/D_v)]/\lambda \qquad (10\text{-}12)$$

In a steady state we do not have any vertical motion on the surface and bottom of the sea. We have, therefore

$$\omega = 0 \quad \text{for} \quad z = 0 \text{ and } h$$

Integrating (10-2) with respect to z from the surface down to the bottom, we have

$$(\partial/\partial x) \int_0^h u \, dz = -w \, \Big|_{z=0}^{z=h} = 0.$$

This means that the integral $\int_0^h u \, dz$ is independent of x, or therefore a constant. But as this integral must vanish directly on the coast or at $x = 0$, we must have

$$\int_0^h u \, dz = 0$$

always. Integrating (10-12) with respect to z from 0 to h and equating the real part of the resulting equation to 0, we have

$$\tau \, (\lambda) = \left\{ (\pi^2 \tau/\rho\omega)/(\lambda^2/2\pi^2) \; [h - PD_v/(P^2 + Q^2) + QD_v/(P^2 + Q^2)] \right\} \; [1 - \cos \, (\lambda \, L/D_h)]/\lambda \qquad (10\text{-}13)$$

This determines the relation between the wind stress τ and the slope of the sea surface induced by the former. Substitution of (10-13) in (10-12) gives u_1 and v_1. Substitutions of u_1 and v_1 thus obtained and (10-13) into (10-4), (10-5) and (10-6) give the horizontal components u and v and the surface slope $\partial \zeta/\partial \xi$. The vertical component w of motion can be derived from the equation of continuity (10-2) as

$$w = -(\partial/\partial x) \int_O{}^z u dz$$

Upwelling in a Deep Sea

When the sea is sufficiently deep and the ratio h/D_v increases indefinitely, we have from (10-13)

$$\tau(\lambda) \to 0$$

and (10-12) becomes

$$\overline{W} = [(i\,\tau\,D_v/A_v)/\sqrt{\lambda^2 + 2\pi^2 i}]\, e^{-\sqrt{\lambda^2 + 2\pi^2 i}\; z/D_v}\, [1 - \cos(\lambda L/D_h)]/\lambda$$

The components of velocity are then given by

$$u(x,z) = (2\pi\,\tau/\rho\omega\; \sin\phi\,D_v) \int_O{}^\infty G(\lambda,z)\, R(\lambda,x)\, d\lambda \qquad (10\text{-}14)$$

$$v(x,z) = (2\pi\,\tau/\rho\omega\, \sin\phi\,D_v) \int_O{}^\infty H(\lambda,z)\, R(\lambda,x)\, d\lambda \qquad (10\text{-}15)$$

$$w(x,z) = (2\pi\,\tau/\rho\omega\, \sin\phi\,D_v)\, (D_h/D_v) \int_O{}^\infty L(\lambda,z)\, S(\lambda,x)\, d\lambda \qquad (10\text{-}16)$$

where

$$G(\lambda,z) = \left\{[Q\cos(Qz/D_v) + P\sin(Qz/D_v)]/(P^2 + Q^2)\right\} e^{-Pz/D_v}$$

$$H(\lambda,z) = \left\{[P\cos(Qz/D_v) + Q\sin(Qz/D_v)]/(P^2 + Q^2)\right\} e^{-Pz/D_v}$$

$$L(\lambda,z) = \left\{2PQ\,[\cos(Qz/D_v)\, e^{-Pz/D_v} - 1] + (P^2 - Q^2)\sin(Qz/D_v)\, e^{-Pz/D_v}\right\}/(P^2 + Q^2)^2$$

$$R(\lambda,x) = \sin(\lambda x/D_h)\, [1 - \cos(\lambda L/D_h)]/\lambda$$

$$S(\lambda,x) = \cos(\lambda x/D_h)\, [1 - \cos(\lambda L/D_h)]/1$$

and P and Q are the real and imaginary parts of $\sqrt{(\lambda^2 + 2\pi^2 i)}$ whose expressions are given by (10-11).

These results show that owing to the effect of the wind blowing parallel to the coast, a vertical circulation in the plane perpendicular to the coast can be expected in addition to a coastal current parallel to the wind direction. The vertical component of this circulation evidently represents the upwelling. However, the upwelling must be ascending currents close to the coast. We sometimes have descending currents on the coast produced by an onshore wind as described by Longard and Banks (13). But such descending currents will be excluded from our discussion because they do not bring cold and nutritious water from deeper layers.

From expressions (10-14), (10-15) and (10-16) for u, v and w, it can be expected that the horizontal velocity of water in this process is approximately D_h/D_v or $\sqrt{(A_h/A_v)}$ times as large as the vertical velocity. This result will be very useful in estimating the approximate speed of upwelling.

If we define a function $\psi(x,z)$ as

$$\psi(x,z) = (2\pi \tau/\rho\omega \sin\phi) \quad \Phi_y(x,z) = (2\pi \tau/\rho\omega \sin\phi) \times \text{real part of}$$

$$i\int_0^\infty \left\{ [(e^{-\sqrt{\lambda^2 + 2\pi^2 i}\, z/D_v} - 1)/(\lambda^2 + 2\pi^2 i)] \right\} \left\{ \sin(\lambda L/D_h)\, [1 - \cos(\lambda L/D_h)]/\lambda \right\} d\lambda \quad (10\text{-}17)$$

we can show that this is the stream function in the plane perpendicular to the coast (xz-plane), and u and w are derived from this formula as

$$u = -\partial\psi/\partial z \qquad\qquad w = +\partial\psi/\partial x$$

so that any curve $\psi(x,z) =$ constant will represent a streamline. If we draw such curves at a constant difference $\psi(x,z)$, the velocity of the current is inversely proportional to the distance between two successive curves.

A Numerical Example

So far the author has elucidated the process of upwelling in a quantitative manner and obtained the expressions for the motion of water produced by a wind blowing parallel to the coast in a belt of finite width. The stream function $\psi(x,z)$ [= $(2\pi\tau/\rho\omega \sin\phi)\ \Phi_y$] can be computed from

Table 10-1
Values of the Stream Function $_y$ (x,z) x 10⁴

z/D$_v$	x/D$_h$								
	0	0.1396	0.2793	0.5585	0.8378	1.1170	1.3963	1.6755	1.9548
0	0	0	0	0	0	0	0	0	0
0.2	0	−175	−313	−449	−461	−465	−455	−425	−310
0.4	0	−277	−499	−725	−748	−750	−734	−682	−496
0.6	0	−321	−578	−843	−867	−866	−848	−788	−576
0.8	0	−334	−600	−871	−891	−887	−869	−808	−589
1.0	0	−332	−598	−865	−890	−873	−856	−797	−579
1.2	0	−329	−592	−852	−863	−853	−837	−782	−568
1.4	0	−327	−587	−844	−853	−843	−827	−773	−561
1.6	0	−326	−585	−840	−848	−837	−823	−769	−558

(10-17) for any distance x/D_h from the coast and any depth z/D_v below the surface. The results of computation for several values of x/D_h and z/D_v are given in Table 10-1 and illustrated by Figure 10-2. The figures in Table 10-1 should be multiplied by

$$(2\pi \, \tau/\rho\omega \, \sin \phi) \times 10^{-4}$$

The values of the stream function are all negative. This means that an upwelling develops close to the coast and that there is a motion of water in an offshore direction in the upper layers of the sea directly below the surface swept by wind. An intense upwelling can be seen within $0.5 D_h$ from the coast and the streamlines go down gradually outside the wind zone, meaning that the process of sinking occurs there.

The fact that the expressions for the velocity components all include $\sin \phi$ in the denominator shows that the lower the latitude the more intense will be the process of upwelling. Perhaps the strong upwelling off the Peruvian coast may be ascribed to this theoretical result.

It is interesting and useful to compute the magnitude of the offshore currents and the velocity of upwelling from the stream function (10-17)

and to compare the result with those formerly estimated from various sources.

It is of course not easy to estimate the magnitude of the coefficients of mixing. The vertical mixing coefficient D_v may be estimated at something like 1000 cgs. If we adopt this value, D_v is about 162 meters for a latitude of 30°N. To estimate A_h is even more difficult. But actual oceanographic observations show that $A_h/A_v = 10^6$ approximately. This means that D_h is just about 1000 times as large as D_v, or 162 kilometers. Furthermore, we do not know much about the width of the coastal wind belt. In this computation, the author tentatively assumed $L = 2.0944 D_h$, that is about twice as large as D_h, or 339 kilometers.

It is still more difficult to estimate the wind velocity of the northwesterlies prevailing off the coast of Southern California in the earlier summer months. The author took $\tau = 1$ cgs. This corresponds to a wind velocity between five and six meters per second. If we consider the upwelling off the coast of Southern California, taking $\phi = 30°N$, we have

$$2\pi \, \tau/\rho\omega \, \sin \phi = 2.37 \times 10^5 \; \sec^{-1}$$

From Table 10-1 we can compute the average velocity between the surface and the layer $0.2 D_v$ deep by

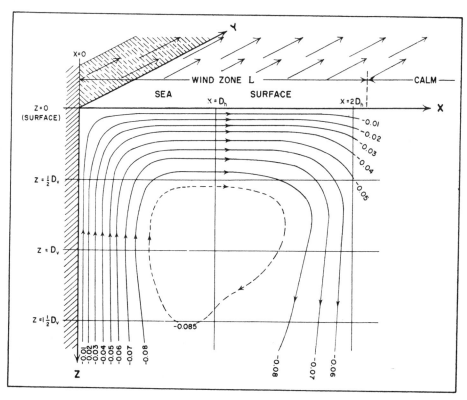

Figure 10-2. Upwelling as induced by a wind parallel to the coast (illustrated by the streamlines in the vertical plane perpendicular to the coast).

$$- (2\pi\, \tau/\rho\omega \sin\phi)\, \Delta\psi/\Delta z = +2.37 \times 10^5 \times 0.045/0.2\, D_v = 3.35 \text{ cm/sec (offshore)}.$$

This is the maximum velocity of the offshore current in the layer between the surface and approximately 32.4 meters.

The maximum upwelling velocity can be estimated in a similar manner, namely, at

$$(2\pi\, \tau/\rho\omega \sin\phi)\, (\Delta\psi/\Delta x) = -2.37 \times 10^5 \times 0.033/0.14\, D_h = -3.14 \times 10^{-3} \text{ cm/sec} = 2.7 \text{ m/day upward}.$$

This speed of upwelling is approximately 80 meters/month. McEwen (3) estimated the speed of upwelling off the coast of Southern California at 10 to 20 meters/month. From Sverdrup's (18) diagrams showing the distribution of isopycnals observed at an interval of 38 days, Saito (14) obtained 2.25 meters/day or about 80 meters/month

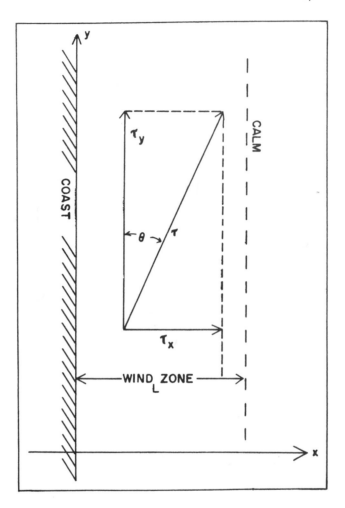

Figure 10-3. Wind making an angle with the coast.

for the upwelling velocity off Southern California. The present result appears to show a speed a little too high but may be suggestive of the order of magnitude of ascending motion in the process of upwelling.

From the diagram in Figure 10-2, we can see that the water mass participating in this process of upwelling comes from the layers from $z = D_y$ to $1.5D_y$ or more. If we take $D_y = 162$ meters, the layers from which the upwelled water originates are located somewhere around 200 meters or deeper. This also agrees with the estimation derived from the results of oceanographic observations off the coast of Southern California (10).

Upwelling Due to a Wind Not Parallel to the Coast

In the foregoing discussion the author has shown the possibility of upwelling induced by a wind parallel to the coast. It may be anticipated that a similar motion can occur in the same vertical plane as a result of an offshore wind, that is, a wind blowing offshore at right angles to the coast. By a study of a combination of these two motions, it should be possible to determine what kind of vertical water motion would correspond to a wind of given direction. In the following lines, a method will be given by which we can determine the most favorable wind direction for upwelling.

Let us examine a wind which is not perpendicular to the coast and divide the wind stress τ into its two components τ_x and τ_y in the x- and y-directions (Figure 10-3). Then the surface condition to be satisfied by u and v should be written, in place of (10-8)

$$-A_v\,(\partial/\partial z)\,(u+iv)\,\Big|_{z\,=0} \quad \left\{ \begin{array}{ll} = \tau_x + i\,\tau_y & (0 \leqslant x \leqslant L) \\[2mm] = 0 & (L < x < \infty) \end{array} \right\}$$

while the dynamical equations, equation of continuity and other boundary conditions remain the same.

In a manner similar to that used in our foregoing discussion, we can integrate the equations of motion. The stream function $\psi(x,z)$ in the plane perpendicular to the coast (xz-plane) and the horizontal components of current velocity are given by

$$\psi\,(w,z) = \text{real part of } [2\pi\,(\tau_x + i\,\tau_y)/\rho\omega \sin \phi] \int_0^\infty [(e^{-\sqrt{\lambda^2 + 2\pi^2 i}\,z/D_v} - 1)/(\lambda^2 + 2\pi^2 i)]$$

$$\left\{ \sin\,(\lambda\,x/D_h)\,[1 - \cos\,(\lambda\,L/D_h)]/\lambda \right\} d\,\lambda$$

and

$$u + iv = [2\pi\,(\tau_x + i\,\tau_y)/\rho\omega \sin \phi] \int_0^\infty [e^{-\sqrt{\lambda^2 + 2\pi^2 i}\,z/D_v}/\sqrt{\lambda^2 + 2\pi^2 i}]$$

$$\left\{ \sin\,(\lambda\,x/D_h)\,[1 - \cos\,(\lambda\,L/D_h)]/\lambda \right\} d\,\lambda$$

or, in real forms

$$\psi(x,z) = (2\pi\,\tau_x/\rho\omega \sin \phi)\,\Phi_x\,(x,z) + (2\pi\,\tau_y/\rho\omega \sin \phi)\,\Phi_y\,(x,z)$$

$$u(x,z) = (2\pi\,\tau_x/\rho\omega \sin \phi\,D_v)\,N\,(x,z) - (2\pi\,\tau_y/\rho\omega \sin \phi\,D_v)\,M\,(x,z)$$

$$v(x,z) = (2\pi\,\tau_x/\rho\omega \sin \phi\,D_v)\,M\,(x,z) + (2\pi\,\tau_y/\rho\omega \sin \phi\,D_v)\,N\,(x,z)$$

Table 10-2
Values of the Stream Function $\psi_x(x,z)$ x 10^4

z/D$_v$	x/D$_h$								
	0	0.1396	0.2793	0.5585	0.8378	1.1170	1.3963	1.6755	1.9548
0	0	0	0	0	0	0	0	0	0
0.2	0	−168	−303	−301	−254	−270	−280	−305	−202
0.4	0	−194	−342	−310	−230	−238	−257	−303	−199
0.6	0	−179	−309	−234	−135	−135	−161	−226	−133
0.8	0	−162	−275	−172	− 68	− 58	− 83	−162	− 85
1.0	0	−152	−256	−138	− 19	− 17	− 43	−129	− 61
1.2	0	−148	−248	−125	− 5	− 3	− 30	−107	− 54
1.4	0	−148	−274	−124	− 4	− 3	− 30	−107	− 54
1.6	0	−147	−248	−126	− 6	− 6 −	− 33	−110	− 57

Table 10-3
Values of the Function $N(x,z)$ x 10^3

z/D$_v$	x/D$_h$								
	0	0.1396	0.2793	0.5585	0.8378	1.1170	1.3963	1.6755	1.9548
0	0	+240	+277	+284	+253	+252	+252	+266	+243
0.2	0	+ 41	+ 64	+ 55	+ 35	+ 31	+ 38	+ 47	+ 22
0.4	0	− 3	− 8	− 28	− 42	− 45	− 42	− 28	− 17
0.6	0	− 10	− 20	− 38	− 47	− 50	− 48	− 39	− 29
0.8	0	− 7	− 14	− 24	− 29	− 30	− 29	− 24	− 20
1.0	0	− 3	− 6	− 11	− 12	− 12	− 12	− 10	− 8
1.2	0	− 1	− 2	− 2	− 3	− 3	− 2	− 2	− 1
1.4	0	0	0	− 1	+ 1	+ 1	+ 1	+ 1	+ 1
1.6	0	0	+ 1	+ 1	+ 2	+ 2	+ 2	+ 1	+ 1

Table 10-4
Values of the Function $M(x,z)$ x 10^3

z/D_v	x/D_h								
	0	0.1396	0.2793	0.5585	0.8378	1.1170	1.3963	1.6755	1.9548
0	0	−98	−174	−248	−255	−243	−263	−235	−174
0.2	0	−70	−127	−186	−191	−182	−197	−175	−129
0.4	0	−34	− 62	− 93	− 96	− 90	− 97	− 85	− 62
0.6	0	−12	− 21	− 31	− 30	− 26	− 30	− 26	− 18
0.8	0	− 2	− 3	− 2	0	+ 3	+ 1	+ 1	+ 1
1.0	0	+ 1	+ 3	+ 6	+ 10	+ 30	+ 9	+ 8	+ 6
1.2	0	+ 1	+ 3	+ 6	+ 8	+ 8	+ 8	+ 6	+ 5
1.4	0	+ 1	+ 2	+ 3	+ 4	+ 4	+ 4	+ 3	+ 2
1.6	0	0	+ 1	+ 1	+ 1	+ 1	+ 1	+ 1	+ 1

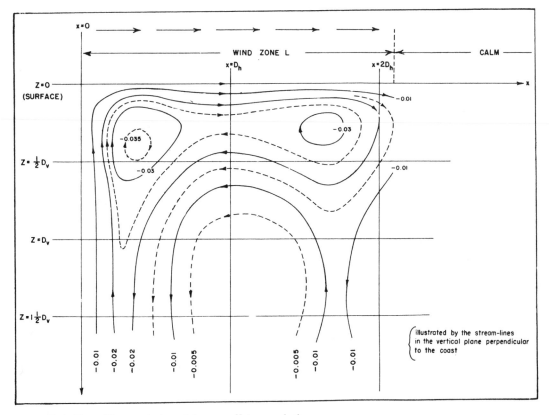

Figure 10-4. Upwelling as induced by an offshore wind.

where

$$\Phi_x(x,z) = \int_0^\infty \left\{ (P^2 - Q^2) \left[e^{-Pz/D_v} \cos(Qz/D_v) - 1 \right] - 2PQ \, e^{-Pz/D_v} \sin(Qz/D_v) \right\} / (P^2 + Q^2)^2$$
$$\times \left\{ \sin(\lambda x/D_h) [1 - \cos(\lambda L/D_h)]/\lambda \right\} d\lambda$$

$$\overline{\otimes}_y(x,z) = \int_0^\infty \left\{ (P^2 - Q^2) \, e^{-Pz/D_v} \sin(Qz/D_v) + 2PQ \left[e^{-Pz/D_v} - 1 \right] \right\} / (P^2 + Q^2)^2$$
$$\times \left\{ \sin(\lambda x/D_h) [1 - \cos(\lambda L/D_h)]/\lambda \right\} d\lambda$$

$$M(x,z) = \int_0^\infty \left\{ [Q \cos(Qz/D_v) + P \sin(Qz/D_v)]/(P^2 + Q^2) \right\} e^{-Pz/D_v}$$
$$\left\{ \sin(\lambda x/D_h) [1 - \cos(\lambda L/D_h)]/\lambda \right\} d\lambda$$

$$N(x,z) = \int_0^\infty \left\{ [P \cos(Qz/D_v) - Q \sin(Qz/D_v)]/(P^2 + Q^2) \right\} e^{-Pz/D_v}$$
$$\left\{ \sin(\lambda x/D_h) [1 - \cos(\lambda L/D_h)]/\lambda \right\} d\lambda$$

Thus the stream function is composed of two components, that is, those due to the wind stress τ_x perpendicular and τ_y parallel to the coast.

If we put $\tau_x = 0$ and $\tau_y = \tau$ we have exactly the foregoing discussion in which the wind blows parallel to the coast.

The component stream functions $\Phi_x(x,z)$ and $\Phi_y(x,z)$ and the functions $M(x,z)$ and $N(x,z)$, which give the horizontal components of velocity, are evaluated numerically against several values of x/D_h and z/D_v, and are compiled in Tables 10-1 to 10-4.

Vertical Circulation Patterns

Figures 10-2 and 10-4 illustrate the contents of Tables 10-1 and 10-2, respectively. They show the streamlines in the vertical circulations induced in the plane perpendicular to the coast (xz-plane) by offshore and longshore winds. Both are clockwise circulations. The values of stream functions are all negative, showing that upwelling exists close to the coast produced by the wind stresses in positive directions of x and y. In each of these cases, the water is upwelled from levels deeper than $z = D_y$, but, as a whole, it appears that the upwelled water originates in deeper levels with a longshore than with an offshore wind. The circulation induced by a longshore wind has been described in detail in the foregoing discussion. But the circulation due to an offshore wind of the same stress will carry a much smaller amount of water than will that due to a longshore wind. Moreover, the former has a rather complicated structure, having two eddies (Figure 10-4) in upper layers, one situated close to the coast, the other near the outer boundary of the wind belt. These circumstances mean that upwelling due to a longshore wind is far more effective in lowering the temperature of the region close to the coast than that induced by an offshore one, since the former brings a much larger amount of colder water to the surface from deeper levels than the latter.

Most Favorable Wind Direction for Upwelling

The amount T of water upwelled through a long horizontal strip 1 centimeter wide and extending from the coast to a distance x will be evaluated by

$$T = \int_0^x \rho \, w \, dx = \rho \int_0^x (\partial \psi / \partial x) \, dx = \rho \, \psi \, (x,z) - \rho \, \psi(0, z) = \rho \, \psi(x, z)$$

since the stream function vanishes along the coastal wall ($x = 0$). Thus the vertical mass transport T will be largest when we take x at a distance at which Φ_x and Φ_y have the largest negative values. By inspection of Tables 10-1 and 10-2 and Figures 10-2 and 10-4, we can estimate

$$T_y = -0.090 \, (2\pi \, \tau_y / \rho \omega \sin \phi) \qquad \text{by longshore wind, and}$$

$$T_x = -0.0355 \, (2\pi \, \tau_x / \rho \omega \sin \phi) \qquad \text{by offshore wind} \qquad\qquad (10\text{-}18)$$

If the wind blows in an offshore direction making an angle θ with the coast (Figure 10-3), we have

$$\tau_x = \tau \sin \theta \qquad\qquad \tau_y = \tau \cos \theta$$

where τ is the absolute magnitude of the wind stress. Then we have from (10-18)

$$T = T_x + T_y = (-0.355 \sin \theta - 0.090 \cos \theta) \, 2\pi \, \tau / \omega \sin \phi$$

This gives an approximate amount of water upwelled to the upper layers per unit length of the coast. The most intense upwelling will therefore take place when $dT/d\theta = 0$, or

$$\tan \theta = -0.0355/-0.090 \qquad \text{or} \qquad \theta = 21°.5$$

providing the magnitude of the wind stress remains the same. This result means that upwelling will be

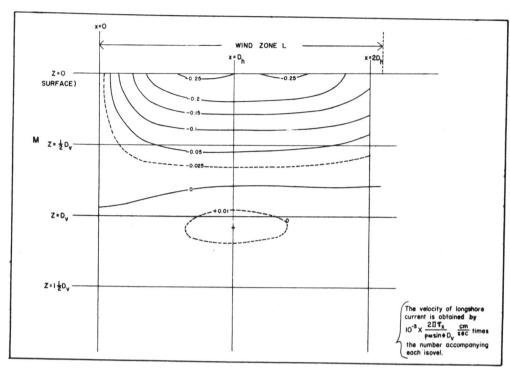

Figure 10-5. Distribution of longshore currents $M(x,z)$ induced by a wind $_x$ parallel to the coast.

most intense when the wind direction makes an angle of 21.5° with the coast line in an offshore direction, that is, when it is slightly deviated offshore from the coast line. This angle of deviation will of course depend upon the width L of the wind belt. But for a value L exceeding D_h, it is understood that θ approaches an asymptotic value not much different from 21.5°. According to Arnold Glaser, upwelling off the coast of Lima, Peru, appears to be most intense when the southerlies blow in a slight offshore direction rather than when they blow parallel to the coast.

In the Gulf of Mexico, we have easterlies almost all year round. This means that we can always expect upwelling along the northern coast of Cuba and the Yucatan Coast of Mexico. We have some evidence of upwelling or existence of colder water off these coasts (15), and we know that many American vessels fish in the waters off

the Yucatan Coast. This fact may suggest the existence of upwelling in this region. According to the surface observations of the Texas A&M Department of Oceanography made recently, there are some indications of colder surface water in the western part of the Gulf of Mexico very close to the coasts. If these conditions really exist, this temperature distribution may be ascribed to upwelling due to the easterly winds prevailing in these latitudes.

Coastal Currents

Recently a number of observations (16, 17, 18 and 19) on the flow of water close or adjacent to coasts have been reported. Many authors have discussed this subject, but they have not yet been able to arrive at a satisfactory explanation of possible causes of coastal or longshore currents. How-

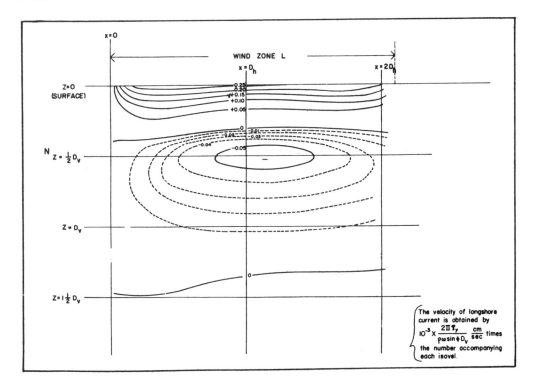

Figure 10-6. Distribution of longshore currents *N(x,z)* induced by an offshore wind ₓ.

ever, most of these currents are ascribed to the action of surf and waves. The longshore component which we have obtained may explain some part of these flows. The expression for this component is given by

$$v(x,z) = (2\pi \tau_x/\rho\omega \sin \phi D_v) N(x,z) - (2\pi \tau_y/\rho\omega \sin \phi D_v) M(x,z) \tag{10-19}$$

where M and N are the functions compiled in Tables 10-3 and 10-4 and illustrated in Figure 10-5 and 10-6. The two terms in the right-hand member of (10-19) show the longshore currents induced by offshore and longshore winds respectively. If the wind is parallel to the coast, we have $\tau_x = 0$ and have

$$v = -(2\pi \tau_y/\rho\omega \sin \phi D_v) M(x,z)$$

Table 10-5
The Angle Between the Wind and Surface Current

x/D_h	0	0.1396	0.2793	0.5585	0.8378	1.1170	1.3963	1.6755	1.9548
Angle	0°	22°.2	32°.2	41°.2	45°.2	44°.3	46°.3	41°.5	35°.7

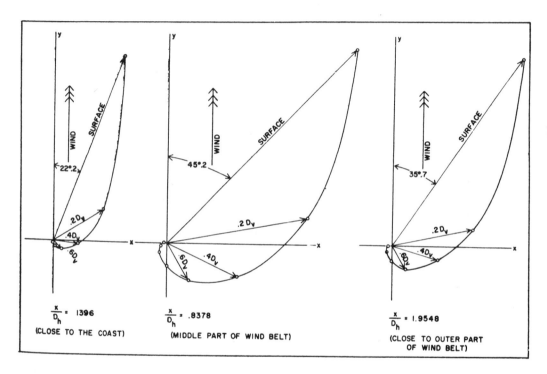

Figure 10-7. Vertical variation of drift currents.

and if it is blowing in an offshore direction perpendicular to the coast, we have

$$v = + \, (2\pi \, \tau_x / \rho \omega \sin \phi \, D_v) \, N \, (x,z)$$

From Tables 10-3 and 10-4 and Figures 10-5 and 10-6 it is possible for us to tell the difference between the two cases. The coastal current induced by a longshore wind is directed leeward as far down as the level $z = 0.9 \, D_v$, but the direction of flow is reversed at deeper levels. On the other hand, the longshore current produced by an offshore wind flows to the right of the wind direction in layers shallower than $z = 0.3D_v$ only. There is a secondary maximum of velocity in the opposite direction at the approximate level $z = 0.5D_v$.

Vertical Variation of Horizontal Currents; Ekman Spiral

The expressions

$$u\,(x,z) = (2\pi\,\tau_x/\rho\omega\,\sin\phi\,D_v)\,N\,(x,z) - (2\pi\,\tau_y/\rho\omega\,\sin\phi\,D_v)\,M\,(x,z) \left.\rule{0pt}{14pt}\right\}$$

$$v\,(x,z) = (2\pi\,\tau_x/\rho\omega\,\sin\phi\,D_v)\,M\,(x,z) + (2\pi\,\tau_y/\rho\omega\,\sin\phi\,D_v)\,N\,(x,z) \left.\rule{0pt}{14pt}\right\}$$

for the horizontal currents show that the angle between the wind and the current induced at a given depth depends only on the distance from the coast and not on the direction of wind. The angle between the wind and the surface current induced by it is computed for several different distances from the coast and compiled in Table 10-5.

Thus the angle between the wind and surface current is near 45° at the middle part of the wind belt but decreases both toward the coast and toward the outer margin of the wind belt. Along the coast there is a very weak current in approximately the same direction as the wind. The vertical variation of currents is practically equal to the Ekman (11) spiral along the median line of the wind belt but is more or less flattened close to the coast and near the outer margin of the wind belt. Some of these spirals are illustrated in Figure 10-7.

Acknowledgments

The author expresses his sincerest appreciation to Dale F. Leipper, who encouraged him in carrying out the present research and in publishing the result during his stay in the Department of Oceanography while Visiting Graduate Professor, 1952-1953, and to R.B. Montgomery of Brown University who has also given many valuable suggestions to the author. He is also much obliged to Dale Leipper, W. Armstrong Price, Robert O. Reid, John Hurt and Arnold Glaser for their valuable suggestions and information; to Richard M. Adams for checking the English in the manuscript; and to George B. Austin who prepared the diagrams. This paper is based in part on work done under the sponsorship of the Office of Naval Research and the Bureau of Ships.

References

1. Thorade, H. 1909. Über die kalifornische Meeresströmung. An. Hydrog. 37:17-34, 63-76.

2. McEwen, G.F. 1912. The distribution of ocean temperatures along the west coast of North America deduced from Ekman's theory of the upwelling of cold water from the adjacent ocean depths. Internatl. Rev. gestamt. Hydrobiol. Hydrogr. 5:243-286.

3. _____. 1929. A mathematical theory of the vertical distribution of temperature and salinity in water under the action of radiation, conduction, evaporation, and mixing due to the resulting convection. Bull. Scripps Inst. Ocean. 2:197-306.

4. Gunther, E.R. 1936. A report on oceanographical investigations in the Peru coastal current. Disc. Rep. 13:109-275.

5. Defant, A. 1936. Das Kaltwasserauftriebsgebiet vor der Küste Sudwestafrikas, Länderkundliche Forschung, Festchrift Norbert Krebs zur Vollendung des 60, Lebensjahres dargebracht, p. 52-66.

6. _____. 1952. Theoretische Überlegungen zum Phonomen des Windstaus und des Auftriebes an ozeanischen Küsten. Deuts. Hydrog. Zs. 5:69-80.

7. Sverdrup, H.U. 1931. Some oceanographic results of the Carnegie's work in the Pacific— The Peruvian Current. Hydrogr. Rev. 8:240-244.

8. _____. 1938. On the process of upwelling. J. Mar. Res. 1:155-164.

9. _____ and Fleming, R.H. 1941. The waters off the coast of Southern California, March to July, 1937. Bull. Scripps Inst. Ocean. 4:261-378.

10. _____ et al. 1942. The oceans. New York: Prentice Hall, p. 501.

11. Ekman, V.W. 1905. On the influence of the earth's rotation of ocean currents. Ark. Mat. Ast. Fysik 2 (11).

12. Takegami, T. 1934. The boundary value problem of the wind current in a lake or a sea. Mem. Kyoto Imper. Univ. (ser. A) 27:305-318.

13. Longard, J.R. and Banks, R.E. 1952. Wind-induced vertical movement of the water on an open coast. Trans. Amer. Geophys. Union 33:377-380.

14. Saito, Yukimasa. 1951. On the velocity of the

vertical flow in the ocean. *J. Inst. Polytech.* (Osaka City Univ., ser. B) 2:1-4.

15. Smith, F.G. Walton et al. 1951. Distribution of vertical water movement calculated from surface drift vectors. *Bull. Mar. Sci. Gulf Carib.* 1:187-195.

16. Putnam, J.A., Munk, W.H. and Traylor, M.A. 1949. The prediction of longshore currents. *Trans. Amer. Geophys. Union* 30:337-345.

17. Shepard, F.P. and Inman, D.L. 1950. Near-shore water circulation related to bottom topography and wave refraction. *Trans. Amer. Geophys. Union* 31:196-212.

18. ___ and Inman, D.L. 1951. Nearshore circulation. *Proc. First Conf. Coastal Eng.*, p. 50-59.

19. Inman, D.L. and Quinn, W.H. 1953. Currents in the surf zone. *Coastal Eng.*, p. 24-36.

11

seiches due to a submarine bank

Formerly the theory of seiches dealt chiefly with the cases of land-locked basins and gulfs. Recently the seiches in a channel have been treated mathematically by the present writer (1).

Some cases of sea-seiches observed on the coasts facing the open ocean have been frequently ascribed to the existence of the shelf between the shore and the deep ocean. This kind of phenomena was discussed by R.A. Harris and called the "shelf seiches" by him.

Thus, seiches can take place in a sea having a coast on one side only. Then the question arises: Is the existence of the coast absolutely essential for seiches? The present chapter deals with this problem and shows that seiches or stationary oscillations can take place in complete absence of the coasts when we suppose a submarine bank.

Consider an ocean unbounded and infinitely wide. Neglect the rotation of the earth and the curvature of the free surface.

Let the depth h of this ocean be given by

$$h = h_0 \left[1 + (x^2/a^2)\right]^{1/2} \qquad (11\text{-}1)$$

where h_0 is the depth at $x = O$ and a the distance from $x = O$ to a point where $h = \sqrt{(2h_0)}$.

Then the bottom configuration is characterized by the existence of a submarine elevation, or a bank, just as seen by this equation. The depth is smallest $=h_0$ above the bank and becomes gradually larger as we leave the bank (see Figure 11-1).

The differential equation for the elevation ζ of the water above the undisturbed level is

$$(\partial^2 \zeta)/(\partial t^2) = g\ (\partial/\partial x)\ [h(x)\ (\partial \zeta/\partial x)]$$

where t is the time and g the acceleration due to gravity.

Assuming the motion to be a simple harmonic oscillation with the period $T = 2\pi/\sigma$ we may put

$$\zeta = U\ (x/a)\ \cos \sigma t$$

where U is independent of t. Then we have from (11-1),

$$d/d\rho\ \left[(1 + \rho^2)^{1/2}\ (dU/d\rho)\right] + (\theta/2)\ U = 0 \qquad (11\text{-}2)$$

where

$$\rho\ =\ x/a\ ,\ \text{and}$$
$$\theta\ =\ (2\sigma^2 a^2/gh_0). \qquad (11\text{-}3)$$

By the substitution $\rho = \sinh 2\xi$ (11-2) further becomes

$$(d^2 U/d\xi^2) + 2\theta \cosh 2\xi \cdot U = 0 \qquad (11\text{-}4)$$

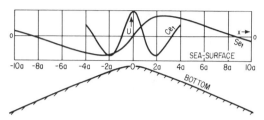

Figure 11-1. Distribution of amplitude in the seiches due to a submarine bank.

Now that the oscillations of water are considered to be the largest immediately above the bank and subside gradually as the distance to it gets larger, we must seek for the solutions of (11-4) which are finite in the range $0 \leqslant \xi < \infty$ and tend to zero as $\xi \to \infty$.

Consider the particular solutions finite for $-\infty < \xi < \infty$ and vanishing at $\xi = \pm \infty$, of the modified Mathieu equation

$$(d^2y/d\xi^2) - (a - 2\theta\cosh 2\xi)y = 0.$$

These are evidently the modified Mathieu functions or the Mathieu functions of an imaginary argument. There are four series of such functions, that is,

$$Ce_{2n}(\xi, \theta), \quad Se_{2n+1}(\xi, \theta), \quad Ce_{2n+1}(\xi, \theta)$$

and $Se_{2n+2}(\xi, \theta)$

where $n = 0, 1, 2$ ...If we arrange these functions in the increasing order of the suffices, we have

$$Ce_0(\xi, \theta), Se_1(\xi, \theta), Ce_1(\xi, \theta), Se_2(\xi, \theta), \ldots$$
$$(11-5)$$

In each of these functions, a is uniquely determined for a given value of θ and is called the characteristic number. Of course, a is different for each of the above functions (11-5) for the same value of θ. Hence, if a is given, θ is particular to each of the modified Mathieu functions above described. a

takes the value $a = 0$ if θ be chosen appropriately, except for $Ce_0(\xi, \theta)$.

Hence, suppose $a = 0$. Then θ takes different values for each of the functions (11-5). Hence, we may write in this case,

$$Ce_0(\xi, \theta_0'), \quad Se_1(\xi, \theta_1), \quad Ce_1(\xi, \theta_1'),$$
$$Se_2(\xi, \theta_2), \ldots \qquad (11-6)$$

Then the equation satisfied by these functions will be the differential equation (11-4). Thus, (11-6) are the solutions of the equation (11-4) which are finite in the range $-\infty < \xi < \infty$ and tend to zero for $\xi = \pm \infty$. The method of determining the values of θ's is described in the next section.

There is no value of θ which makes $a = 0$ for Ce_0 except $\theta_0' = 0$, which corresponds to the motionless state.*

Hence, the gravest mode is the oscillation proportional to $Se_1(\xi, \theta_1)$. Using the substitution

$$Se_{2n+1}(\xi, \theta_{2n+1})$$
$$= \sum_{r=0}^{\infty} B_{2r+1}^{2n+1}(\theta) \sinh(2r+1)\xi,$$

we obtain, by the theory of Mathieu functions, the period-equation:

$$\theta - 1 + (\theta^2/3^2 -) (\theta^2/5^2 -) (\theta^2/7^2 -), \ldots$$
$$= 0\dagger$$

the value of θ_1, is given by the smallest root of this equation, or by $\theta_1 = 0.9080163$.

The period of oscillation for this mode is, by (11-3),

*When there is a boundary at a certain distance from the bank, this mode will be an oscillation with a finite period.
†For the process of reducing this period-equation, refer to the theory of Mathieu functions. See, for example, E.L. Ince, Tables of the Elleptic-cylinder Functions, Proc. Roy. Soc. Edinburgh, Vol. LII, Part IV, No. 22, 1932. Pp. 355-423.

Table 11-1. Distribution of Amplitude U.

First mode (Se_1)				Second mode (Ce_1)			
x/a	U	x/a	U	x/a	U	x/a	U
0.0	0.0000	5.0	0.5751	0.0	1.0000	2.5	−0.5880
0.2	0.1335	5.2	0.5441	0.1	0.9813	2.6	−0.5224
0.4	0.2597	5.4	0.5123	0.2	0.9265	2.7	−0.4520
0.6	0.3735	5.6	0.4800	0.3	0.8391	2.8	−0.3781
0.8	0.4723	5.8	0.4473	0.4	0.7244	2.9	−0.3018
1.0	0.5555	6.0	0.4143	0.5	0.5889	3.0	−0.2241
1.2	0.6242	6.2	0.3811	0.6	0.4392	3.1	−0.1462
1.4	0.6795	6.4	0.3479	0.7	0.2822	3.2	−0.0688
1.6	0.7229	6.6	0.3147	0.8	0.1239	3.3	0.0072
1.8	0.7560	6.8	0.2816	0.9	−0.0305	3.4	0.0810
2.0	0.7801	7.0	0.2487	1.0	−0.1766	3.5	0.1521
2.2	0.7961	7.2	0.2162	1.1	−0.3110	3.6	0.2197
2.4	0.8052	7.4	0.1839	1.2	−0.4310	3.7	0.2834
2.6	0.8083	7.6	0.1521	1.3	−0.5357	3.8	0.3428
2.8	0.8061	7.8	0.1207	1.4	−0.6233	3.9	0.3975
3.0	0.7992	8.0	0.0898	1.5	−0.6936	4.0	0.4473
3.2	0.7883	8.2	0.0595	1.6	−0.7466		
3.4	0.7739	8.4	0.0298	1.7	−0.7827		
3.6	0.7563	8.6	0.0007	1.8	−0.8028		
3.8	0.7360	8.8	−0.0284	1.9	−0.8076		
4.0	0.7134	9.0	−0.0561	2.0	−0.7985		
4.2	0.6888	9.2	−0.0832	2.1	−0.7766		
4.4	0.6624	9.4	−0.1095	2.2	−0.7432		
4.6	0.6345	9.6	−0.1350	2.3	−0.6998		
4.8	0.6054	9.8	−0.1598	2.4	−0.6476		
		10.0	−0.1838				

$$T_1 = (2\pi a)/(\sqrt{\theta_1/2gh_0})$$

$$= 6.5936 \; a/(\sqrt{gh_0}).$$

The second mode is that proportional to $Ce_1(\xi, \theta_1')$. By the substitution

$$Ce_{2n+1}(\xi, \theta'_{2n+1})$$

$$= \sum_{r=0}^{\infty} B_{2r+1}^{2n+1}(\theta) \cosh(2r+1)\xi,$$

we obtain the period-equation

$$- \theta' - 1 + \theta'^2/(3^2 -) \ (\theta'^2)/(5^2 -) \ (\theta'^2)$$
$$/(7^2 -), \ldots \ = \ 0 \ ;$$

the smallest root of this equation is what we require. It is $\theta_1' = 7.51361$, giving $T_2 = 3.2417$ $a/\sqrt{(gh_0)}$.

A sequence of oscillations proportional to Se_2, Ce_2, etc., are easily obtained in a similar way. Reference should again be made to the theory of Mathieu functions.

For computing the distribution of amplitude in each mode, we have only to solve the differential equation (11-4) with the given value of $\mathcal{C}$ and suitable initial values of U and $dU/d\rho$. To carry out this process, the numerical integration is most effective.

Now, if we put

$$U = U_1/(1 + \rho^2)^{1/4}$$

we have the equation for U_1:

$$\frac{d^2 U_1}{d\rho^2}$$

$$+ \left[\frac{\sigma/2}{(1 + \rho^2)^{1/2}} + \frac{1/4}{(1 + \rho^2)} - \frac{3/4}{(1 + \rho^2)^2} \right] U_1$$

$$= 0.$$

This equation can be integrated by Störmer's method. The functions U are approximately equal to the circular functions for ρ small. Hence, the constant factors were determined in analogy to the circular functions, that is, $U(0) = 0$, $U'(0) =$ $\sqrt{(\theta/2)}$ for odd functions and $U(0) = 1$, $U'(0) = 0$ for even functions. The first three or four values of U_1 were computed by the Runge-Kutta method. Then the values of U_1 were derived and further integration was carried out by Störmer's method.[*] The results for Se_1 and Ce_1 are compiled in Table 11-1.

In a rectangular canal of lengths $2l$ and uniform depth h_{01} the period of oscillation is given by

$$T_n = 4l/n\sqrt{(gh_0)}$$

where $n = 1, 2, 3,\ldots$.When l becomes indefinitely large, every period of oscillation also increases indefinitely. But when the depth is not uniform as mentioned in our investigation, only the fundamental mode is of an infinitely long period, and other modes are all of finite lengths of period. This is a very queer conclusion but nevertheless true.

Acknowledgments

In conclusion the author expresses his best thanks to Gakuzyutsu Kinko Kwai for the grant given for the present research and to Dr. T. Okada, director of the Imperial Marine Observatory, Kobe, Japan, who kindly read this paper before printing.

References

1. Hidaka, K. 1935. Seiches in a channel. *Memoirs of the Imp. Mar. Obs.* (5)4:327-358.

[*]Here we used Störmer's method modified by H. Levy, See H. Levy and E.A. Baggott, *Numerical Studies in Differential Equations*, London, 1934, p. 158. The error is about $[(\Delta\rho)^6/240] \ (d \ U_1/d\rho^6)$.

12

a theory
of shelf seiches

The theory of seiches due to a submarine bank given in the previous chapter leads us to the theory of the very interesting case of shelf seiches.

We consider a straight coast with a vertical wall. Take this coast for $x = 0$ and suppose the depth h of the ocean is given by

$$h = h_O \left[1 + (x^2/a^2)\right]^{1/2}$$

where h_O is the depth immediately before the coast and a is the distance from the shore to the isobath of $\sqrt{2}h_O$.

The equations of motion and the conditions to be satisfied are the same as in the case of bank seiches, except for an additive condition that the motion of the water perpendicular to the coast is nil, that is, $(\partial \zeta / \partial x)_{x=0} = 0$.

This at once leads us to impose an additive condition $(dU/d\rho)_{\rho=0} = 0$.

Now it is evident that the modified Mathieu functions which satisfy the condition 12-1 are the Ce functions.

The gravest mode is the oscillation proportional to Ce_1. This is the same as the second mode in the bank seiches. The period of oscillation is given by $T_1 = 3.214\, a/\sqrt{(gh_O)}$.

The second mode is the oscillation proportional to Ce_2. We have for this mode

$$\theta_2' = 21.29863$$

and

$$T_2 = 1.9254\, [a/\sqrt{(gh_O)}]\,.$$

Further modes can be, of course, obtained in a similar manner. Reference should be made to theory of Mathieu functions.

The distribution of amplitude for these two gravest modes was found by the method described in the previous chapter. The results are given in Table 12-1 and Figure 12-1.

Table 12-1
Distribution of Amplitude U.

First mode				Second mode			
x/a	U	x/a	U	x/a	U	x/a	U
0.0	1.0000	2.0	-0.7985	0.0	1.0000	2.0	0.5600
0.1	0.9813	2.1	-0.7766	0.1	0.9474	2.1	0.6665
0.2	0.9265	2.2	-0.7432	0.2	0.7964	2.2	0.7392
0.3	0.8391	2.3	-0.6998	0.3	0.5668	2.3	0.7773
0.4	0.7244	2.4	-0.6476	0.4	0.2868	2.4	0.7818
0.5	0.5889	2.5	-0.5880	0.5	-0.0115	2.5	0.7548
0.6	0.4392	2.6	-0.5224	0.6	-0.2972	2.6	0.6995
0.7	0.2822	2.7	-0.4520	0.7	-0.5445	2.7	0.6197
0.8	0.1239	2.8	-0.3781	0.8	-0.7346	2.8	0.5200
0.9	-0.0305	2.9	-0.3018	0.9	-0.8566	2.9	0.4052
1.0	-0.1766	3.0	-0.2241	1.0	-0.9070	3.0	0.2800
1.1	-0.3110	3.1	-0.1462	1.1	-0.8887	3.1	0.1493
1.2	-0.4313	3.2	-0.0688	1.2	-0.8096	3.2	0.0176
1.3	-0.5357	3.3	0.0072	1.3	-0.6807	3.3	-0.1110
1.4	-0.6233	3.4	0.0810	1.4	-0.5149	3.4	-0.2327
1.5	-0.6936	3.5	0.1521	1.5	-0.3256	3.5	-0.3443
1.6	-0.7466	3.6	0.2197	1.6	-0.1261	3.6	-0.4430
1.7	-0.7827	3.7	0.2834	1.7	0.0716	3.7	-0.5267
1.8	-0.8028	3.8	0.3428	1.8	0.2572	3.8	-0.5938
1.9	-0.8076	3.9	0.3975	1.9	0.4221	3.9	-0.6436
		4.0	0.4473			4.0	-0.6751

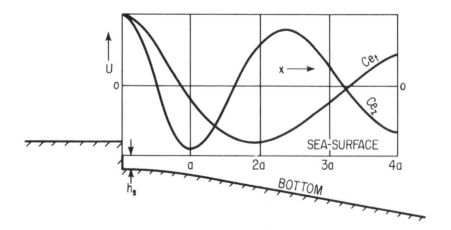

Figure 12-1. Distribution of amplitude in the shelf seiches.

Acknowledgements

In conclusion the author expresses his best thanks to the *Nippon Gakuzyutsu Shinko Kwai* for the grant given for the present research and to Dr. T. Okada, director of the Imperial Marine Observatory, Kobe, Japan, who read this paper before printing.

13

ocean currents, upwelling and sinking in a zonal ocean

Introduction

In 1958, the present author published a dynamical calculation of the equatorial current system of the Pacific (1) under the assumption that lateral mixing is taking place in addition to vertical mixing. The calculation was two-dimensional because the equatorial current system of the Pacific can be approximately regarded as a stationary flow without appreciable variation in east-west direction. Flow patterns in a meridional section were studied. In that paper, however, the authors could not succeed in calculating the upwelling around the equator.

Afterward the author published a different method enabling him to derive the vertical flow in addition to the horizontal. The principle of this method was published in 1961 (2). A slight modification was recently made on this method and given in this paper.

Numerical calculation will be possible although it is not simple.

The two-dimensional dynamical equations were solved for a zonal flow in the equatorial region of a meridional section of oceans. Expressions for horizontal and vertical components of velocity were derived by assuming a balance between the pressure gradient, Coriolis and frictional forces. The frictional forces consist of terms of both vertical and lateral mixing. No non-linear terms are not taken into account.

Three components of current velocity can be derived from the observational data-density distribution and wind stresses on the sea surface.

Theory

Suppose a meridional section passing through the equator. Take the x-axis positive eastward, the y-axis northward and z-axis vertically downward from the undisturbed level surface. Near the equator the east-west variation in velocity and pressure gradients is generally much smaller than the meridional variation, so we can neglect the term $\partial^2/\partial x^2$ compared with $\partial^2/\partial y^2$. Then the steady-state equations of motion are as follows:

$$\left.\begin{array}{l} A_l\,(\partial^2 u/\partial y^2) + A_z(\partial^2 u/\partial z^2) + 2\omega\sin\phi\rho v = \partial p/\partial x; \\[2mm] A_l\,(\partial^2 v/\partial z^2) + A_z\,(\partial^2 v/\partial z^2) - 2\omega\sin\phi\rho u = \partial p/\partial y, \end{array}\right\}$$

$$(13\text{-}1)$$

where u and v are the x- and y-components of velocity, respectively, p is the pressure, ρ the density of sea water, A_l and A_z are the coefficients of horizontal and vertical mixing, respectively, ω is the angular velocity of the earth's rotation and ϕ is the latitude connected with y by a relation $y = R\phi$ where R is the radius of the earth. The conditions to be satisified at the sea surface are

$$z = 0: \quad -A_z \, (\partial u/\partial z) = \tau_x; \quad -A_z \, (\partial v/\partial z) = \tau_y$$

where τ_x and τ_y are the components of wind stress and generally functions of y or ϕ. Further let us use the bottom conditions:

$$z = h: \quad (\partial u/\partial z) = (\partial v/\partial z) = 0$$

which means that there is no stress at the bottom $z = h$.

If we differentiate both equations of (13-1) with respect to z and combine the two by a complex expression for the vertical shears,

$$(\partial u/\partial z) + i \, (\partial v/\partial z) = W \, (y, z), \tag{13-2}$$

we shall have

$$A_f(\partial^2 W/\partial y^2) + A_z \, (\partial^2 W/\partial z^2) - i \cdot 2\omega \sin\phi\rho W = g \, [(\partial\rho/\partial x) + i(\partial\rho/\partial y)] \tag{13-3}$$

with the conditions

$$z = 0: \quad W = -(\tau x + i\tau y/A_z) \tag{13-4}$$

and

$$z = h: \quad W = 0. \tag{13-5}$$

In order to solve equation (13-3) subject to the conditions (13-4) and (13-5), we suppose that there are stations all spaced at an equal interval Δy on a meridian and replace in (13-3) the second derivative of $\overline{W}$ with respect to y with a finite difference, or

$$\partial^2 W_k/\partial y^2 = [W_{k+1} - 2W_k + W_{k-1} \, /(\Delta y)^2]$$

where $\overline{W}$ and $(\partial^2 \overline{W}_k)/(\partial y)^2$ represent $\overline{W}$ and its second derivative at a particular station (kth station) on the meridian under consideration, and $\overline{W}_{k-1}$ and $\overline{W}_{k+1}$ refer to the $(k-1)$th and $(k+1)$th stations which are located by a spacing Δy to the north and south of kth station respectively. Then equation (13-3) becomes the following set of m simultaneous equations:

$$A_z \frac{d^2 W_1}{dz^2} + A_l \frac{W_N - 2W_1 + W_2}{(\Delta y)^2} - i \cdot 2\omega \sin\phi_1 \rho W_1 = g \left(\frac{\partial \rho}{\partial x} + i\frac{\partial \rho}{\partial y} \right)_1 ,$$

$$A_z \frac{d^2 W_2}{dz^2} + A_l \frac{W_1 - 2W_2 + W_3}{(\Delta y)^2} - i \cdot 2\omega \sin\phi_2 \rho W_2 = g \left(\frac{\partial \rho}{\partial x} + i\frac{\partial \rho}{\partial y} \right)_2 ,$$

$$A_z \frac{d^2 W_3}{dz^2} + A_l \frac{W_2 - 2W_3 + W_4}{(\Delta y)^2} - i \cdot 2\omega \sin\phi_3 \rho W_3 = g \left(\frac{\partial \rho}{\partial x} + i\frac{\partial \rho}{\partial y} \right)_3 ,$$

$$\cdots\cdots$$

$$A_z \frac{d^2 W_m}{dz^2} + A_l \frac{W_{m-1} - 2W_m + W_s}{(\Delta y)^2} - i2\omega \sin\phi_m \rho W_m = g \left(\frac{\partial \rho}{\partial x} + i\frac{\partial \rho}{\partial y} \right)_m$$

$$(13\text{-}6)$$

where $\overline{W}_N$ and $\overline{W}_S$ are the values of $\overline{W}$ at the northern and southern boundaries of the meridional section, and $((\partial \rho/\partial x) + i(\partial \rho/\partial y))$, $((\partial \rho/\partial x) + i(\partial \rho/\partial y))_2$,——, are the values of $(\partial \rho/\partial x) + i(\partial \rho/\partial y)$ at m stations located in between.

Multiply both sides of (13-6) by a set of m undetermined multipliers l, il_2, $i^2 l_3$,——, $i^{k-1} lk$, ——,$i^{m-1} lm$ respectively and add together; then we have

$$A_z \frac{d^2}{dz^2} (l_1 W_1 + il_2 W_2 + i^2 l_3 W_3 + \cdots + i^{m-1} l_m W_m) +$$

$$+ \frac{A_l}{(\Delta y)^2} \left[\left\{ -(2 + i \cdot 2\omega \sin\phi_1 \frac{(\Delta y)^2}{A_l} \rho) l_1 + il_2 \right\} W_1 \right.$$

$$+ \left\{ l_1 - (2 + i \cdot 2\omega \sin\phi_2 \frac{(\Delta y)^2}{A_l} \rho) il_2 + i^2 l_3 \right\} W_2$$

$$+ \left\{ il_2 - (2 + i \cdot 2\omega \sin\phi_3 \frac{(\Delta y)^2}{A_l} \rho) i^2 l_3 + i^3 l_4 \right\} W_3$$

$+ \ldots$ continued

+ . . . continued

$$+ \left\{ i^{m-2}l_{m-1} - (2 + i\cdot 2\omega\sin\phi_m \frac{(\Delta y)^2}{A_l}\rho)\, i^{m-1}l_m \right\}\; W_m$$

$$= -\frac{A_l}{(\Delta y)^2}\,(l_1 W_N + i^{m-1}l_m W_s) + g\left(\frac{\partial\rho}{\partial x} + i\,\frac{\partial\rho}{\partial y}\right)_1 \cdot l_2$$

$$+ g\left(\frac{\partial\rho}{\partial x} + i\,\frac{\partial\rho}{\partial y}\right)_2 \cdot il_2 + \cdots + g\left(\frac{\partial\rho}{\partial x} + i\,\frac{\partial\rho}{\partial y}\right)_m \cdot i^{m-1}l_m .$$

$$(13\text{-}7)$$

Here we introduce an unknown ξ which satisfies the following equations:

$$-(2 + i\cdot 2\omega\sin\phi_1 \cdot (\Delta y)^2/A_l\rho)l_1 + il_2 = -(2 + i\xi)l_1 ,$$

$$l_1 - (2 + i\cdot 2\omega\sin\phi_2 \cdot (\Delta y)^2/A_l\rho)il_2 + i^2 l_3 = -(2 + i\xi)il_2 ,$$

$$il_2 - (2 + i\cdot 2\omega\sin\phi_3\,(\Delta y)^2/A_l\rho)i^2 l_3 + i^3 l_4 = -(2 + i\xi)i^2 l_3 ,$$

$$i^{m-2}l_{m-1} - (2 + i\cdot 2\omega\sin\phi_m\,(\Delta y)^2/A_l\rho)i^{m-1}l_m = -(2 + i\xi)i^{m-1}lm$$

or

$$\{\xi - 2\omega\sin\phi_1\,(\Delta y)^2/A_l\rho\}\; l_1 + l_2 = 0,$$

$$-l_1 + \{\xi - 2\omega\sin\phi_2(\Delta y)^2/A_l\,\rho\}l_2 + l_3 = 0,$$

$$-l_2 + \{\xi - 2\omega\sin\phi_3\,(\Delta y)^2/A_l\,\rho\}l_3 + l_4 = 0,$$

$$-l_{m-1} + \{\xi - 2\omega\sin\phi_m\,(\Delta y)^2/A_l\,\rho\}\, lm = 0,$$

$$(13\text{-}8)$$

Eliminating l_1, l_2, l_3, ——lm from these m equations, we shall have an algebraic equation of mth order in ξ. This equation has m roots ξ_1, ξ_2, ξ_3,——ξ_m. For each of these roots, or ξ_j (j=1, 2, 3, ...,m), we have a set of multipliers l_{1j}, l_{2j}, l_{3j},——, l_{mj}.

Then equation (13-7) takes the following form.

$$(d^2 V_j/dz^2) - (Al/Az)\,[2 + i\xi/(\Delta y)^2]\, V_j = F_j(z)$$

$$(13\text{-}9)$$

where

$$V_j = l_{1j} W_1 + i l_{2j} W_2 + i^2 l_{3j} W_3 + \cdots + i^{m-1} l_{mj} W_m$$

$$= \sum_{k=1}^{m} i^{k-1} l_{kj} W_k \tag{13-10}$$

and

$$F_j(z) = (Al/Az)\{l_{ij} W_n + i^{m-1} l_{mj} W_s/(\Delta y)^2\} + 1/Az \sum_{k=1}^{m} i^{k-1} l_k g\{\partial\rho/\partial x + i (\partial\rho/\partial y)\}_k \tag{13-11}$$

The conditions to be satisfied by V_j can easily be derived as

$$z = 0 : V_j = -\overline{\tau_j}/A_z \tag{13-12}$$

where

$$\tau_j = \sum_{k=1}^{m} i^{k-1} l_{kj} (\tau_x + i \tau_y)_k$$

and

$$z = h : V_j = 0. \tag{13-13}$$

If we solve the equation (13-9) subject to the conditions (13-12) and (13-13), we have

$$V_j(z) = -\overline{\tau j}/Az \, \sinh\lambda j \, (h-3)/\sinh\lambda_j h - \int_O^h F_j (s) K_j(z,s) \, ds \tag{13-14}$$

where

$$K_j(z,s) = \sinh\lambda j(h-3)\sinh\lambda js/\lambda j\sinh\lambda jh \quad (s<z) = \sinh\lambda jz\sinh\lambda j(h-s)/\lambda j\sinh\lambda jh \quad (s>z)$$

and

$$\lambda_j = +\sqrt{(Ae/Az \quad 2+i\xi j/(\Delta y)^2)} .$$

According to the result of numerical solution of the equations (13-8), the quantities l_{kj} have largest values on the diagonal ($k = j$) of a matrix

$|l_{kj}|$ and decrease rapidly as we go away from the latter, or the larger the value of $|k - j|$, the smaller l_{kj} becomes.

This means that the term

$$-A_l/A_z \left\{ l_{1j}\, W_N + i^{m-1}\, l_{mj}\, W_S/(\Delta y)^2 \right\}$$

in (13-11) has very little influence on the result except at a few stations closest to the northern and southern boundaries of the meridional section. For this reason, we may put $W_n = W_s = 0$ in (13-11) and use a simpler expression of $Fj\,(z)$, or

$$F_j(z) = 1/Az \sum_{k=1}^{m} i^{k-1} l_{kij} \cdot g\, [\partial\rho/\partial x) + i\, (\partial\rho/\partial y)]_k,$$

provided that we avoid applying the result to a few stations closest to the northern and southern boundaries.

Computation of Horizontal Components of Velocity

By inversion of the equations (13-10), we shall have

$$i^{k-1}\, W_k = \sum_{j=1}^{m} \bar{l}_{j,k}\, V_j$$

$$(k = 1, 2, \cdots, m)$$

This becomes further

$$W_k = \sum_{j=1}^{m} (\bar{l}_{jk}/i^{k-1})\, V_j \qquad (13\text{-}15)$$

Integrating (13-15) with respect to z from $z = z$ to $z = h$, it follows:

$$u_k + iv_k = -\sum (\bar{l}_{jk}/i^{k-1}) \int_z^h V_j\,(s)\,ds$$

where $V_j(z)$ is given by (13-10). The real and imag-

inary parts of this expression represent the velocity components u and v at the kth station.

Computation of Upwelling

Now we have, replacing k by $k + 1$ in (13-15),

$$W_{k+1} = \sum_{j=1}^{m} \frac{\bar{l}_{j,\,k+1}}{i^k}\, V_j\,(z)$$

Now the value of $\partial W/\partial y$, at the mid-point between kth and $(k + 1)$th stations, or $\partial W_{k+\frac{1}{2}}/\partial y$ is

$$\partial W_{k+\frac{1}{2}}/\partial y = W_{k+1} - W_k/\Delta y$$

$$= \sum_{j=1}^{m} \left\{ (\bar{l}_{j,\,k+1}/i^k) - (\bar{l}_{j,k}/i^{k-1}) \right\} V_j/\Delta y$$

very accurately. However, since

$$W = \frac{(\partial U}{\partial z)} + \frac{(\partial V}{\partial z)}$$

by (13-2), the above equation becomes

$$(\partial^2 U_{k+\frac{1}{2}}/\partial y\,\partial z) + i(\partial^2 V_{k+\frac{1}{2}}/\partial y\,\partial z)$$

$$= \sum_{j=1}^{m} \left(\bar{l}_{j,k+1}/i^k \right) - \left(\bar{l}_{j,k}/i^{k-1} \right) V_j(z)/\Delta y.$$

Taking the real part only, we have

$$(\partial^2 v_{k+\frac{1}{2}}/\partial y\,\partial z) = \text{imaginary part of}$$

$$\sum_{j=1}^{m} \left(\bar{l}_{jk}/i^{k-1} \right) - \left(\bar{l}_{j,k+1}/i^k \right) \bar{V}_j(z)/\Delta y \qquad (13\text{-}16)$$

where $V_j(z)$ is given by (13-14).

From the equation of continuity we have, for a place not far from the equator,

$$\partial v / \partial y = -\partial w / \partial z \qquad\qquad\qquad\qquad\qquad (13\text{-}17)$$

where w is the vertical velocity component counted positive downward. Substituting (13-17) in the left-hand side of (13-16), we have

$$d^2 W_{k+1/2}(z) \, / \, (dz^2) \; = \; \text{imaginary part of}$$

$$\sum_{j=1}^{m} \; \left(\overline{l}_{j,k+1} / i^k \right) - \left(\overline{l}_{j,k} / i^{k-1} \right) V_j(z) \, / \, (\Delta y) \; .$$

If we integrate this equation subject to the conditions $w = 0$ for $z = 0$ and $z = h$, we shall have

$$W_{k+1/2}(z) \; = \; \text{imaginary part of}$$

$$\sum_{j=1}^{m} \; \left\{ (\overline{l}_{j,k+1} / i^k) - (\overline{l}_{j,k}/i^{k-1}) \right\} \; (1 \, / \, \Delta y) \; \cdot \; \int_0^h V_j(s) \, G\,(z,s) \, ds$$

where

$$G(z,s) = (h-z)s/h \;\; (s < z) \Big\}$$
$$\qquad = z(h\text{-}s)/h \;\; (s > z) \Big\}$$

Since $V_j(z)$ is given by (13-14), we have

$$\int_0^h V_j(s)\,G(z,s)\,ds = \int_0^h \left[-\overline{\tau} i / A z \right\} \; \sinh \lambda_j \, (h-z)/\sin\lambda_j h \right\} \; -\int_0^h F_j(\sigma)\,K_j(s,\sigma) \Big] \; G\,(z,s)\,ds$$

$$= -\overline{\tau}_i / A z \int_0^h \sinh\lambda_i \, (h-s)/\sinh\lambda_j h \;\; G\;(z,s)\,ds \, - \, 1/\lambda_j^2 \int_0^h F_j\,(s)\,M_j\,(z,s)\,ds$$

where

$$M_j(z,s) = \lambda_j^2 \int_0^h G\;(z,\sigma)\,K_j\,(0,s)\,d\sigma \; .$$

Now that we have

$$\int_o^h \sinh\lambda_j \ (h-s) \ G \ (z,s) \ ds$$

$$= \int_O^z \sinh\lambda_j \ (h-s) \ (h-z)s/h \ ds + \int_z^h \sinh\lambda_j \ (h-s) \ z(h-s)/h \ ds$$

$$= \sinh\lambda_j h/\lambda_j^2 \ \left\{ h-z/h \ - \ \sinh\lambda_j \ (h-z)/\sinh\lambda_j h \right\}$$

and $M_j(z, s)$ becomes for $s < z$ as

$$M_j(z,s) = \lambda_j^2 \int_O^s (h-z)\sigma/h \cdot \sinh\lambda_j \ (h-s) \ \sinh\lambda_j \ \sigma \ / \ \lambda_j^2 \ \sinh\lambda_j \ h \ d\sigma$$

$$+ \lambda_j^2 \int_s^z (h-z)\sigma/h \cdot \sinh\lambda_j \ s \ \sinh\lambda_j \ (h-\sigma) \ / \ \lambda_j^2 \ \sinh\lambda_j h \ d\sigma$$

$$- \lambda_j^2 \int_z^h z(h-\sigma)/h \cdot \sinh\lambda_j \ s \ \sinh\lambda_j \ (h-\sigma) \ /\lambda_j^2 \ \sinh\lambda_j \ h \ d\sigma$$

$$= z(h-s)/h - \sinh\lambda_j \ (h-z) \ \sinh\lambda_j h/\lambda_j\sinh\lambda_j h \ \ (s < z) \ .$$

Similar expression will be obtained for $M_j(z,s)$ when $s > z$. Thus we have

$$1/\Delta y \int_O^h V_j(s) \ G \ (z,s) \ ds$$

$$= - \bar{\tau}_j \ / \ Az\Delta y\lambda_j^2 \ \left\{ h-z/h \ - \ \sinh\lambda_j \ (h-z)/\sinh\lambda_j h \right\}$$

$$- g/Az\Delta y\lambda_j^2 \int_O^h \sum_{k=1}^m i^{k-1} \ l_{kj} \ (\partial\rho/\partial x + i \ \partial\rho/\partial y)_k , \ M_j \ (z,s) \ ds$$

where $M_j(z,s)$ is a symmetric kernel of the form:

$$M_j(z, s) = \frac{(h-z)s}{h} \ - \ \frac{\sinh\lambda_0 \ (h-z)\sinh\lambda_0 - s}{\lambda_j \sinh\lambda_j h} \ \ (s < z)$$

$$= \frac{z \ (h-s)}{h} \ - \ \frac{\sinh\lambda_0 z \sinh\lambda_0 \ (h-s)}{\lambda_j \sinh\lambda_j h} \ \ (s > z).$$

Thus the upwelling velocity $-w_{k+\frac{1}{2}}(z)$ will be, since w is counted positive downwards, given by

$$-W_{k+\frac{1}{2}}(z) \quad \sum_{j=1}^{m} (\bar{l}_{j,k+1}/i^k - \bar{l}_{j,k}/i^{k-1}) \cdot [\bar{\tau}_j/Az\Delta y\lambda_j^2 \left\{ h-z/h - \sinh\lambda_j (h-z)/\sinh\lambda_j h\right\}$$

$$+ g/Az\Delta y\lambda_j^2 \cdot \int_o^h (\partial\bar{\rho}/\partial x + i\partial\bar{\rho}/\partial y) \cdot M_j(z,s)\, ds\,]$$

where

$$\bar{\tau}_j = \sum_{k=1}^{m} i^{k-1} l_{kj} (\tau_x + i\tau_y)_k$$

and

$$g\left\{(\partial\bar{\rho}/\partial x) + i(\partial\bar{\rho}/\partial y)\right\} j = \sum_{k=1}^{m} i^{k-1} l_{kj} g (\partial\rho/\partial x + i \partial\rho/\partial y)_k.$$

References

1. Hidaka, Koji and Nagata, Y. 1958. The dynamic computation of the equatorial current system of the Pacific, with special application to the equatorial undercurrent, *Geophys. Journ. Royal Astron. Soc.* 1:198-207.
2. Hidaka, Koji. 1961. Equatorial upwelling and sinking in a zonal ocean with lateral mixing. *Geophys. Journ. Royal Astron. Soc.* 4:359-371.

14

water oscillations in a rotating, rectangular basin of variable depth

Introduction

The tidal oscillation in rotating water was first solved by Lord Kelvin (1) and by Lord Rayleigh for the case of slow rotation (2). The cases where the depth of the basin decreases from the center to the boundaries after a paraboloidal law were first solved by Horace Lamb for a circular basin (3). For an elliptic basin approximate solution was worked out by S. Goldstein (4) and the problem was solved in a complete form by G.R. Goldsbrough (5). Particularly, Lamb and Goldsbrough showed that the paraboloidal law of depth gives simpler solutions mostly consisting of polynomials of finite orders.

The oscillations of water in a nonrotating basin with a paraboloidal law of depth was solved by the present author (6). The solution was obtained in an infinite double series of zonal harmonics. Now the case of a rectangular, rotating basin whose depth decreases from the center to the edges will be discussed in this chapter. The solution is given again in an infinite double series of zonal harmonics but in more complicated form due to an addition of rotational terms. A numerical solution was worked out for several gravest modes. Electronic computation will give a complete solution of all possible modes.

Theory

Take the x-, y- and z-axes positive eastward, northward and downward, respectively. Let the rectangular basin under consideration be bounded by $x = \pm a$, $y = \pm b$, so that the lengths of four edges are $2a$ and $2b$, respectively. The law of depth is given by

$$h = h_0 \left(1 - x^2/a^2\right)\left(1 - y^2/b^2\right),$$

where h_0 is depth at the center ($x = 0$: $y = 0$) of the basin. The depth is *nil* along the four edges of the basin.

The equations of motion of small oscillations are

$$\frac{\partial u}{\partial t} = 2\omega v - g\frac{\partial \zeta}{\partial x},$$

$$\frac{\partial v}{\partial t} = 2\omega u - g\frac{\partial \zeta}{\partial y}, \tag{14-1}$$

while the equation of continuity is

$$\frac{\partial \zeta}{\partial t} + \frac{\partial}{\partial x}(hu) + \frac{\partial}{\partial y}(hv) = 0 \tag{14-2}$$

153

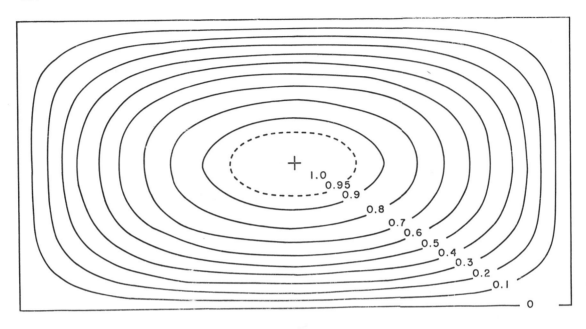

Figure 14-1. Rectangular basin ($b = 1/2\ a$) with isobaths.

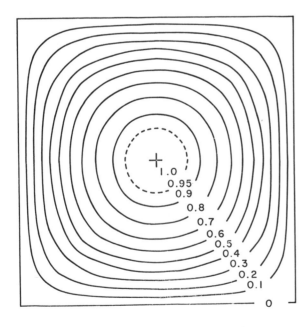

Figure 14-2. Square basin ($b = a$) with isobaths.

where u and v are the velocity components in x- and y-directions respectively, ζ is the elevation of water above a certain level ($z = 0$), t is the time, g is the acceleration of gravity and ω is the angular velocity of the rotating basin.

Assuming that the motion of water is a simple harmonic oscillation with a frequency σ, we may put

$$u\,,\,v\,,\,\zeta \propto e^{i\sigma t} \qquad (14\text{-}3)$$

Substituting (14-3) in (14-1) and (14-2) dropping the time factor, we have after some modification

$$u = \frac{g}{\sigma^2 - 4\omega^2}\left[i\sigma\,\frac{\partial \zeta}{\partial x} + 2\omega\,\frac{\partial \zeta}{\partial y}\right].$$

$$v = \frac{g}{\sigma^2 - 4\omega^2}\left[-2w\frac{\partial \zeta}{\partial x} + i\sigma\,\frac{\partial \zeta}{\partial y}\right], \qquad (14\text{-}4)$$

$$i\sigma\zeta + \frac{\partial}{\partial x}\,(hu) + \frac{\partial}{\partial y}\,(hv) = 0 \qquad (14\text{-}5)$$

The boundary conditions to be satisfied at the edges $x = \pm a$ and $y = \pm b$ are simply that ζ, u and v must be finite along these four edges where the depth is *nil.*

Again, substituting (14-4) in (14-5) we have, on writing

$$\left. \begin{array}{l} \zeta = \zeta_1 + i\zeta_2 \\ \\ u = u_1 + u_2, \ v = v_1 + iv_2 \end{array} \right\} \qquad (14\text{-}6)$$

and

$$\sigma/2\omega = f,$$

$$\frac{\partial}{\partial x}\left(h\frac{\partial \zeta_1}{\partial x}\right) + \frac{\partial}{\partial y}\left(h\frac{\partial \zeta_1}{\partial y}\right) + \frac{4\omega^2}{g}(f^2 - 1)\zeta_1 + \frac{1}{f}\ \frac{\partial}{\partial x}\left(h\frac{\partial \zeta_2}{\partial y}\right) - \frac{\partial}{\partial y}\left(h\frac{\partial \zeta_2}{\partial x}\right) = 0 \qquad (14\text{-}7)$$

and

$$\frac{\partial}{\partial x}\left(h\frac{\partial \zeta_2}{\partial x}\right) + \frac{\partial}{\partial y}\left(h\frac{\partial \zeta_2}{\partial y}\right) + \frac{4\omega^2}{g}(f^2 - 1)\zeta_2 - \frac{1}{f}\ \frac{\partial}{\partial x}\left(h\frac{\partial \zeta_1}{\partial y}\right) - \frac{\partial}{\partial y}\left(h\frac{\partial \zeta_1}{\partial x}\right) = 0 \qquad (14\text{-}8)$$

Further, let

$$x/a = \xi \ , y/b = \eta$$

and

$$4\omega^2 ab/gh_0 = \beta;$$

then the two equations (14-7) and (14-8) will be reduced to

$$\frac{b}{a}\ \frac{\partial}{\partial \xi}(1 - \xi^2)(1 - \eta^2)\frac{\partial \zeta_1}{\partial \xi} + \frac{a}{b}\ \frac{\partial}{\partial \eta}(1 - \xi^2)(1 - \eta^2)\frac{\partial \zeta_2}{\partial \eta}$$
$$+ \beta\,(f^2 - 1)\,\zeta_2 - \frac{1}{f}\ \frac{\partial}{\partial \xi}(1 - \xi^2)(1 - \eta^2)\frac{\partial \zeta_1}{\partial \eta} - \frac{\partial}{\partial \eta}(1 - \xi^2)(1 - \eta^2)\frac{\partial \zeta_1}{\partial \xi} = 0 \qquad (14\text{-}9)$$

and

$$\frac{b}{a}\ \frac{\delta}{\delta \xi}(1 - \xi^2)(1 - \eta^2)\frac{\delta \zeta_2}{\delta \xi} + \frac{a}{b}\ \frac{\delta}{\delta \eta}(1 - \xi^2)(1 - \eta^2)\frac{\delta \zeta_2}{\delta \eta}$$
$$+ \beta\ (f^2 - 1)\ \zeta_2 - \frac{1}{f}\ \frac{\delta}{\delta \xi}(1 - \xi^2)(1 - \eta^2)\frac{\delta \zeta_1}{\delta \eta}\left\{\frac{\delta}{\delta \eta}(1 - \xi^2)(1 - \eta^2)\frac{\delta \zeta_1}{\delta \xi}\right\} = \qquad (14\text{-}10)$$

Integration in Double Series of Zonal Harmonics

In order to solve these two simultaneous equations, we assume

$$\zeta_1 = \sum_{m,n} A_{mn} P_m(\xi) \cdot P_n(\eta) \qquad (14\text{-}11)$$

and

$$\zeta_2 = \sum_{m,n} B_{m+1,\,n-1} P_{m+1}(\xi) \cdot P_{n-1}(\eta)$$

$$(14\text{-}12)$$

where $P_m(\xi)$ and $P_n(\eta)$ are zonal harmonics of orders m and n.

The boundary conditions will be satisfied by the series (14-11) and (14-12) because zonal harmonics remain finite at $\xi = \pm 1$, $\eta = \pm 1$. Substitution of (14-11) and (14-12) in (14-9) and (14-10) gives

$$\sum A_{mn}\left\{\frac{b}{a}\frac{d}{d\xi}(1-\xi^2)\frac{aP_m}{d\xi}\cdot(1-\eta^2)P_m(\eta)\right.$$

$$+\frac{a}{b}(1-\xi^2)P_m(\xi)\cdot\frac{d}{d\eta}(1-\eta^2)\frac{dP_n}{d\eta}+\beta(f^2-1)P_m(\xi)P_n(\eta)\bigg\}$$

$$+\frac{1}{f}\sum_{m,n}B_{m+1,\,n-1}\cdot\frac{d}{d\xi}(1-\xi^2)P_{m+1}\cdot(1-\eta^2)\frac{dP_n}{d\eta} \qquad (14\text{-}13)$$

$$-(1-\xi^2)\frac{dP_{m+1}}{d\xi}\cdot\frac{d}{d\eta}(1-\eta^2)P_{m-1}(\eta)\bigg\}=0$$

and

$$\sum_{m,n}B_{m+1,\,n-1}\left\{\frac{b}{a}\frac{d}{d\xi}(1-\xi^2)\frac{dP_{m+1}}{d\xi}(1-\eta^2)P_{n-1}\right.$$

$$+\frac{a}{b}(1-\xi^2)P_{m+1}\cdot\frac{d}{d\eta}(1-\eta^2)\frac{dP_{n-1}}{d\eta}+\beta(f^2-1)\cdot P_{m+1}(\xi)\cdot P_{n-1}(\eta)$$

$$-\frac{1}{f}\sum_{m,n}A_{mn}\left\{\frac{d}{d\xi}(1-\xi^3)P_m\cdot(1-\eta^2)\frac{dP_n}{d\eta}\right. \qquad (14\text{-}14)$$

$$-(1-\xi^2)\frac{dP_n}{d\xi}\cdot\frac{d}{d\eta}(1-\eta^2)P_n\bigg\}=0$$

By the virtue of the following formulas associated with zonal harmonics:

$$\frac{d}{d\xi}(1-\xi^2)\frac{dP_m}{d\xi} = -m(m+1)P_m(\xi) \tag{14-15}$$

$$\left.\begin{array}{l} (1-\xi^2)P_n + 1(\xi) = -\dfrac{(m-1)m}{(2m-1)(2m+1)}P_{m-2}(\xi) \\[3mm] + 1\dfrac{m^2}{(2m-1)(2m+1)} - \dfrac{(m+1)^2}{(2m+1)(2m+3)} P_m(\xi) - \dfrac{(m+1(m+2)}{(2m+1)(2m+3)} P_{m+2}(\xi) \end{array}\right\} \tag{14-16}$$

$$\frac{d}{d\xi}(1-\xi^2)P_m(\xi) = -\frac{(m+1)(m+2)}{2m+1}P_{m+1} + \frac{(m-1)m}{2m+1}P_{m-1} \tag{14-17}$$

and

$$(1-\xi^2)\frac{dP_m}{d\xi} = -\frac{m(m+1)}{2m+1}(P_{m+1} - P_{m-1}) \tag{14-18}$$

the factors $(1-\xi^2)$ and $(1-\eta^2)$ can be eliminated from (4-13) and (4-14). The resulting sequence equations for $A_{m,n}$ and $B_{m+1,n-1}$ becomes

$$m(m+1)\cdot\frac{(n+1)(n+2)}{(2n+3)(2n+5)}\cdot\frac{b}{a} A_{m,m+2} + m(m+1)\cdot\frac{(n-1)n}{(2n-3)(2n-1)}\cdot\frac{b}{a}\cdot A_{m,n-2}$$

$$+\frac{(m+1)(m+2)}{(2m+3)(2m+5)}\cdot n(n+1)\frac{a}{b}\cdot A_{m+2,n} + \frac{(m-1)m}{(2m-3)(2m-1)}\cdot n(n+1)\cdot\frac{a}{b}\cdot A_{m-2,n}$$

$$+\left[\beta(f^2-1)-m(m+1)\left\{1-\frac{n^2}{(2n-1)(2n+1)}-\frac{(n+1)^2}{(2n+1)(2n+3)}\right\}\cdot\frac{b}{a}\right.$$

$$\left. -\left\{1-\frac{m^2}{(2m-1)(2m+1)}-\frac{(m+1)^2}{(2m+1)(2m+3)}\right\}\cdot n(n+1)\cdot\frac{a}{b}\right]\cdot A_{m,n}$$

$$+\frac{1}{f}\cdot\left\{\frac{m(m+1)}{2m-1}\cdot\frac{(n-1)\cdot n}{2n-1} - \frac{(m-1)\cdot m}{2m-1}\cdot\frac{n(n+1)}{2n-1}\right\}\cdot B_{m-1,n-1}$$

$$-\frac{1}{f}\left\{\frac{m(m+1)}{2m-1}\cdot\frac{(n+1)(n+2)}{2n+3} - \frac{(m-1)\cdot m}{2m-1}\cdot\frac{n(n+1)}{2n+3}\right\}\cdot B_{m-1,n+1}$$

$$-\frac{1}{f}\cdot\left\{\frac{m(m+1)}{2m+3}\cdot\frac{(n-1)\cdot n}{2n-1} - \frac{(m+1)(m+2)}{2m+3}\cdot\frac{n(n+1)}{2n-1}\right\}\cdot B_{m+1,n-1}$$

$$+\frac{1}{f}\left\{\frac{m(m+1)}{2m+3}\cdot\frac{(n+1)(n+2)}{2n+3} - \frac{(m+1)(m+2)}{2m+3}\cdot\frac{n(n+1)}{2n+3}\right\}\cdot B_{m+1,n+1} = 0 \tag{14-19}$$

In these equations, the suffices are so arranged that if the suffices in A for ξ-expansion is odd numbers, those in η must be even and *vice versa*.

This is clear from the properties of the recurrence equations (14-15), (14-16), (14-17) and (14-18). Thus, these two sets of sequence equations will determine the ratio $A_{01} : A_{03} : ---,$ $A_{21} : A_{23} : ---, B_{10} : B_{12} : ---, B_{30} : B_{32} :$ $---$. The principle used by Kelvin, Darwin and Hough (stated in Reference 7) in the dynamical theory of tides will apply for proving the convergency of the series (14-11) and (14-12) in connection with the sequence equations (14-19) and (14-20).

Approximate Solutions for Gravest Modes

The gravest modes of oscillations will be those in which the terms:

$$P_0(\xi)P_1(\eta) \quad \text{or} \quad P_1(\xi)P_0(\eta) \qquad (14\text{-}20)$$

predominate. Hence, the expressions

$$\zeta_1 = A_{01}\, P_0(\xi)\, P_1(\eta) \qquad (14\text{-}21)$$

and

$$\zeta_2 = B_{10}\, P_1(\xi)\, P_0(\eta) \qquad (14\text{-}22)$$

will give approximate solutions. For more accurate expressions for these modes, it may be assumed that

$$\zeta_1 = A_{01}\, P_0(\xi)\, P_1(\eta) + A_{03}\, P_0(\xi)\, P_3(\eta)$$
$$+ A_{21}\, P_2(\xi)\, P_1(\eta), \qquad (14\text{-}23)$$

and

$$\zeta_2 = B_{10}\, P_1(\xi)\, P_0(\eta) + B_{12}\, P_1(\xi)\, P_2(\eta)$$
$$+ B_{20}\, P_2(\xi)\, P_0(\eta). \qquad (14\text{-}24)$$

In order to learn the primary nature of oscillations, the approximate solutions of the forms (14-21) and (14-22) are very convenient. The sequence equations (14-19) and (14-20) in this case reduce to the following two equations:

$$\left\{ \beta\,(f^2 - 1) - \frac{4}{3}\,\frac{a}{b} \right\} A_{01} + \frac{4}{3}\,\frac{1}{f}\, B_{10} = 0$$

$$(14\text{-}25)$$

and

$$\frac{4}{3} \cdot \frac{1}{f}\, A_{01} + \left\{ \beta\,(f^2 - 1) - \frac{4}{3}\,\frac{b}{a} \right\} \beta_{10}$$

$$= 0 \qquad (14\text{-}26)$$

Eliminating A_{01} and B_{10} between (14-25) and (14-26), it follows that

$$\left\{ \beta\,(f^2 - 1) - \frac{4}{3}\,\frac{a}{b} \right\} \left\{ \beta\,(f^2 - 1) - \frac{4}{3}\,\frac{b}{a} \right\}$$

$$- \left(\frac{4}{3} \right)^2 \frac{1}{f^2} = 0 \qquad (14\text{-}27)$$

or

$$\left\{ f^4 - \left(1 + \frac{4}{3}\,\frac{\dfrac{a}{b} + \dfrac{b}{a}}{\beta} \right) f^2 \right.$$

$$\left. + \left(\frac{4}{3\beta} \right)^2 \right\} \cdot (f^2 - 1) = 0. \qquad (14\text{-}28)$$

Equation (14-28) gives three pairs of real roots

$$f^2 = \frac{1 + \dfrac{4}{3} \dfrac{\dfrac{a}{b} + \dfrac{b}{a}}{\beta} \pm \sqrt{\left(1 + \dfrac{4}{3} \dfrac{\dfrac{a}{b} + \dfrac{b}{a}}{\beta}\right)^2 - \left(\dfrac{8}{3\beta}\right)^2}}{2} \tag{14-29}$$

and

$$f^2 = 1 \tag{14-30}$$

The root $f^2 = 1$ gives only trivial solutions and does not give an oscillations as is clearly seen from equations (14-7) and (14-8). So this case will be excluded from further discussions.

From (14-29), it follows that

$$\beta(f^2 - 1) - \frac{4}{3}\frac{a}{b} = \pm \sqrt{\frac{\left\{\beta + \dfrac{4}{3}\left(\dfrac{a}{b} + \dfrac{b}{a}\right)\right\}^2 - \dfrac{8}{3}}{2}} - \beta - \frac{4}{3}\left(\frac{a}{b} - \frac{b}{a}\right) \tag{14-31}$$

and

$$\beta(f^2 - 1) - \frac{4}{3}\frac{b}{a} = \pm \sqrt{\frac{\left\{\beta + \dfrac{4}{3}\left(\dfrac{a}{b} + \dfrac{b}{a}\right)\right\}^2}{2}} - \beta + \frac{4}{3}\left(\frac{a}{b} - \frac{b}{a}\right) \tag{14-32}$$

Equation (14-27) gives, assuming $f > o$ in advance,

$$\frac{4}{3} \cdot \frac{1}{f} = + \sqrt{\left\{\beta(f^2 - 1) - \frac{4}{3}\frac{a}{b}\right\} \cdot \left\{\beta(f^2 - 1) - \frac{4}{3}\cdot\frac{b}{a}\right\}} \tag{14-33}$$

Further substituting (14-31), (14-32) and (14-33) in (14-25), it follows that ($f > o$).

$$B_{10} = \pm \sqrt{\left| \frac{\sqrt{\left\{\beta + \dfrac{4}{3}\left(\dfrac{a}{b} + \dfrac{b}{a}\right)\right\}^2 - \left(\dfrac{8}{3}\right)^2} - \beta - \dfrac{4}{3}\left(\dfrac{a}{b} - \dfrac{b}{a}\right)}{\sqrt{\left\{\beta + \dfrac{4}{3}\dfrac{a}{b} + \dfrac{b}{a}\right\}^2 - \dfrac{8}{3}^2} - \beta + \dfrac{4}{3}\dfrac{a}{b} - \dfrac{b}{a}}} \right|} \cdot A_{01} \tag{14-34}$$

where |———| under the radical means the absolute value.

If we assume $f > o$, the sign of the right-hand member of (14-33) will be negative so that there will be no change in (14-34). Now that it follows from (14-3) and (14-6) that

$$\zeta = (\zeta_1 + i\,\zeta_2) \cdot e^{i\sigma t}$$

taking the real part only,

$$\zeta = \zeta_1 \cos\sigma t - \zeta_2 \sin\sigma t .$$

Substituting from (14-21) and (14-22), the elevation of water surface ζ is given by

$$\zeta(x, y, t,) = A_{01} P_0 \left(\frac{x}{a}\right) \cdot P_1 \left(\frac{y}{b}\right) \cos\sigma t - B_{10} P_1 \left(\frac{x}{a}\right) \cdot P_0 \left(\frac{y}{b}\right) \sin\sigma t . \tag{14-35}$$

Since zonal harmonics of oth and 1st order are given by $P_0(x) = 1$ and $P_1(x) = X$ respectively, (14-35) becomes

$$\zeta(x, y, t) = A_{01} \frac{y}{b} \cos\sigma t - B_{10} \cdot \frac{x}{a} \sin\sigma t . \tag{14-36}$$

This expression signifies that the water surface in these modes are approximately *a plane* rotating about the vertical axis with a frequency

$$f = \frac{\sigma}{2\omega} \quad .$$

Finally substitution from (14-34) gives

$$\zeta(x, y, t) = A_{01} \lfloor \frac{y}{b} \cos\sigma t$$

$$\pm \sqrt{ \left| \frac{ \pm\sqrt{\left\{\beta + \frac{4}{3}\left(\frac{a}{b} + \frac{b}{a}\right)\right\}^2 - \left(\frac{8}{3}\right)^2} - \beta - \frac{4}{3}\left(\frac{a}{b} - \frac{b}{a}\right) }{ b\sqrt{\left\{\beta + \frac{4}{3}\left(\frac{a}{b} + \frac{b}{a}\right)\right\}^2 - \left(\frac{8}{3}\right)^2} - \beta + \frac{4}{3}\left(\frac{a}{b} - \frac{b}{a}\right) } \right| } \cdot \sin|\sigma| t \Bigg]$$

where A_{01} is an arbitrary constant, the frequencies corresponding to this mode being

$$f = + \sqrt{\frac{1 + \dfrac{4}{3}\dfrac{\frac{a}{b} + \frac{b}{a}}{\beta} \pm \sqrt{\left(1 + \dfrac{4}{3}\cdot\dfrac{\frac{a}{b} + \frac{b}{a}}{\beta}\right)^2 - \left(\dfrac{8}{3\beta}\right)^2}}{2}}$$

Co-Tidal Lines and Co-Range Lines

Co-tidal lines along which high water takes place at the same time is given by

$$\partial\zeta/\partial t = 0.$$

From (14-36) it follows that

$$-A_{01}\cdot\frac{y}{b}\ \sin\sigma t - B_{10}\frac{x}{a}\ \cos\sigma t = 0$$

or

$$\tan\sigma t + \frac{B_{10}}{A_{01}}\ \cdot\left(\frac{x}{a}\Big/\frac{y}{b}\right) = 0\ .$$

This is the equation of co-tidal lines.

In the oscillation of water in a rotating basin, these co-tidal lines rotate either in the same direction with the rotation of the basin or in an opposite direction to the rotation of the basin. In the former case, the oscillation is called a "positive wave," while the wave rotating in an opposite direction to the rotation of the basin is a "negative wave." In the present discussion, all frequencies f being assumed positive $(f > o)$ in advance, the frequency f larger than 1 in (14-29) gives a negative wave, while the frequency smaller than 1 corresponds to a "positive wave."

Co-range lines are the curves on which the amplitude of oscillation or tidal range is a constant. The expression (14-36) can be transformed in the form:

$$\zeta(x, y, t) = \sqrt{\frac{x^2}{(a/B_{10})^2} + \frac{y^2}{(b/A_{01})^2}} \, \cos\left\{\sigma t - \tan^{-1}\left(\frac{B_{10}}{A_{01}} \cdot \frac{b}{a} \cdot \frac{x}{y}\right)\right\}$$

Thus the co-range lines are given by a system of ellipses:

$$\frac{x^2}{(a/B_{10})^2} + \frac{y^2}{(b/A_{01})^2} = C^2$$

where C^2 is any positive constant.

Numerical Result

Numerical result was worked out for a square basin ($b = a$) and a rectangular basin ($b = \frac{1}{2}a$). For these two types of basins, computation was made for

$$\beta = \frac{4\omega^2 ab}{gh_0} = 1, 2, 5 \text{ and } 10,$$

specifying different depth h_0. For $\beta = 0$, we have the solutions for nonrotating basins discussed in detail by the author (6).

Three roots of the equation (14-27) or (14-28) or the frequencies, given by (14-29) and (14-30), are compiled in Table 14-1.

Comparison between the corresponding frequencies for these two types of basins shows that the smaller frequencies are larger in a square basin than in a rectangular basin by 3 to 10%, while the ratios are reversed for the larger frequencies. These ratios approach to unity as β increases indefinitely.

Square Basins

From (14-34), it follows that for square basins ($b = a$)

$$B_{10} = \mp A_{01}$$

always for the frequencies given by (14-29), or

Table 14-1

Frequencies, $f = \dfrac{a}{zw}$

$\beta = 1$	$\beta = 2$	$\beta = 5$	$\beta = 10$
	Square Basin ($b = a$)		
0.758306	0.457427	0.218795	0.119139
1.758306	1.457427	1.218795	1.119139
	Rectangular Basin ($b = \frac{1}{2}a$)		
0.677379	0.422650	0.209329	0.116056
1.968374	1.577345	1.273910	1.148852

$$f = + \left\{\frac{1 + \dfrac{8}{3\beta} \pm \sqrt{\left(1 + \dfrac{8}{3\beta}\right)^2 - \left(\dfrac{8}{3\beta}\right)^2}}{2}\right\}$$

The corresponding expressions for ζ, the surface contours are given by

$$\zeta(x, y, t) = A_{01}(y\cos\sigma t - x\sin\sigma t.)$$

The co-tidal lines are given by

$$\tan\sigma t = \frac{x}{y} \, \frac{x}{y},$$

while co-range lines are given by concentric circles

$$x^2 + y^2 = C^2$$

where C^2 is a positive constant. The co-tidal lines indicate that the water surface, being approximately a plane, rotates at a uniform angular velocity about the vertical axis in the same direction as the rotation of the basin (positive wave) for a frequency f smaller than 1, or for the first root of each pair of the upper half of Table 14-1, the motion corresponding to the frequencies larger than 1 gives an approximately *plane* surface rotating in an opposite direction to the rotation of the basin.

These modes correspond to the case $s = 1$ and $n = 3$ in Lamb's solution (3) for a circular basin whose depth distribution is given by

$$h = h_0 \left(1 - n^2/a^2\right)$$

where a is the radius of the basin and h_0 is the depth at the center of the basin. In his case the surface of the water is a perfect plane. The solution for ζ is given by

$$\zeta = A_s \frac{r}{a}^s F\left(\alpha,\ \beta,\ \gamma,\ \frac{r^2}{a^2}\right) e^i \left(\sigma t + s\theta + \epsilon\right)$$

where

$$\alpha = \frac{1}{2} n + \frac{1}{2} s,\ \beta = 1 + \frac{1}{2} s - \frac{1}{2} n,\ \partial = s + 1$$

and a is a constant. The frequencies equation is given by a cubic:

$$\frac{(6^2 - 4\omega^2)\, a^2}{gh_0} - \frac{4\omega s}{\sigma} = n(n-2) - s^2$$

Rectangular Basins

In a rectangular basin ($b = \frac{1}{2}a$), the expression for ζ becomes

$\beta = 1.$ (*a*) $f = 0.677379$ (positive wave)

$$\zeta(x, y, t) = \frac{y}{b} \cdot \cos\sigma t - 1.629674\ \frac{x}{a}\ \sin\sigma t,$$

(*b*) $f = 1.968373$ (negative wave)

$$\zeta(x, y, t) = \frac{y}{b}\ \cos\sigma t + 0.306808\ \frac{x}{a}\ \sin\sigma t.$$

$\beta = 2.$ (*a*) $f = 0.422646$ (positive wave)

$$\zeta(x, y, t) = \frac{y}{b}\ \cos\sigma t - 1.366126\ \frac{x}{a}\ \sin\sigma t,$$

(*b*) $f = 1.577350$ (negative wave)

$$\zeta(x, y, t) = \frac{y}{b}\ \cos\sigma t + 0.366025\ \frac{x}{a}\ \sin\sigma t.$$

$\beta = 5.$ (*a*) $f = 0.209330$ (positive wave)

$$\zeta(x, y, t) = \frac{y}{b}\ \cos\sigma t - 1.169247\ \frac{x}{a}\ \sin\sigma t,$$

(*b*) $f = 1.273910$ (negative wave)

$$\zeta(x, y, t) = \frac{y}{b}\ \cos\sigma t + 0.427626\ \frac{x}{a}\ \sin\sigma t.$$

$\beta = 10.$ (*a*) $f = 0.116056$ (positive wave)

$$\zeta(x, y, t) = \frac{y}{b}\ \cos\sigma t - 1.090808\ \frac{x}{a}\ \sin\sigma t,$$

(*b*) $f = 1.148853$ (negative wave)

$$\zeta(x, y, t) = \frac{y}{b}\ \cos\sigma t + 0.458367\frac{x}{a}\ \sin\sigma t.$$

If we take the two oscillations of different frequencies for $\beta = 1$ for representatives because they are extreme cases, the co-tidal lines are given by

$\beta = 1$ (*a*) $f = 0.677379$ (positive wave)

$$\tan\sigma t = -1.629674\,([x]/[a] \,/\, [y]/[b])$$

and

(*b*) $f = 1.968373$ (negative wave)

$$\tan\sigma t = +0.306808\,([x]/[a] \,/\, [y]/[b]),$$

while the co-range lines will be given by a group of ellipses,

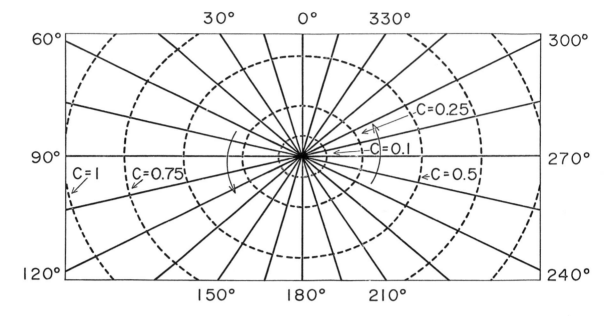

Figure 14-3. Co-tidal and co-range lines for = 1, f = o.677379 (positive wave).

$\beta = 1$ (a) f = 0.677379

$$(x^2)/[(a/1.629674)^2] + (y^2)/[(a/2)^2] = C^2$$

(b) f = 1.968373

$$(x^2)/[(a/0.306808)^2] + (y^2)/[(a/2)^2] = C^2,$$

where c^2 is an arbitrary positive constant.

If we take $ot = 0°$, $15°$, $30°$, $45°$, ____ $345°$ and $360°$, we can make clear how the co-tidal lines rotate in the same or in an opposite direction as the rotation of the basin. They are shown in Figures 14-3 and 14-4, together with the co-range lines in an arbitrary unit. The case $\beta = 1$, $f = 0.677379$ gives co-tidal lines sweeping at approximately uniform angular velocity in a counterclockwise direction, while the co-range lines are ellipses of small eccentricity. On the contrary, the case $\beta = 1$, $f = 1.968373$ gives a clockwise rotation with slower angular velocity around $0°$ and $180°$ and more rapid around $90°$ and $270°$, while the co-

range lines are elongated ellipses with the major axis parallel to the longer edges of the basin.

For other cases $\beta = 2, 5$ and 10 the cotidal and co-range lines fall between these two extreme cases.

More Exact Numerical Solutions

When more accurate numerical solutions are required, two more terms will be added to each of the expressions ζ_1 and ζ_2 as shown in (14-23) and (14-24), or

$$\zeta_1 (x, y, t) = A_{01} P_0 (\xi) P_1 (\eta)$$
$$+ A_{03} P_0 (\xi) P_3 (\eta) + A_{21} P_2 (\xi) P_1(\eta)$$

and

$$\zeta_2 (x, y, t) = B_{10} P_1 (\xi) P_0 (\eta)$$
$$+ B_{12} P_1 (\xi) P_2 (\eta) + B_{30} P_3 (\xi) P_0 (\eta)$$

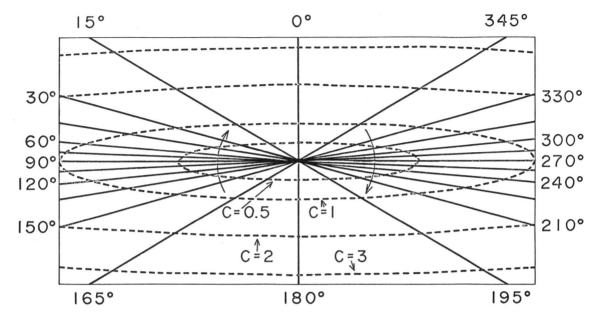

Figure 14-4. Co-tidal and co-range lines for = 1.968373 (negative wave).

In this case, three pairs of sequence equations will be derived from (14-19) and (14-20). They are as follows:

$$\beta\ (f^2-1)\ -\frac{4}{3}\frac{a}{b}\ \ A_{01}+\ \frac{4}{15}\ \frac{a}{b}\ A_{21}\ +\frac{4}{3}\cdot\ \frac{1}{f}\ B_{10}\ -\ \frac{4}{15}\cdot\frac{1}{f}B_{12}=0_1$$

$$\beta\ (f^2-1)\ -\ 8\frac{a}{b}\ \ \ A_{03}\ +\frac{8}{5}\cdot\ \frac{1}{f}B_{12}=0,$$

$$\frac{4}{3}\frac{a}{b}\ A_{01}\ +\ \ \beta\ (f^2-1)\ -(\frac{12}{5}\cdot\frac{b}{a}+\ \frac{20}{21}\ \frac{a}{b})\ \ \ A_{21}\ +\frac{4}{3}\cdot\ \frac{1}{f}B_{10}$$

$$+\ \frac{32}{15}\ \cdot\ \frac{1}{f}\ B_{12}\ -\ \frac{24}{7}\ -\frac{1}{f}B_{30}=0,$$

$$-\frac{4}{3}\cdot\ \frac{1}{f}\ A_{01}\ +\frac{4}{15}\cdot\ \frac{1}{f}\ A_{21}\ +\ \ \beta\ (f^2-1)-\frac{4}{3}\cdot\ \frac{b}{a}\ \ B_{10}$$

$$+\ \frac{4}{15}\ \cdot\ \frac{b}{a}B_{12}=0$$

$$\frac{4}{3}\cdot\ \frac{1}{f}\ A_{01}\ -\ \frac{24}{7}\ \cdot1\frac{1}{f}\ A_{03}\ +\ \frac{32}{15}\ \cdot\ \frac{1}{f}\ A_{21}\ +\frac{4}{3}\cdot\ \frac{b}{a}B_{10}$$

$$+\ \ \beta\ (f^2-1)-\ \frac{20}{21}\ \cdot\ \frac{b}{a}+\ \frac{12}{5}\ \cdot\ \frac{a}{b}\ B_{12}=0,$$

$$-\frac{8}{5}\cdot\ \frac{1}{f}\ A_{21}\ +\ \ \beta\ (f^2-1)-\ 8\ \frac{b}{a}B_{30}=0$$

Table 14-2,

Frequencies f and Coefficients in the Expansions

$$\zeta_1 = \sum_{m,n} A_{m,n}\, P_m\, \frac{x}{a}\, P_m\, \frac{y}{a} \quad \text{and} \quad \zeta_2 = \sum_{m,n} B_{m+1,n-1}\, P_{m+1}\, \frac{x}{a}\, P_{n-1}\, \frac{y}{b}$$

Square Basin (b = a)

β =	1		2		5		10	
f =	0.766207	1.974999	0.472446	1.564525	0.235722	1.262121	0.134373	1.140649
A_{01} =	1.000000	1.000000	1.000000	1.000000	1.000000	1.000000	1.000000	1.000000
A_{03} =	-0.018564	0.352870	-0.077058	0.325383	-0.246295	0.302857	-0.478398	0.295117
A_{21} =	-0.074814	0.224754	-0.217389	-1.624101	-0.461633	-1.202956	-0.715909	-1.047173
B_{10} =	1.000000	-1.000000	1.000000	-1.000000	1.000000	-1.000000	1.000000	-1.000000
B_{12} =	-0.074790	2.224754	-0.217389	1.624101	-0.461633	1.202956	-0.715907	1.047173
B_{30} =	-0.018570	-0.352870	-0.077058	-0.325385	-0.246295	-0.302864	-0.478398	-0.295117

Rectangular Basin (b = ½a)

β =	1		2		5		10	
f =	0.684035	2.106409	0.433953	1.641064	0.222823	1.294749	0.130908	1.157886
A_{01} =	1.000000	1.000000	1.000000	1.000000	1.000000	1.000000	1.000000	1.000000
A_{03} =	-0.020620	0.007708	-0.065430	-0.000422	-0.197792	-0.018796	-0.396366	-0.021254
A_{21} =	0	-1.077270	-0.107825	-0.795183	-0.113253	-0.621376	-0.442864	-0.493595
B_{10} =	1.622770	-0.283745	1.352419	-0.346005	1.148634	-0.416298	1.045776	-0.461763
B_{12} =	-0.145739	0.127485	-0.312744	-0.005460	-0.433888	-0.191923	-0.837614	-0.193694
B_{30} =	0	-1.453314	-0.070697	-1.263055	-0.216759	-1.224258	-0.391422	-1.150193

Eliminating A_{01}, A_{03}, A_{21}, B_{10}, B_{12} and B_{30} out of these six equations, the following frequency equation (14-37) will be obtained for f.

The solution of equation (14-37) gives the results shown in Table 14-2, which comprises more exact frequencies and corresponding coefficients A_{01}, A_{03}, A_{21}, B_{10}, B_{12} and B_{30} for the same square and rectangular basins whose approximate solutions have been discussed in equations (12-2)-(12-5). A comparison of this table with Table 14-1 enables us to notice that the values of f here are slightly larger than the former ones. However, because of the slight differences amounting to only less than 10%, we may estimate that the errors involved in these second approximations will be so small that the latter values for the frequencies may be regarded as correct enough. Moreover, the fact that A_{03}, A_{21}; B_{12} and B_{30} are small compared with A_{01} and B_{10} in most modes shows that the second approximations are fairly accurate (excluding the cases $b = \frac{1}{2}a$, $\beta = 1$, $f = 2.106409$; $\beta = 2$, $f = 1.641064$; $\beta = 5$, $f = 1.294749$ and $\beta = 10$, $f = 1.157886$).

$$h = h_0 \left[1 - \frac{x^2}{a^2} \right] \left[1 - \frac{y^2}{b^2} \right]$$

These considerations suggest that the solutions given in (14-3), (14-4) and (14-5) may be used as close approximations to the oscillations of water in rotating basins whose depth varies according to a paraboloidal law:

Although there are two waves with the same frequencies and opposite signs, it can be shown that they are identical because only the signs of B's change when those of f change, while A's remains unaltered. The direction of the rotation of waves must be decided only after drawing co-tidal lines (5, 8). Evaluation of a complete set of frequencies and wave patterns by an electronic computer will not be difficult.

Acknowledgments

The author is much indebted to Dr. K. Kajiura, professor at the Earthquake Research Institute, the University of Tokyo, for his kind advice in explaining the nature of the oscillations of the rotating basin under consideration. He also expresses his deep appreciation to Misses Mariko Karasawa, Fuyumi Wada and Midori Aoki for tedious numerical computations of the algebraic equations of higher order for estimating the frequencies of the oscillations.

References

1. Thomson, Sir William. 1879. On gravitational oscillations of rotating water. *Proc. Roy. Soc. Edinburgh.* Reprinted in *Phil. Mag.* 10: 97-109.
2. Lord Rayleigh. 1903. On the vibration of a rectangular sheet of rotating liquid. *Phil. Mag.* 6: 297-301, Paper 5: 93.
3. Lamb, Horace. 1932. *Hydrodynamics.* 6th Edition No. 212: 326-328.
4. Goldstein, S. 1928. *Tidal motion in rotating elliptic basins.* Monthly Notices of Royal Astronomical Society of London, Geophysical Supplement 2: 44-56.
5. Goldsbrough, G.R. 1936. The tidal oscillations in an elliptic basin of variable depth-II. *Roy. Soc. London Proc.* A-884: 12-32.
6. Hidaka, Koji. 1937. Tidal oscillations in a rectangular basin of variable depth (3rd paper) (Problem of water oscillations in various types of basins and canals-Part II) *Memoirs of the Imperial Marine Observatory* 4 (3): 259-278.
7. Darwin, Sir George H. 1886. On the dynamical theory of the tides of long period. *Proc. Roy. Soc. London A-41,* 337-342. Paper 1, pp. 366-371.
8. Goldsbrough, G.R. 1915. The dynamical theory of the tides in a polar basin. *Proc. London Math. Soc. Ser.* 2 (14), Part I: 31-66.

part 6: austin h. church

Matrix Algebra

Axial Numbers

Beam Numbers

Finite Element Method

Professor Austin H. Church received an M.E. degree from Cornell University in 1928, and a Sc.M. degree in physics from New York University in 1934.

From 1928 to 1931 he was an engineer with the Westinghouse E. and M. Company at South Philadelphia; from 1931 to 1937 an instructor of Machine Design at the Cooper Union in New York City; from 1937 to 1940 an engineer working on special problems for the De Laval Steam Turbine Company in Trenton. He has been in the Mechanical Engineering Department of New York University from 1940 to date teaching various courses in the area of solid mechanics. He was Chairman of the Department from 1946 through 1963; and 1957 through 1958 was on leave to work on vibration problems for the General Engineering Laboratory of the General Electric Company in Schenectady; and for the academic year 1963 through 1964 was a Visiting Professor of Mechanical Engineering at Duke University.

He is a Fellow of the American Society of Mechanical Engineers and a member of various other organizations. In addition to many consulting experiences and technical papers, he is the author of "Centrifugal Pumps and Blowers"; revised Guillet's "Kinematics of Machines"; and the author of "Mechanical Vibrations". The first two of these efforts have Spanish editions.

15 elements of structural analysis

matrix algebra

Introduction

The design and analysis of underwater structures depends upon solving a series of linear simultaneous equations, and this can be a lengthy, tedious procedure. In the 1850s Sylvester, Hamilton and Cayley devised a method of handling sets of equations by formulating them into arrays or matrices, and developed the basic principles for manipulating them. While this was helpful, it did not appreciably decrease the amount of labor involved, so the method was not generally used until the digital computer was developed about a hundred years later.

Because the computer is rapid and hence cheaper to use, can do repetitive calculations without errors, and can be programmed to give eight or sixteen figure accuracy, it has been found to be a powerful tool in applying matrix methods. The result has been that today matrix principles are widely used in all fields of engineering.

Many texts regarding matrices are available on all levels, ·and more are being released constantly. It is not feasible to list them all (1-9).

The first step in applying matrix methods is to understand the basic principles of matrix algebra which are given in this chapter. It is hoped that this elementary review of matrix manipulation and its application to simple structural analysis will be helpful to the uninitiated. In this effort rigorous proofs will be omitted and the emphasis will be on achieving a physical understanding of the principles involved. Those interested in pursuing these topics further are referred to the texts cited in the References.

Matrices

A *matrix* is an array of numbers or symbols arranged in rows and columns, whereas a *vector* is a single row or column of numbers or symbols. To illustrate, consider the set of simultaneous equations:

$$
\begin{aligned}
3\,x_1 + 2\,x_2 + x_3 &= 9 \\
x_1 - 2\,x_2 + x_3 &= -1 \quad (15\text{-}1) \\
2\,x_1 + x_2 - 2\,x_3 &= 11
\end{aligned}
$$

which could be expressed in more general terms as

$$
\begin{aligned}
a_{11}\,x_1 + a_{12}\,x_2 + a_{13}\,x_3 &= y_1 \\
a_{21}\,x_1 + a_{22}\,x_2 + a_{23}\,x_3 &= y_2 \quad (15\text{-}2) \\
a_{31}\,x_1 + a_{32}\,x_2 + a_{33}\,x_3 &= y_3 \,.
\end{aligned}
$$

The subscripts of the a_{ij} coefficients are written so that i designates the number of the row and j the number of the column. These coefficients can be written as a matrix $\mathbf{A}$:

$$A = \begin{bmatrix} a_{11} & a_{12} & a_{13} \\ a_{21} & a_{22} & a_{23} \\ a_{31} & a_{32} & a_{33} \end{bmatrix} \quad \text{or} \quad \begin{bmatrix} 3 & 2 & 1 \\ 1 & -2 & 1 \\ 2 & 1 & -2 \end{bmatrix}$$

There are two column vectors, indicated by curly brackets, in equation (15-1) $\{x_1 \ x_2 \ x_3\}$; and $\{y_1 \ y_2 \ y_3\}$ or $\{9 \ -1 \ 11\}$, designated as **x** and **y**. Hence, the set of equations can be written in matrix form as

$$\mathbf{A} \ \mathbf{x} \ = \ \mathbf{y}. \tag{15-3}$$

It is important in understanding equation (15-3) to recognize that it is expressing equation (15-1) or (15-2), i.e., that the coefficients a_{ij} are multiplied by x_i successively and the products added to give y_i.

In this example the matrix **A** is square, i.e., it has the same number of rows and columns; but it also could be rectangular. In general, the *order* of a matrix is given as m by n, where m is the number of rows and n is the number of columns.

The principal diagonal of a matrix is defined as those elements whose subscripts i and j are equal. Thus, in equation (15-1) the principal diagonal is 3, −2, −2; or in equation (15-2) it is a_{11}, a_{22}, a_{33}.

A *scalar matrix* is one in which all of the elements on the principal diagonal have the same value and all other elements are zero. This corresponds to a single number. If all of the elements on the principal diagonal are unity, it is a *unit matrix* **I**, corresponding to the number 1. If all of the elements in a matrix are zero, it is known as a *null matrix*.

A matrix is said to be *symmetric* if $a_{ij} = a_{ji}$. Thus,

$$\begin{bmatrix} 1 & 3 & -2 \\ 3 & 0 & 4 \\ -2 & 4 & 5 \end{bmatrix}$$

is a symmetric matrix, and it may be noted that corresponding elements are symmetrically placed relative to the principal diagonal.

In manipulating matrices it is frequently necessary to transpose a matrix; that is, interchange the rows and columns, which corresponds to making $a_{ij} = a_{ji}$. This operation is indicated by using a superscript T or a prime. Referring to equation (15-1):

$$\mathbf{A}^T \ = \ \mathbf{A}' \ = \ \begin{bmatrix} 3 & 1 & 2 \\ 2 & -2 & 1 \\ 1 & 1 & -2 \end{bmatrix} \tag{15-4}$$

It may be noted that the transpose of a symmetric matrix is the same as the original matrix.

When dealing with large matrices it is frequently desirable to condense them into smaller matrices by partitioning them. Thus,

$$\left[\begin{array}{ccc|cc} 2 & 3 & 1 & 5 & 8 \\ 1 & 4 & 5 & 9 & 6 \\ 1 & 2 & 3 & 1 & 3 \\ \hline 0 & 0 & 0 & 1 & 2 \\ 0 & 0 & 0 & 3 & 4 \end{array}\right] = \left[\begin{array}{c|c} A & B \\ \hline C & D \end{array}\right] \tag{15-5}$$

where

$$A = \begin{bmatrix} 2 & 3 & 1 \\ 1 & 4 & 5 \\ 1 & 2 & 3 \end{bmatrix} \ ; B = \begin{bmatrix} 5 & 8 \\ 9 & 6 \\ 1 & 3 \end{bmatrix} ;$$

$$C = \begin{bmatrix} 0 & 0 & 0 \\ 0 & 0 & 0 \end{bmatrix} \ ; D = \begin{bmatrix} 1 & 2 \\ 3 & 4 \end{bmatrix}$$

Partitioning is usually clarified by using dashed lines as shown in equation (15-5).

Determinants

The determinant $|A|$ of a coefficient matrix is a square array of numbers or symbols which when expanded gives a number that can be used in solving the set of equations. If $|A|$ equals zero the equations cannot be solved, and the determinant is said to be *singular*.

The expansion of second and third order determinants is covered in elementary algebra courses. Thus,

$$|A| = \begin{vmatrix} a_{11} & a_{12} \\ a_{21} & a_{22} \end{vmatrix} = a_{11}\,a_{22} - a_{12}\,a_{21}$$

and

$$|A| = \begin{vmatrix} a_{11} & a_{12} & a_{13} \\ a_{21} & a_{22} & a_{23} \\ a_{31} & a_{32} & a_{33} \end{vmatrix}$$

$$= \begin{array}{l} a_{11}a_{22}a_{33} + a_{12}a_{23}a_{31} + a_{13}a_{21}a_{32} \\ -a_{13}a_{22}a_{31} - a_{11}a_{23}a_{32} - a_{12}a_{21}a_{33} \end{array}$$

It is *not* possible to extend this procedure to expand higher order determinants, but many ways have been devised to do this. They will be considered later after some of the basic properties of determinants have been discussed.

A few rules governing the manipulation and evaluation of determinants will be considered for the determinant

$$|A| = \begin{vmatrix} a & b \\ c & d \end{vmatrix} = ad - bc$$

1. Transposing $|A|$ does not change its value.

$$|A^T| = \begin{vmatrix} a & c \\ b & d \end{vmatrix} = ad - bc \ .$$

2. Interchanging any two rows (or columns) changes the sign of $|A|$:

$$|B| = \begin{vmatrix} c & d \\ a & b \end{vmatrix} = bc - ad \ ;$$

$$|C| = \begin{vmatrix} b & a \\ d & c \end{vmatrix} = bc - ad.$$

3. If a row (or column) is changed by adding (or subtracting) the corresponding elements of another row (or column) times a constant, the value of the determinant is unchanged.

$$|B| = \begin{vmatrix} a-kc & b-kd \\ c & d \end{vmatrix}$$

$$= ad - kcd - bc + kcd$$

4. If all the elements of a row (or column) are multiplied by a constant, the value of the determinant is multiplied by that constant.

$$|B| = \begin{vmatrix} ka & kb \\ c & d \end{vmatrix}$$

$$= kad - kbc = k\,|A|.$$

If all the elements of a determinant of order n are multiplied by a constant k then the value of the determinant is $k^n |A|$.

5. If corresponding elements of a row (or column) of a determinant are linear combinations of another row (or column) the value of the determinant is zero. Conversely, if the value of a determinant is zero, then at least one row *and* one column are linearly dependent on other rows and columns. Two examples involving third order determinants are

$$|A| = \begin{vmatrix} a & b & c \\ ka & kb & kc \\ d & e & f \end{vmatrix}$$

$$= \begin{matrix} akbf + bkcd + kace \\ -akbf - bkcd - kace \end{matrix} = 0$$

and

$$|B| = \begin{vmatrix} a & b & c \\ a+d & b+e & c+f \\ d & e & f \end{vmatrix}$$

$$= \begin{matrix} aef + abf + bdf + bcd + cde + ace \\ -aef - abf - bdf - bcd - cde - ace \end{matrix}$$

$$= 0.$$

6. If a determinant contains a row (or column) of zeros its value is zero. Thus,

$$|A| = \begin{vmatrix} a & 0 \\ c & 0 \end{vmatrix} = 0 - 0 = 0.$$

The *rank* of a determinant (or a matrix) is the order of the largest nonsingular determinant. Thus, referring to the examples of Rule 5, the rank of these determinants is two, as the determinant of the entire matrix is zero, but the determinant of

$$\begin{vmatrix} a & b \\ d & e \end{vmatrix}$$

which is of the second order, is not.

One method of expanding large determinants is the *method of minors*. The *minor* of an element of a determinant is the determinant that remains when the row and column of the element are deleted. The *cofactor* of an element a_{ij} is $(-1)^{i+j}$ times the minor of the element. The value of a determinant equals the sum of the products of the elements in any row (or column) multiplied by the corresponding cofactor. To illustrate, it is possible to expand the determinant $|A|$ by using the minors of the top row:

$$|A| = \begin{vmatrix} 3 & 1 & -1 & 1 \\ 0 & 3 & 1 & 2 \\ 1 & 4 & 2 & 1 \\ 5 & -1 & -3 & 5 \end{vmatrix}$$

Then

$$|A| = 3(-1)^{1+1} \begin{vmatrix} 3 & 1 & 2 \\ 4 & 2 & 1 \\ -1 & -3 & 5 \end{vmatrix}$$

+ . . . continued

+ . . . continued

$$+ 1(-1)^{1+2} \begin{vmatrix} 0 & 1 & 2 \\ 1 & 2 & 1 \\ 5 & -3 & 5 \end{vmatrix}$$

$$+ (-1)(-1)^{1+3} \begin{vmatrix} 0 & 3 & 2 \\ 1 & 4 & 1 \\ 5 & -1 & 5 \end{vmatrix}$$

$$+ 1(-1)^{1+4} \begin{vmatrix} 0 & 3 & 1 \\ 1 & 4 & 2 \\ 5 & -1 & -3 \end{vmatrix}$$

or $|A| = 3(-2) - 1(-26) - 1(-42) - 1(18) = 44$.

The same value would be obtained if any other row or column were used.

As mentioned previously, there are many other ways of expanding large determinants. Space does not permit their inclusion here but they can be found in the references cited.

Determinants can be used to solve the set of equations given as equation (15-1) by use of *Cramers Rule*. That is, the value of x_j can be found by replacing the j'th column of the $|A|$ determinant by the y vector to obtain determinant $|B|$. Then, $x_j = |B| / |A|$. To illustrate using equation (15-1):

$$x_3 = \frac{\begin{vmatrix} 3 & 2 & 9 \\ 1 & -2 & -1 \\ 2 & 1 & 11 \end{vmatrix}}{\begin{vmatrix} 3 & 2 & 1 \\ 1 & -2 & 1 \\ 2 & 1 & -2 \end{vmatrix}} = \frac{-44}{22} = -2.$$

The values of x_1 and x_2 could be found in a similar manner.

It might be noted that if $|A| = 0$, all values of x_j would equal infinity, which shows that the equations cannot be solved if the determinant of the coefficient matrix is singular.

Matrix Algebra

Two matrices are equal only if all of their elements a_{ij} are identical.

Addition or *subtraction* of matrices is performed by adding or subtracting corresponding elements. These operations can be performed in any order; i.e., they are *commutative*.

$$D = A - B + C = C + A - B.$$

$$D = \begin{bmatrix} 4 & -2 \\ 3 & 5 \end{bmatrix} - \begin{bmatrix} 1 & 0 \\ 2 & -3 \end{bmatrix}$$

$$+ \begin{bmatrix} 5 & 4 \\ 3 & -2 \end{bmatrix} = \begin{bmatrix} 8 & 2 \\ 4 & 6 \end{bmatrix}$$

If a matrix is multiplied by a constant, each element is multiplied by that constant.

$$3 \begin{bmatrix} 1 & 2 \\ 3 & 4 \end{bmatrix} = \begin{bmatrix} 3 & 6 \\ 9 & 12 \end{bmatrix} .$$

This is equivalent to three successive additions of the matrix. It should be noted that this rule is not the same as Rule 4 given for determinants.

Multiplication of two matrices is accomplished by summing the product of the elements in turn of any row of the first matrix and corresponding elements in any column of the second matrix. This process can be done only if the matrices are conformable for multiplication; i.e., the number of elements in each row of the first matrix must equal the number of elements in each column of the second matrix.

$$\begin{bmatrix} 2 & 1 \\ 3 & 2 \end{bmatrix} \cdot \begin{bmatrix} 1 & 2 & 4 \\ 3 & -1 & 2 \end{bmatrix}$$

$$= \begin{bmatrix} 2(1)+1(3) & 2(2)+1(-1) & 2(4)+1(2) \\ 3(1)+2(3) & 3(2)+2(-1) & 3(4)+2(2) \end{bmatrix}$$

$$= \begin{bmatrix} 5 & 3 & 10 \\ 9 & 4 & 16 \end{bmatrix} .$$

In general multiplication is not commutative; i.e., $\mathbf{A} \cdot \mathbf{B} \neq \mathbf{B} \cdot \mathbf{A}$.

$$\mathbf{A} = \begin{bmatrix} 0 & 2 \\ 1 & 3 \end{bmatrix} ; \qquad \mathbf{B} = \begin{bmatrix} 1 & 2 \\ 3 & 4 \end{bmatrix}$$

$$\mathbf{A} \cdot \mathbf{B} = \begin{bmatrix} 6 & 8 \\ 10 & 14 \end{bmatrix}$$

but

$$\mathbf{B} \cdot \mathbf{A} = \begin{bmatrix} 2 & 8 \\ 4 & 18 \end{bmatrix}$$

It is possible that the product of two matrices can yield a null matrix without either of the product matrices being null.

$$\mathbf{A} = \begin{bmatrix} 0 & 1 \\ 0 & 2 \end{bmatrix} ; \qquad \mathbf{B} = \begin{bmatrix} 3 & 2 \\ 0 & 0 \end{bmatrix} ;$$

$$\mathbf{A} \cdot \mathbf{B} = \begin{bmatrix} 0 & 0 \\ 0 & 0 \end{bmatrix} ;$$

but

$$\mathbf{B} \cdot \mathbf{A} = \begin{bmatrix} 0 & 7 \\ 0 & 0 \end{bmatrix} .$$

If several matrices are to be multiplied the operations may be performed in any order. Thus, if $\mathbf{D} = \mathbf{A} \cdot \mathbf{B} \cdot \mathbf{C}$, it is possible to premultiply $\mathbf{C}$ by $\mathbf{B}$ and then premultiply the product by $\mathbf{A}$; i.e., work from right to left or to postmultiply $\mathbf{A}$ by $\mathbf{B}$ and then postmultiply this product by $\mathbf{C}$; i.e., work from left to right. If the multiplication is to be done by hand, it may be simpler to work in a particular direction as illustrated in the example below.

Given:

$$\mathbf{A} = \begin{bmatrix} 2 \\ -1 \\ 4 \end{bmatrix} ; \quad \mathbf{B} = [5 \ 2 \ -3] ;$$

$$C = \begin{bmatrix} 1 \\ 0 \\ 2 \end{bmatrix}$$

Working from left to right:

$$\begin{bmatrix} 2 \\ -1 \\ 4 \end{bmatrix} \quad [5 \; 2 \; -3]$$

$$= \begin{bmatrix} 10 & 4 & -6 \\ -5 & -2 & 3 \\ 20 & 8 & -12 \end{bmatrix} \begin{bmatrix} 1 \\ 0 \\ 2 \end{bmatrix} = \begin{bmatrix} -2 \\ 1 \\ -4 \end{bmatrix} .$$

Working from right to left:

$$[5 \; 2 \; -3] \begin{bmatrix} 1 \\ 0 \\ 2 \end{bmatrix} = [-1] \quad ;$$

$$\begin{bmatrix} 2 \\ -1 \\ 4 \end{bmatrix} \quad [-1] \quad = \begin{bmatrix} -2 \\ 1 \\ -4 \end{bmatrix} .$$

The transpose of the product of several matrices equals the product of the transposes of these matrices taken in the reverse order; i.e., $[\mathbf{A} \cdot \mathbf{B} \cdot \mathbf{C}]^T = \mathbf{C}^T \mathbf{B}^T \mathbf{A}^T$. Using the example just given, the transpose of $[\mathbf{ABC}]$ from the previous example is

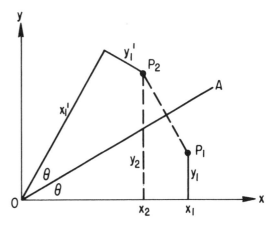

Figure 15-1. Rotation of point P_1 to P_2 about line OA.

$[-2 \; 1 \; -4]$. Taking the transpose of each matrix gives $\mathbf{C}^T \mathbf{B}^T \mathbf{A}^T$ or

$$[1 \; 0 \; 2] \begin{bmatrix} 5 \\ 2 \\ -3 \end{bmatrix} [2 - 1 \; 4] = [-2 \; 1 \; -4] .$$

Geometrical Interpretation

The definitions and rules regarding matrices and determinants tend to be somewhat confusing. They can be clarified to some extent by relating them to the position of a point in a plane. The position of the point can be specified by a vector

$$\begin{bmatrix} x \\ y \end{bmatrix}$$

where x and y are distances measured from an origin O.

Assume that a point P_1 is initially at a position x_1 and y_1, and that it is desired to rotate it to a point P_2 at x_2 and y_2 about a line OA making an angle θ with the x axis as shown in Figure 15-1.

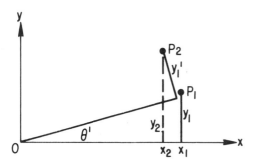

Figure 15-2. Rotation of point P_1 to P_2 about origin O through the angle O.

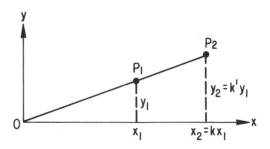

Figure 15-3. Displacement of point P_1 to P_2 along a radial line through origin O.

From the trigonometry of the figure it can be seen that

$$x_2 = x_1 \cos 2\theta + y_1 \sin 2\theta$$

$$y_2 = x_1 \sin 2\theta - y_1 \cos 2\theta.$$

These equations can be put into matrix form as

$$\begin{bmatrix} x_2 \\ y_2 \end{bmatrix} = \begin{bmatrix} \cos 2\theta & \sin 2\theta \\ \sin 2\theta & -\cos 2\theta \end{bmatrix} \begin{bmatrix} x_1 \\ y_1 \end{bmatrix}.$$

$$(15-6)$$

If θ is zero the matrix is

$$\begin{bmatrix} 1 & 0 \\ 0 & -1 \end{bmatrix}$$

and the point is rotated about the x axis. If θ is 90° the matrix is

$$\begin{bmatrix} -1 & 0 \\ 0 & 1 \end{bmatrix}$$

and the point is rotated about the y axis.

Another example is to assume that it is desired to rotate the point P_1 through an angle θ' about the origin 0 as shown in Figure 15-2.

From the trigonometry of Figure 15-2 it can be seen that

$$x_2 = x_1 \cos \theta' - y_1 \sin \theta'$$

$$y_2 = x_1 \sin \theta' + y_1 \cos \theta'$$

These equations can be put into matrix form as

$$\begin{bmatrix} x_2 \\ y_2 \end{bmatrix} = \begin{bmatrix} \cos \theta' & -\sin \theta' \\ \sin \theta' & \cos \theta' \end{bmatrix} \begin{bmatrix} x_1 \\ y_1 \end{bmatrix} \quad (15-7)$$

Another common type of matrix results if the point P_1 is "stretched" or moved along a radial line passing through the origin 0 as shown in Figure 15-3. It may now be observed that $x_2 = k\,x_1$ and $y_2 = k'\,y_1$ or

$$\begin{bmatrix} x_2 \\ y_2 \end{bmatrix} = \begin{bmatrix} k & 0 \\ 0 & k' \end{bmatrix} \begin{bmatrix} x_1 \\ y_1 \end{bmatrix}.$$

Each of these examples represent rather special cases, so it is of interest to see the result of applying a general matrix such as

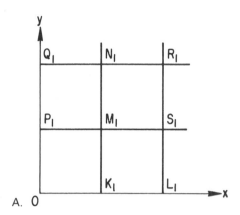

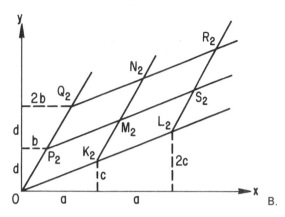

Figure 15-4. Displacement of a grid of squares due to the application of a general matrix.

$$\begin{bmatrix} a & b \\ c & d \end{bmatrix}$$

to the square grid of points shown in Figure 15-4A. Then

$$\begin{bmatrix} x_2 \\ y_2 \end{bmatrix} = \begin{bmatrix} a & b \\ c & d \end{bmatrix} \begin{bmatrix} x_1 \\ y_1 \end{bmatrix}$$

or

$$x_2 = a\,x_1 + b\,y_1$$
$$y_2 = c\,x_1 + d\,y_1\;.$$

The result is the distorted grid shown as Figure 15-4B, and it may be observed that an original square such as $O\,P_1\,M_1\,K_1$ is transformed into the parallelogram $O\,P_2\,M_2\,K_2$. Hence, the general matrix produces horizontal stretches a and b plus vertical stretches c and d. These correspond to strains in an element due to loading. The same matrix can be applied to problems in fluid flow, heat transfer and electric fields.

It is, of course, possible to determine the values of the matrix elements a, b, c and d from experiments on a grid subjected to loading when desirable.

Several important conclusions can be drawn from these examples:

1. Perhaps the most fundamental one is that a matrix is a mathematical instruction indicating a transformation. Thus, it is similar to a division sign or an integral sign that indicates a mathematical operation.

2. The first example indicates the movement of a point about a line. If this operation is repeated twice, the point will return to its original position. Hence, the square of the matrix of equation (15-6) should yield a unit matrix; and it does give

$$\begin{bmatrix} \cos^2 2\theta + \sin^2 2\theta & \cos 2\theta \sin 2\theta - \sin 2\theta \cos 2\theta \\ \sin 2\theta \cos 2\theta - \cos 2\theta \sin 2\theta & \cos^2 2\theta + \sin^2 2\theta \end{bmatrix}$$

$$= \begin{bmatrix} 1 & 0 \\ 0 & 1 \end{bmatrix}$$

3. Geometrical interpretation can be used to show that, in general, the multiplication of matrices $A\cdot B$ does not equal $B\cdot A$. Consider the matrices

$$A = \begin{bmatrix} 1 & 0 \\ 0 & -1 \end{bmatrix} \text{and } B = \begin{bmatrix} 0 & 1 \\ 1 & 0 \end{bmatrix}.$$

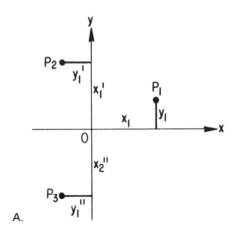

From equation (15-6) matrix **A** represents the rotation of a point P about the x axis for $\theta = 0$. From equation (15-7) matrix **B** represents the rotation of the point about the origin for $\theta' = 90°$. If the point is first rotated about the origin 0 and then about the x axis as shown in Figure 15-5A successive positions are P_1, P_2 and P_3 and the matrix equation is

$$\begin{bmatrix} x_3 \\ y_3 \end{bmatrix} = A \cdot B \begin{bmatrix} x_1 \\ y_1 \end{bmatrix}$$

$$= \begin{bmatrix} 1 & 0 \\ 0 & -1 \end{bmatrix} \cdot \begin{bmatrix} 0 & -1 \\ 1 & 0 \end{bmatrix} \begin{bmatrix} x_1 \\ y_1 \end{bmatrix}$$

$$= \begin{bmatrix} 0 & -1 \\ -1 & 0 \end{bmatrix} \begin{bmatrix} x_1 \\ y_1 \end{bmatrix} = \begin{bmatrix} -y_1 \\ -x_1 \end{bmatrix}$$

A.

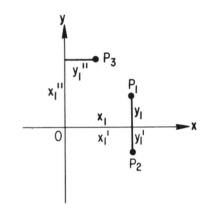

B.

Figure 15-5. Geometric demonstration that A • B ≠ B • A.

If the point is first rotated about the x axis and then about the origin 0 as shown in Figure 15-5B, the matrix equation is

$$\begin{bmatrix} x_3 \\ y_3 \end{bmatrix} = B \cdot A \begin{bmatrix} x_1 \\ y_1 \end{bmatrix} = \begin{bmatrix} 0 & -1 \\ 1 & 0 \end{bmatrix} \begin{bmatrix} 1 & 0 \\ 0 & -1 \end{bmatrix} \begin{bmatrix} x_1 \\ y_1 \end{bmatrix}$$

$$= \begin{bmatrix} 0 & 1 \\ 1 & 0 \end{bmatrix} \begin{bmatrix} x_1 \\ y_1 \end{bmatrix} = \begin{bmatrix} y_1 \\ x_1 \end{bmatrix}$$

It will be noted that $A \cdot B \neq B \cdot A$ and the position of point P_3 is different in the two cases.

4. These geometrical interpretations could be extended to higher order matrices and the basic principles would still apply, but they would be more difficult to visualize.

Figures 15-4A and B show that the general matrix converts a square into a parallelogram. The ratio of the area of the parallelogram to that of the square is a constant g, which depends on the size of the elements of the general matrix.

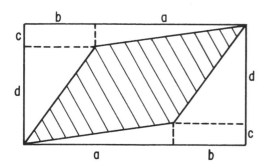

Figure 15-6. Distortion of a unit square due to application of a general matrix.

Using the matrix of Figure 15-4, the original unit square is converted to the crosshatched parallelogram of Figure 15-6. It may be observed that this parallelogram is enclosed in a rectangle having an area of (a+b)(c+d). The uncrosshatched area of the rectangle is 2 bc + ac + bd; hence, the area of the parallelogram is (a+b)(c+d) − 2bc − ac − bd = ad − bc. Hence, g = (ad − bc)/1, and it will be recognized that the numerator of this is the determinant of the general matrix, or

$$\begin{vmatrix} a & b \\ c & d \end{vmatrix}.$$

If two matrices U and V are multiplied, operation U may change the area g times, while operation V may change the area h times. Hence, the product of U·V would change the area g·h times, and |U|·|V| = |UV|. The process is commutative, and the change in area can be positive or negative. The determinant of the stretch matrix is

$$\begin{vmatrix} k & 0 \\ 0 & k \end{vmatrix} = k \cdot k';$$

and it may be observed from Figure 15-3 that the k′ term changes the vertical length k′ times, while the k term changes the horizontal length k times. Hence, the total area change is k·k′.

Rotations carry the point around without changing or producing any area, so the determinants of the first two matrices of this section are unity.

Matrix Inversion

The section on Matrix Algebra did not include material on matrix division since it cannot be done directly. The process is accomplished by finding an inverse A^{-1} of the divisor matrix A such that

$$A^{-1}A = AA^{-1} = I. \qquad (15\text{-}8)$$

Therefore, the problem is reduced to finding the inverse of a matrix that will satisfy equation (15-8). Methods of finding this inverse will now be discussed. However, there are two important restrictions concerning inverses, namely that (1) only a square matrix has an inverse, and (2) a singular matrix (one whose determinant is zero) has no inverse.

Terms Only on the Principal Diagonal

If the matrix to be inverted contains terms only on the principal diagonal, it is a simple matter to find the inverse. It is only necessary to find the reciprocals of each of these terms.

$$A = \begin{bmatrix} a_{11} & 0 & 0 \\ 0 & a_{22} & 0 \\ 0 & 0 & a_{33} \end{bmatrix};$$

$$\text{then } A^{-1} = \begin{bmatrix} \dfrac{1}{a_{11}} & 0 & 0 \\ 0 & \dfrac{1}{a_{22}} & 0 \\ 0 & 0 & \dfrac{1}{a_{33}} \end{bmatrix}. \qquad (15\text{-}9)$$

If these two matrices are multiplied (in any order) it will be found that the product yields a unit matrix in accordance with equation (15-8).

General Case

The inverse A^{-1} of a general matrix A can be found by first determining the cofactor matrix A^c and then dividing each element of the transposed cofactor matrix by the determinant of A. Thus

$$A^{-1} = [A^c]^T/|A| \qquad (15\text{-}10)$$

The application of equation (15-10) to find the inverse of a second order matrix is to interchange the elements on the principal diagonal and to change the sign of the terms on the cross diagonal. Then each element is divided by the determinant of the original matrix.

If

$$A = \begin{bmatrix} a_{11} & a_{12} \\ a_{21} & a_{22} \end{bmatrix} \text{ then } |A| = a_{11}a_{22} - a_{12}a_{21}$$

and

$$A^{-1} = 1/|A| \begin{bmatrix} a_{22} & -a_{12} \\ -a_{21} & a_{11} \end{bmatrix} \qquad (15\text{-}11)$$

Equation (15-11) can be shown to be valid by inserting it into equation (15-8):

$$A \cdot A^{-1} = \begin{bmatrix} a_{11} & a_{12} \\ a_{21} & a_{22} \end{bmatrix} \frac{1}{(a_{11}a_{22} - a_{12}a_{21})} \begin{bmatrix} a_{22} & -a_{12} \\ -a_{21} & a_{11} \end{bmatrix}$$

$$= \frac{1}{a_{11}a_{22} - a_{12}a_{21}} \begin{bmatrix} a_{11}a_{22} - a_{12}a_{21} & -a_{11}a_{12} + a_{12}a_{11} \\ a_{21}a_{22} - a_{22}a_{21} & -a_{21}a_{12} + a_{11}a_{22} \end{bmatrix} = \begin{bmatrix} 1 & 0 \\ 0 & 1 \end{bmatrix} = I.$$

While the same procedure can be illustrated for third order matrices using symbols, the expressions get quite lengthy and cumbersome; therefore it is more feasible to illustrate it with a numerical example.

$$\text{Given: } A = \begin{bmatrix} 2 & -1 & 3 \\ 0 & 4 & -2 \\ 1 & -3 & 5 \end{bmatrix}; A^c = \begin{bmatrix} 14 & -2 & -4 \\ -4 & 7 & 5 \\ -10 & 4 & 8 \end{bmatrix}$$

$$|A| = 40 + 2 + 0 - 12 - 12 - 0 = 18.$$

By Equation (15-10):

$$A^{-1} = \frac{1}{18} \begin{bmatrix} 14 & -4 & -10 \\ -2 & 7 & 4 \\ -4 & 5 & 8 \end{bmatrix}.$$

The value of A^{-1} can be checked by substituting it in equation (15-8):

$$A \cdot A^{-1} = \begin{bmatrix} 2 & -1 & 3 \\ 0 & 4 & -2 \\ 1 & -3 & 5 \end{bmatrix} 1/18 \begin{bmatrix} 14 & -4 & -10 \\ -2 & 7 & 4 \\ -4 & 5 & 8 \end{bmatrix}$$

$$= \begin{bmatrix} 1 & 0 & 0 \\ 0 & 1 & 0 \\ 0 & 0 & 1 \end{bmatrix} = I.$$

This method of finding inverses is not practical for matrices whose order is higher than three, as it becomes too cumbersome. A variety of methods are available for finding inverses of large matrices (1, 2, 3, 4, 7, 10). Only two of these will be given here.

Gauss Elimination Method

As was pointed out in Rule 3 of determinants, the value of a matrix is not changed if the elements of any row are multiplied by a constant and added to another row of elements.

In the Gauss Elimination Method the matrix A_1 to be inverted is set up beside a unit matrix A_2 of the same order. Then, by a series of operations or transformations of the type just described, matrix A_1 is transformed into a unit matrix, and these same operations are also applied to the second matrix A_2. Matrix A_2 then becomes A^{-1}, or the inverse of A_1.

This is illustrated by a numerical example. To clarify the procedure each pair of matrices is designated by a different letter for each successive transformation. The operations are specified on the left side and the corresponding transformations are given on the right.

$$A_1 = \begin{bmatrix} -2 & 3 & 1 \\ -5 & 4 & 2 \\ 3 & 0 & -1 \end{bmatrix}; A_2 = \begin{bmatrix} 1 & 0 & 0 \\ 0 & 1 & 0 \\ 0 & 0 & 1 \end{bmatrix}$$

Transformation 1

$$b_{1j} = a_{1j}/a_{11}$$
$$b_{2j} = a_{2j} - b_{1j}a_{21}$$
$$b_{3j} = a_{3j} - b_{1j}a_{31}$$

$$B_1 = \begin{bmatrix} 1 & -3/2 & -1/2 \\ 0 & -7/2 & -1/2 \\ 0 & -9/2 & 1/2 \end{bmatrix}; B_2 = \begin{bmatrix} -1/2 & 0 & 0 \\ -5/2 & 1 & 0 \\ 3/2 & 0 & 1 \end{bmatrix}$$

Transformation 2

$$c_{1j} = b_{1j} - c^{2j}b_{12}$$
$$c_{2j} = b_{2j}/b_{22}$$
$$c_{3j} = b_{3j} - c_{2j}b_{32}$$

$$C_1 = \begin{bmatrix} 1 & 0 & -2/7 \\ 0 & 1 & 1/7 \\ 0 & 0 & -1/7 \end{bmatrix}; C_2 = \begin{bmatrix} 4/7 & -3/7 & 0 \\ 5/7 & -2/7 & 0 \\ -12/7 & 1/7 & 1 \end{bmatrix}$$

Transformation 3

$$d_{1j} = c_{1j} - d_{3j} c_{13}$$
$$d_{2j} = c_{2j} - d_{3j} c_{23}$$
$$d_{3j} = c^{3j}/c_{33}$$

$$D_1 = \begin{bmatrix} 1 & 0 & 0 \\ 0 & 1 & 0 \\ 0 & 0 & 1 \end{bmatrix} ; D_2 = \begin{bmatrix} 4 & -3 & -2 \\ -1 & 1 & 1 \\ 12 & -9 & -7 \end{bmatrix}$$

Checking by equation (15-8):

$$A_1 \cdot D_2 = I.$$

$$\begin{bmatrix} -2 & 3 & 1 \\ -5 & 4 & 2 \\ 3 & 0 & -1 \end{bmatrix} \begin{bmatrix} 4 & -3 & -2 \\ -1 & 1 & 1 \\ 12 & -9 & -7 \end{bmatrix}$$

$$= \begin{bmatrix} 1 & 0 & 0 \\ 0 & 1 & 0 \\ 0 & 0 & 1 \end{bmatrix} = I.$$

This method can be applied to a matrix **M** of any order but entails one serious restriction: there cannot be a zero on the principal diagonal. It will be observed that the first step in Transformation k is to divide the row containing the element m_{kk} by m_{kk}. If m_{kk} is zero it is apparent that the procedure "blows up" since the terms on that row would be infinite.

It is seldom in mechanics problems that there is a zero on the principal diagonal; but if it should occur, it is desirable to have an alternate procedure available. One such alternate is the possibility of partitioning the given matrix (2, pp. 81-5; 4, pp. 102-5).

Inversion by Partitioning

A large matrix **S** of order n, which may have zeros on the principal diagonal, can be partitioned into four smaller matrices:

$$S = \begin{bmatrix} A & B \\ \hline C & D \end{bmatrix} \qquad (15\text{-}12)$$

The only restriction is that the submatrices **A** and **D** must be square. Assume that the inverse of **S** is

$$S^{-1} = \begin{bmatrix} K & | & L \\ \hline M & | & N \end{bmatrix} \qquad (15\text{-}13)$$

where submatrices **K** and **N** are of the same order as submatrices **A** and **D**.

Applying equation (15-8) to these matrices:

$$A \cdot K + B \cdot M = I \qquad\qquad A \cdot L + B \cdot N = O$$

$$C \cdot K + D \cdot M = O \qquad\qquad C \cdot L + D \cdot N = I$$

Matrix operations applied to these equations gives

$$K = (A - B D^{-1} C)^{-1}$$

$$N = (D - C A^{-1} B)^{-1}$$

$$L = -A^{-1} B N$$

$$M = -D^{-1} C K$$

If the calculations are to be done by hand, time and effort can be reduced by converting these equations to

$$N = (D - C A^{-1} B)^{-1} \qquad M = -N C A^{-1}$$

$$L = -A^{-1} B N \qquad\qquad K = A^{-1} - A^{-1} B M$$

or to

$$K = (A - B D^{-1} C)^{-1} \qquad L = K B D^{-1}$$

$$M = - D^{-1} C K \qquad N = D^{-1} - D^{-1} C L$$
$$(15\text{-}14)$$

since a fewer number of inversions are required.

Example: Given matrix **S** with two zeros on the principal diagonal:

$$S = \begin{bmatrix} 1 & 0 & 1 & \vline & 1 & 0 \\ 2 & 0 & 0 & \vline & 0 & 2 \\ 0 & 1 & 0 & \vline & 0 & 1 \\ \hline 1 & 0 & 1 & \vline & 2 & 0 \\ 0 & 2 & 0 & \vline & -1 & 1 \end{bmatrix}$$

Using equation (15-14):

$$D^{-1} = 1/2 \begin{bmatrix} 1 & 0 \\ 1 & 2 \end{bmatrix}$$

$$K = \begin{bmatrix} -1 & 1/2 & -2 \\ -1 & 0 & -1 \\ 3 & -1/2 & 2 \end{bmatrix} \qquad N = \begin{bmatrix} 1 & 0 \\ -1 & -1 \end{bmatrix}$$

$$L = \begin{bmatrix} 1 & 1 \\ 1 & 1 \\ -2 & -1 \end{bmatrix} \qquad M = \begin{bmatrix} -1 & 0 & 0 \\ 1 & 0 & 2 \end{bmatrix}$$

It might be desirable for the reader to verify these values by solving the various equations. Assembling the values into equation (15-13):

$$S^{-1} = \begin{bmatrix} -1 & 1/2 & -2 & \vline & 1 & 1 \\ -1 & 0 & 2 & \vline & 1 & 1 \\ 3 & -1/2 & 2 & \vline & -2 & -1 \\ \hline -1 & 0 & 0 & \vline & 1 & 0 \\ 1 & 0 & 2 & \vline & -1 & -1 \end{bmatrix}.$$

which can be checked with the aid of equation (15-8).

If it should be possible to partition matrix **S** so that submatrices **B** and **C** are null in equation (15-12), then equation (15-9) could be applied to give

$$S^{-1} = \begin{bmatrix} A^{-1} & 0 \\ 0 & D^{-1} \end{bmatrix} \qquad (15\text{-}15)$$

Example:

$$\text{Given: } S = \begin{bmatrix} 1 & 0 & 0 & \vline & 0 & 0 \\ 0 & 2 & 1 & \vline & 0 & 0 \\ 1 & -2 & 0 & \vline & 0 & 0 \\ \hline 0 & 0 & 0 & \vline & 1 & 2 \\ 0 & 0 & 0 & \vline & -1 & 0 \end{bmatrix}$$

$$= \begin{bmatrix} A & 0 \\ 0 & D \end{bmatrix}$$

Then: $A^{-1} = \begin{bmatrix} 1 & 0 & 0 \\ 1/2 & 0 & -1/2 \\ -1 & 1 & 1 \end{bmatrix}$

$D^{-1} = \begin{bmatrix} 0 & -1 \\ 1/2 & 1/2 \end{bmatrix}$.

Inserting these submatrices into equation (15-15) gives

$$S^{-1} = \left[\begin{array}{ccc|cc} 1 & 0 & 0 & 0 & 0 \\ 1/2 & 0 & -1/2 & 0 & 0 \\ -1 & 1 & 1 & 0 & 0 \\ \hline 0 & 0 & 0 & 0 & -1 \\ 0 & 0 & 0 & 1/2 & 1/2 \end{array}\right] .$$

Applying equation (15-10) to matrix A gives

$$A^{-1} = 1/22 \begin{bmatrix} 3 & 5 & 4 \\ 4 & -8 & -2 \\ 5 & 1 & -8 \end{bmatrix} .$$

Putting this in equation (15-16) gives

$$x = 1/22 \begin{bmatrix} 3 & 5 & 4 \\ 4 & -8 & -2 \\ 5 & 1 & -8 \end{bmatrix} \begin{bmatrix} 9 \\ -1 \\ 11 \end{bmatrix}$$

$$= \begin{bmatrix} 3 \\ 1 \\ -2 \end{bmatrix} .$$

Solution of Simultaneous Equations

As a simple illustration of the application of matrix algebra, suppose it is desired to solve the set of nonhomogenous equations given by equation (15-1):

$$A \cdot x = y \qquad (15\text{-}3)$$

Premultiplying both sides of equation (15-3) by A^{-1} gives:

$$A^{-1} A x = A^{-1} y$$

$$x = A^{-1} y \qquad (15\text{-}16)$$

References

1. Gere, J.W. and W. Weaver. 1965. *Matrix algebra for engineers.* New York: Van Nostrand.
2. Hohn, F.E. 1958. *Elementary matrix algebra.* New York: Macmillan.
3. Frazer, R.A., W. J. Duncan and A.R. Collar. 1957. *Elementary matrices.* Cambridge: Cambridge University Press.
4. Faddeeva, V.N. 1959. *Computational methods of linear algebra.* New York: Dover.
5. Martin, H.C. 1966. *Introduction to matrix methods of structural analysis.* New York: McGraw-Hill.
6. Sawyer, W.W. 1955. *Prelude to mathematics.* Baltimore: Penguin.
7. Pipes, L.A. 1963. *Matrix methods for engineering.* Englewood Cliffs: Printice-Hall.
8. Pestel, E.G. and F.A. Leckie. 1963. *Matrix methods in elastomechanics.* New York: McGraw-Hill.

9. Rubinstein, M.F. 1966. *Matrix computer analysis of structures.* Englewood Cliffs: Prentice-Hall.

10. Argyris, J.H. and S. Kelsey. 1960. *Energy theorems and structural analysis.* London: Butterworth.

16 elements of structural analysis

axial numbers

Formulation of Matrices

Having some familiarity with the principles of matrix algebra, the next step is to consider methods of formulating the matrices needed to determine the response of structures due to external loading. This will be done in turn for each of two general types of systems based on the number of their degrees of freedom, or the number of coordinates needed to describe their displacement.

Three basic conditions must be satisfied when writing the equations for any linear structure. These are that:

1. The loads (axial forces, bending moments and torques) acting at any point must be in equilibrium.
2. The deformations at any point in the structure must be compatible.
3. The deformations must be proportional to the applied loads (Hooke's Law).

There are two basic approaches of writing the equations that can be put into matrix form to determine the response of a member. In discussing these the word "force" (F) will be taken in a generic sense and could be a force, a torque or a bending moment; the word "deformation" (δ) is the response to that force and could be a deformation in the form of a distance or an angle.

One basic approach is the *stiffness method* in which the force acting at a point is expressed in terms of unknown deformations of the system. This can be written in matrix form as

$$\{F\} = [K] \ \{\delta\} \qquad (16\text{-}1)$$

where $\{F\}$ is a column vector of forces acting on member; $\{\delta\}$ is a column vector of the unknown deformations; and the $[K]$ matrix is made up of elements k_{ij} specifying the load at i required to maintain a unit deformation at j when all other deformations are zero. Note that k_{ij} has the units of pound/inch. Thus, equation (16-1) represents a set of simultaneous force equations such as

$$
\begin{aligned}
F_1 &= k_{11}\,\delta_1 + k_{12}\,\delta_2 + k_{13}\,\delta_3 \\
F_2 &= k_{21}\,\delta_1 + k_{22}\,\delta_2 + k_{23}\,\delta_3 \\
F_3 &= k_{31}\,\delta_1 + k_{32}\,\delta_2 + k_{33}\,\delta_3.
\end{aligned}
$$

A second basic approach is the *flexibility method* in which the deformation at a point is expressed in terms of unknown external forces acting on the system. This can be expressed in matrix form as

$$\{\delta\} = [C] \ \{F\} \qquad (16\text{-}2)$$

where the column vector $\{\delta\}$ and $\{F\}$ are those given previously, and the $[C]$ matrix is made up of

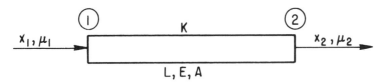

Figure 16-1. Loads and deflections of axially loaded bar.

elements c_{ij} specifying the deflection at i due to a unit load at j and assumes that all other loads on the system are zero. Note that c_{ij} has the units of inch/pound. Thus, equation (16-2) represents a set of simultaneous displacement equations such as

$$\delta_1 = c_{11} F_1 + c_{12} F_2 + c_{13} F_3$$
$$\delta_2 = c_{21} F_1 + c_{22} F_2 + c_{23} F_3$$
$$\delta_3 = c_{31} F_1 + c_{32} F_2 + c_{33} F_3.$$

If equation (16-1) is premultiplied by K^{-1} we obtain $K^{-1}\{F\} = K^{-1} K \{\delta\}$; or $\{\delta\} = K^{-1} \{F\}$. Comparing this with equation (16-2) we note that they are identical, hence $C = K^{-1}$, or $K = C^{-1}$, thus C and K are duals of each other.

Of these two basic approaches, the stiffness method is generally preferred because its matrices are better conditioned, superposition of two matrices is done by direct addition, and redundancy of the system causes no difficulty. Because of its widespread use to solve structural problems the stiffness matrix will be greatly emphasized here.

For some applications, *transfer matrices* have advantages (1). Here both basic approaches are combined in a single matrix. In general, the top rows contain flexibility terms and the bottom rows stiffness terms.

Members subjected to a pure axial load or to a pure torque have similar responses since the deformation is one dimensional; i.e., an axial displacement u, or an angle of twist ϕ.

In discussing members subjected to axial loads, the following symbols are commonly used. A force acting parallel to the x axis is X, and the corresponding deflection is u; a force acting parallel to the y axis is Y, and the corresponding deflec-

tion is v. Plus values of force and displacement are taken to be to the right and upward.

A single bar of area A, length L and modulus of elasticity E is shown in Figure 16-1. For equilibrium: $X_1 + X_2 = 0$. Now,

$$E = \frac{s}{\epsilon} = \frac{XL}{Au} \qquad (16\text{-}3)$$

since the unit stress $s = X/A$, and the unit strain $\epsilon = u/L$. The spring scale of the bar k, or ratio of X to u, from equation (16-3) is EA/L.

If the bar is held fixed at point 1, i.e., $u_1 = 0$, then $X_2 = -X_1 = ku_2 = (EA/L) u_2$. If the bar is held fixed at point 2, then $X_1 = -X_2 = ku_1 = (EA/L) u_1$. Combining these equations for a general case gives

$$X_1 = k u_1 - k u_2$$
$$X_2 = - k u_1 + k u_2.$$

These relationships can be written in matrix form to give the stiffness matrix K of the member.

$$\begin{bmatrix} X_1 \\ X_2 \end{bmatrix} = EA/L \begin{bmatrix} 1 & -1 \\ -1 & 1 \end{bmatrix} \begin{bmatrix} u_1 \\ u_2 \end{bmatrix} = \begin{bmatrix} k & -k \\ -k & k \end{bmatrix} \begin{bmatrix} u_1 \\ u_2 \end{bmatrix}$$

$$(16\text{-}4)$$

Bars in Combination

Assume that three bars act in *series* as shown in Figure 16-2 and are subjected to the axial loads X_1, X_2, X_3 and X_4. It is desired to determine the stiffness matrix K for this combination.

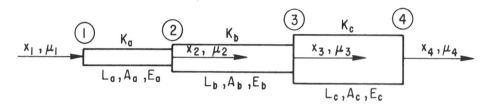

Figure 16-2. Axially loaded bars acting in series.

Each bar has its own spring scale k, where $k = X/u = E\,A/L$. Applying the basic definition for k_{ij}, a table can be set up for the unit displacement of each point in turn and the corresponding forces found to maintain that displacement. Thus,

$u_1\,u_2\,u_3\,u_4$	X_1	X_2	X_3	X_4
+1 0 0 0	$k_a u_1$	$-k_a u_2$	0	0
0+1 0 0	$-k_a u_1$	$(k_a+k_b)u_2$	$-k_b u_3$	0
0 0+1 0	0	$-k_b u_3$	$(k_b+k_c)u_3$	$-k_c u_4$
0 0 0+1	0	0	$-k_c u_3$	$k_c u_4$

The last four columns of this table form the stiffness matrix K of the system. It could also be obtained by the direct addition of the stiffness matrices of equation (16-4) of each member:

$$
\begin{bmatrix}
k_a & -k_a & 0 & 0 \\
-k_a & k_a & 0 & 0 \\
0 & 0 & 0 & 0 \\
0 & 0 & 0 & 0
\end{bmatrix}
+
\begin{bmatrix}
0 & 0 & 0 & 0 \\
0 & k_b & -k_b & 0 \\
0 & -k_b & k_b & 0 \\
0 & 0 & 0 & 0
\end{bmatrix}
+
\begin{bmatrix}
0 & 0 & 0 & 0 \\
0 & 0 & 0 & 0 \\
0 & 0 & k_c & -k_c \\
0 & 0 & -k_c & k_c
\end{bmatrix}
$$

$$
K =
\begin{bmatrix}
k_a & -k_a & 0 & 0 \\
-k_a & k_a+k_b & -k_b & 0 \\
0 & -k_b & k_b+k_c & -k_c \\
0 & 0 & -k_c & k_c
\end{bmatrix}
$$

Hence the matrix equation is

$$
\begin{bmatrix}
X_1 \\ X_2 \\ X_3 \\ X_4
\end{bmatrix}
=
\begin{bmatrix}
k_a & -k_a & 0 & 0 \\
-k_a & k_a+k_b & -k_b & 0 \\
0 & -k_b & k_b+k_c & -k_c \\
0 & 0 & -k_c & k_c
\end{bmatrix}
\begin{bmatrix}
u_1 \\ u_2 \\ u_3 \\ u_4
\end{bmatrix}
$$

If three bars are placed in *parallel* between two rigid plates, as shown in Figure 16-3, it may be noted that only two forces act. The deformation of all the bars must be the same to satisfy compatibility, and the total load is the sum of the spring scales for a unit deflection; i.e., $k_{eq.} = k_a + k_b + k_c$. Hence the stiffness equation in matrix form is

$$
\begin{bmatrix}
X_1 \\ X_2
\end{bmatrix}
=
\begin{bmatrix}
k_a+k_b+k_c & -(k_a+k_b+k_c) \\
-(k_a+k_b+k_c) & k_a+k_b+k_c
\end{bmatrix}
\begin{bmatrix}
u_1 \\ u_2
\end{bmatrix}
$$

Stiffness Matrix Properties

The stiffness matrix has some very interesting properties which can be used to shorten the labor in deriving it for complex situations, or are useful in checking the derivation. These are

1. The matrix is symmetric, that is, $k_{ij} = k_{ji}$. This reciprocity relation is known as Maxwell's Relationship and is based on the fact that the work done by force F_i acting through the displacement caused by the force F_j equals the work done by F_j acting through the displace-

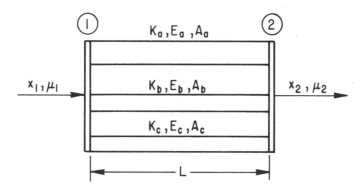

Figure 16-3. Axially loaded bars acting in parallel.

ment caused by F_i. This can be illustrated by referring to Figure 16-2. Assume that point 1 is held fixed and that only forces X_2 and X_4 are applied to the system, i.e., $X_3 = 0$.

If only force X_2 is applied, $u_2 = X_2/k_{22}$, and u_4 equals X_2/k_{24}. If only force X_4 is applied, $u_2 = X_4/k_{42}$ and $u_4 = X_4/k_{44}$.

Now if force X_2 is applied and then followed by force X_4, the total work done on the system is $1/2 \ X_2 \ (X_2/k_{22}) + X_2 \ (X_4/k_{42}) + 1/2 \ X_4 \ (X_4/k_{44})$. If the force X_4 is applied and then followed by X_2, the total work done on the system is $1/2 \ X_4(X_4/k_{44}) + X_4(X_2/k_{24}) + 1/2 \ X_2(X_2/k_{22})$. The total work done on the system must be the same regardless of the order of loading it; therefore these two equations must be equal. Cancelling out like terms gives $k_{24} = k_{42}$. This can be generalized to $k_{ij} = k_{ji}$.

This relationship is valuable for checking or can save time and effort since the value of only half of the elements of the matrix need be calculated.

2. The sum of the elements in any column equal zero. This is based on the principle of equilibrium of the forces acting at a point.
3. The sum of the elements in any row equals zero.

4. All of the terms on the principal diagonal are positive.
5. The value of the determinant of this matrix is zero, hence it has no inverse. This is caused by the lack of boundary conditions so that a "rigid-body" motion is possible. This will be discussed in more detail in the next chapter.

Matrix Solution of a Structure for Given Boundary Conditions

The matrix equation (16-5) is complete but may have to be reordered to care for the boundary conditions that could exist for the system shown in Figure 16-2. Thus points 1 or 4 (or both) could be held fixed, so it is desirable to establish a general method that can be used for these possibilities. It will be found that this same general procedure can also be used for the case of beams.

To be specific, suppose that the points 1 and 4 in Figure 16-2 are fixed, and that external loads are applied at points 2 and 3.

The first step is to reorder the equations by putting the known forces (and unknown displacements) at the top; and the reaction loads (X_1 and X_4) with the known displacements (boundary conditions) at the bottom. This means that rows of the matrix and the two vectors must be reordered, and at the same time the columns must also be reordered so that the various elements will be mul-

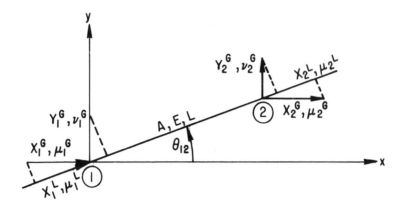

Figure 16-4. One member of loaded pin-jointed truss.

tiplied by the correct displacement term. The reordered matrix is now

$$
\begin{bmatrix} X_2 \\ X_3 \\ \hline X_1 \\ X_4 \end{bmatrix} = \begin{bmatrix} k_a+k_b & -k_b & -k_a & 0 \\ -k_b & k_b+k_c & 0 & -k_c \\ \hline -k_a & 0 & k_a & 0 \\ 0 & -k_c & 0 & k_c \end{bmatrix} \begin{bmatrix} u_2 \\ u_3 \\ u_1 \\ u_4 \end{bmatrix}
$$

$$
= \begin{bmatrix} K_1 & K_2 \\ \hline K_3 & K_4 \end{bmatrix} \begin{bmatrix} u_2 \\ u_3 \\ u_1 \\ u_4 \end{bmatrix}. \quad (16\text{-}6)
$$

The matrix equation can then be partitioned as shown by the dashed lines, so that the stiffness matrix K is composed of the submatrices K_1, K_2, K_3 and K_4.

Then the reordered matrix equation (16-6) can be reduced to several smaller equations.

$$
\begin{bmatrix} X_2 \\ X_3 \end{bmatrix} = K_1 \begin{bmatrix} u_2 \\ u_3 \end{bmatrix}; \text{ or } \begin{bmatrix} u_2 \\ u_3 \end{bmatrix} = K_1^{-1} \begin{bmatrix} X_2 \\ X_3 \end{bmatrix}. \quad (16\text{-}7)
$$

By equation (15-11):

$$
K_1^{-1} = \frac{1}{k_a k_b + k_a k_c + k_b k_c} \begin{bmatrix} k_b+k_c & k_b \\ k_b & k_a+k_b \end{bmatrix}
$$

Then equation (16-7) becomes

$$
\begin{bmatrix} u_2 \\ u_3 \end{bmatrix} = \frac{1}{k_a k_b + k_a k_c + k_b k_c} \begin{bmatrix} (k_b+k_c)X_2 + k_b X_3 \\ k_b X_2 + (k_a+k_b)X_3 \end{bmatrix}. \quad (16\text{-}8)
$$

If these deformations are put in the δ vector of equation (16-6), the reactions can be found from

$$
\begin{bmatrix} X_1 \\ X_4 \end{bmatrix} = K_3 \begin{bmatrix} u_2 \\ u_3 \end{bmatrix}; \text{ or } \begin{bmatrix} X_1 \\ X_4 \end{bmatrix} = \begin{bmatrix} -k_a & u_2 \\ -k_c & u_3 \end{bmatrix}. \quad (16\text{-}9)
$$

It might be desirable to assume numerical values for this problem and verify that equations (16-8) and (16-9) satisfy the laws of statics.

Trusses

A common type of structure is a pin-jointed truss that is made up of a number of axially loaded members. These members are generally inclined to each other so that it is convenient to use two systems of coordinates. The first, or "local" coordinate system uses the axis of the member as the x axis as developed previously. The second, or "global" coordinate system refers to the x and y coordinates of the entire truss.

Intuition indicates that it should be possible to convert from local to global coordinates by employing the proper trigonometric relationships; and it will now be shown that this can be done with a transformation matrix $[T]$. In this development, local forces and deformations will have the superscript L, while the corresponding global values will have the superscript G. Figure 16-4 shows one member of a pin-pointed truss inclined at an angle θ with the global axes x and y. The left end is subjected to forces X_1^G and Y_1^G with resulting deformations u_1^G and V_1^G, while the corresponding forces and deformations at the right end are subscripted by the number 2. Since the member can carry only an axial load the local forces acting on the member are X_1^L and X_2^L; and the local deformations are u_1^L and u_2^L; i.e., $v_1^L = v_2^L = Y_1^L = Y_2^L = 0$. The angle θ is always measured in a counterclockwise direction from a horizontal line to the right of the origin as in mathematics.

The matrix equation, based on local coordinates, can be written by expanding equation (16-4) to give

$$
\begin{bmatrix} X_1 \\ Y_1 \\ X_2 \\ Y_2 \end{bmatrix}^L = \frac{AE}{L} \begin{bmatrix} 1 & 0 & -1 & 0 \\ 0 & 0 & 0 & 0 \\ -1 & 0 & 1 & 0 \\ 0 & 0 & 0 & 0 \end{bmatrix} \begin{bmatrix} u_1 \\ v_1 \\ u_2 \\ v_2 \end{bmatrix}^L \tag{16-10}
$$

or

$$
\{F^L\} = [K^L] \ \{\delta^L\}. \tag{16-11}
$$

From Figure 16-4 the following equations can be written:

$$
X_1^L = X_1^G \cos\theta + Y_1^G \sin\theta
$$

$$
X_2^L = X_2^G \cos\theta + Y_2^G \sin\theta
$$

$$
Y_1^L = 0 = X_1^G \sin\theta + Y_1^G \cos\theta
$$

$$
Y_2^L = 0 = -X_2^G \sin\theta + Y_2^G \cos\theta \tag{16-12}
$$

and

$$
u_1^L = u_1^G \cos\theta + v_1^G \sin\theta
$$

$$
u_2^L = u_2^G \cos\theta + v_2^G \sin\theta
$$

$$
v_1^L = 0 = -u_1^G \sin\theta + v_1^G \cos\theta
$$

$$
v_2^L = 0 = -u_2^G \sin\theta + v_2^G \cos\theta \tag{16-13}
$$

Equation (16-12) can be put into matrix form as

$$
\{F^L\} = [T] \ \{F^G\} \tag{16-14}
$$

or

$$
\begin{bmatrix} X_1 \\ Y_1 \\ X_2 \\ Y_2 \end{bmatrix}^L = \begin{bmatrix} \cos\theta & \sin\theta & 0 & 0 \\ -\sin\theta & \cos\theta & 0 & 0 \\ 0 & 0 & \cos\theta & \sin\theta \\ 0 & 0 & -\sin\theta & \cos\theta \end{bmatrix} \begin{bmatrix} X_1 \\ Y_1 \\ X_2 \\ Y_2 \end{bmatrix}^G
$$

$$\tag{16-15}$$

In a similar manner equation (16-13) can be written as

$$
\{\delta^L\} = [T] \ \{\delta^G\} \tag{16-16}
$$

$$
\begin{bmatrix} u_1 \\ v_1 \\ u_2 \\ v_2 \end{bmatrix}^L =
\begin{bmatrix}
\cos\theta & \sin\theta & 0 & 0 \\
-\sin\theta & \cos\theta & 0 & 0 \\
0 & 0 & \cos\theta & \sin\theta \\
0 & 0 & -\sin\theta & \cos\theta
\end{bmatrix}
\begin{bmatrix} u_1 \\ v_1 \\ u_2 \\ v_2 \end{bmatrix}^G
$$

$$(16\text{-}17)$$

It may be observed that the transformation matrix $[T]$ is the same in equations (16-15) and (16-17); and when partitioned, it can be written as

$$
[T] = \begin{bmatrix} A & 0 \\ 0 & A \end{bmatrix}
$$

where

$$
A = \begin{bmatrix} \cos\theta & \sin\theta \\ -\sin\theta & \cos\theta \end{bmatrix}
$$

By equation (15-9), its inverse is

$$
T^{-1} = \begin{bmatrix} A^{-1} & 0 \\ 0 & A^{-1} \end{bmatrix}
$$

and by equation (15-11)

$$
A^{-1} = \frac{1}{\cos^2\theta + \sin^2\theta} \begin{bmatrix} \cos\theta & -\sin\theta \\ \sin\theta & \cos\theta \end{bmatrix}
$$

Since the value of the determinant $|A|$ is unity, it may be noted that $A^{-1} = A^T$, hence:

$$
T^{-1} = T^T \qquad\qquad (16\text{-}18)
$$

Substituting equations (16-14) and (16-16) into equation (16-11) gives

$$
T\,F^G = K^L\,T\,\delta^G \qquad\qquad (16\text{-}19)
$$

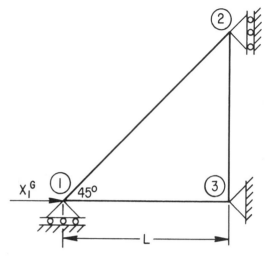

Figure 16-5. Truss of Example 1.

Premultiplying equation (16-19) by T^{-1} and using equation (16-18) gives $T^{-1}\,T\,F^G = T^{-1}\,K^L\,T\delta^G = T^T K^L T \delta^G$. Hence:

$$
K^G = T^T\,K^L\,T . \qquad\qquad (16\text{-}20)
$$

$$
\begin{bmatrix}
\cos\theta & -\sin\theta & 0 & 0 \\
\sin\theta & \cos\theta & 0 & 0 \\
0 & 0 & \cos\theta & -\sin\theta \\
0 & 0 & \sin\theta & \cos\theta
\end{bmatrix}
\begin{bmatrix}
1 & 0 & -1 & 0 \\
0 & 0 & 0 & 0 \\
-1 & 0 & 1 & 0 \\
0 & 0 & 0 & 0
\end{bmatrix}
$$

$$
\begin{bmatrix}
\cos\theta & \sin\theta & 0 & 0 \\
-\sin\theta & \cos\theta & 0 & 0 \\
0 & 0 & \cos\theta & \sin\theta \\
0 & 0 & -\sin\theta & \cos\theta
\end{bmatrix}
$$

$$
K^G = \begin{bmatrix}
\cos^2\theta & \cos\theta\sin\theta & -\cos^2\theta & -\cos\theta\sin\theta \\
\cos\theta\sin\theta & \sin^2\theta & -\cos\theta\sin\theta & -\sin^2\theta \\
-\cos^2\theta & -\cos\theta\sin\theta & \cos^2\theta & \cos\theta\sin\theta \\
-\cos\theta\sin\theta & -\sin^2\theta & \cos\theta\sin\theta & \sin^2\theta
\end{bmatrix}
$$

With these equations the procedure to be followed in determining the truss global stiffness matrix would be to first make a sketch of the truss with the various pin joints numbered, the external loads indicated and the boundary conditions specified. Then the local stiffness matrix K^L of each member, in turn, can be found with the aid of equation (16-10). These matrices can then be rotated into position to obtain the corresponding global stiffness matrix by the use of equation (16-20). The value of θ to be used is found by considering the member as a vector going from i to j i.e., from lower to higher joint number.

After the global stiffness matrix of each member has been found, the corresponding elements can be added directly to obtain the truss global matrix.

This global matrix can then be partitioned in accordance with equation (16-6) and solved for the deformations with equation (16-7), and the reactions determined by equation (16-9).

The procedure just outlined will be illustrated by two examples. In these examples it will be assumed that all of the members have the same cross-sectional area A and the same modulus of elasticity E.

Example 1

The truss shown in Figure 16-5 is subjected to the single external force X_1^G. The boundary conditions indicated on the sketch are $v_1^G = u_2^G = u_3^G = v_3^G = 0$. It is desired to find the deformations u_1^G and v_2^G due to the load X_1^G in terms of L, E, A and X_1^G.

The first step is to find the local stiffness matrix K_{ij}^L of each member so that the corresponding elements can be added to obtain the overall global stiffness matrix K^G of the truss.

Member 1-2: $\theta_{12} = 45°$; length $= \sqrt{2}\,L$; $\cos\theta = 1/\sqrt{(2)}$; $\sin\theta = 1/\sqrt{(2)}$. Local stiffness matrix, Equation (16-10):

$$\begin{bmatrix} X_1 \\ Y_1 \\ X_2 \\ Y_2 \end{bmatrix}^L = \frac{AE}{\sqrt{(2)}L} \begin{bmatrix} 1 & 0 & -1 & 0 \\ 0 & 0 & 0 & 0 \\ -1 & 0 & 1 & 0 \\ 0 & 0 & 0 & 0 \end{bmatrix} \begin{bmatrix} u_1 \\ v_1 \\ u_2 \\ v_2 \end{bmatrix}^L$$

The corresponding global stiffness matrix by equation (16-20) is

$$\begin{bmatrix} X_1 \\ Y_1 \\ X_2 \\ Y_2 \end{bmatrix}^L = \frac{AE}{2\sqrt{(2)}L} \begin{bmatrix} 1 & 1 & -1 & -1 \\ 1 & 1 & -1 & -1 \\ -1 & -1 & 1 & 1 \\ -1 & -1 & 1 & 1 \end{bmatrix} \begin{bmatrix} u_1 \\ v_1 \\ u_2 \\ v_2 \end{bmatrix}^L$$

Member 1-3: $\theta_{13} = 0°$; length $= L$; $\cos\theta = 1$; $\sin\theta = 0$.

$$\begin{bmatrix} X_1 \\ Y_1 \\ X_3 \\ Y_3 \end{bmatrix}^L = \frac{AE}{L} \begin{bmatrix} 1 & 0 & -1 & 0 \\ 0 & 0 & 0 & 0 \\ -1 & 0 & 1 & 0 \\ 0 & 0 & 0 & 0 \end{bmatrix} \begin{bmatrix} u_1 \\ v_1 \\ u_3 \\ v_3 \end{bmatrix}^L$$

$$\begin{bmatrix} X_1 \\ Y_1 \\ X_3 \\ Y_3 \end{bmatrix}^G = \frac{AE}{L} \begin{bmatrix} 1 & 0 & -1 & 0 \\ 0 & 0 & 0 & 0 \\ -1 & 0 & 1 & 0 \\ 0 & 0 & 0 & 0 \end{bmatrix} \begin{bmatrix} u_1 \\ v_1 \\ u_3 \\ v_3 \end{bmatrix}^G$$

Member 2-3: $\theta_{23} = 270°$; length $= L$; $\cos\theta = 0$; $\sin\theta = -1$.

$$\begin{bmatrix} X_2 \\ Y_2 \\ X_3 \\ Y_3 \end{bmatrix}^L = \frac{AE}{L} \begin{bmatrix} 1 & 0 & -1 & 0 \\ 0 & 0 & 0 & 0 \\ -1 & 0 & 1 & 0 \\ 0 & 0 & 0 & 0 \end{bmatrix} \begin{bmatrix} u_2 \\ v_2 \\ u_3 \\ v_3 \end{bmatrix}^L$$

$$
\begin{bmatrix} X_2 \\ Y_2 \\ X_3 \\ Y_3 \end{bmatrix}^G = \frac{AE}{L} \begin{bmatrix} 0 & 0 & 0 & 0 \\ 0 & 1 & 0 & -1 \\ 0 & 0 & 0 & 0 \\ 0 & -1 & 0 & 1 \end{bmatrix} \begin{bmatrix} u_2 \\ v_2 \\ u_3 \\ v_3 \end{bmatrix}^G
$$

Adding corresponding elements of the member global matrices K_{ij}^G gives the global matrix K^G of the entire truss.

$$
\begin{bmatrix} X_1 \\ Y_1 \\ X_2 \\ Y_2 \\ X_3 \\ Y_3 \end{bmatrix}^G = \frac{AE}{L} \begin{bmatrix} \frac{1+1}{2\sqrt{2}} & \frac{1}{2\sqrt{2}} & \frac{-1}{2\sqrt{2}} & \frac{-1}{2\sqrt{2}} & -1 & 0 \\ \frac{1}{2\sqrt{2}} & \frac{1}{2\sqrt{2}} & \frac{-1}{2\sqrt{2}} & \frac{-1}{2\sqrt{2}} & 0 & 0 \\ \frac{-1}{2\sqrt{2}} & \frac{-1}{2\sqrt{2}} & \frac{1}{2\sqrt{2}} & \frac{1}{2\sqrt{2}} & 0 & 0 \\ \frac{-1}{2\sqrt{2}} & \frac{-1}{2\sqrt{2}} & \frac{1}{2\sqrt{2}} & \frac{1+1}{2\sqrt{2}} & 0 & -1 \\ -1 & 0 & 0 & 0 & 1 & 0 \\ 0 & 0 & 0 & -1 & 0 & 1 \end{bmatrix} \begin{bmatrix} u_1 \\ v_1 \\ u_2 \\ v_2 \\ u_3 \\ v_3 \end{bmatrix}^G
$$

The next step is to reorder this truss global matrix, putting the unknown deflections on the top two lines and the boundary conditions on the lower four lines. This then becomes

$$
\begin{bmatrix} X_1 \\ Y_2=0 \\ \hline Y_1 \\ X_2 \\ X_3 \\ Y_3 \end{bmatrix}^G = \frac{AE}{L} \left[\begin{array}{cc|cccc} \frac{1+2\sqrt{2}}{2\sqrt{2}} & \frac{-1}{2\sqrt{2}} & \frac{1}{2\sqrt{2}} & \frac{-1}{2\sqrt{2}} & -1 & 0 \\ \frac{-1}{2\sqrt{2}} & \frac{1+2\sqrt{2}}{2\sqrt{2}} & \frac{-1}{2\sqrt{2}} & \frac{1}{2\sqrt{2}} & 0 & -1 \\ \hline \frac{1}{2\sqrt{2}} & \frac{-1}{2\sqrt{2}} & \frac{1}{2\sqrt{2}} & \frac{-1}{2\sqrt{2}} & 0 & 0 \\ \frac{-1}{2\sqrt{2}} & \frac{1}{2\sqrt{2}} & \frac{-1}{2\sqrt{2}} & \frac{1}{2\sqrt{2}} & 0 & 0 \\ -1 & 0 & 0 & 0 & 1 & 0 \\ 0 & -1 & 0 & 0 & 0 & 1 \end{array} \right] \begin{bmatrix} u_1 \\ v_2 \\ \hline v_1 \\ u_2 \\ u_3 \\ v_3 \end{bmatrix}^G
$$

This matrix is then partitioned as shown by the dashed lines in accordance with equation

(16-6). The inverse of the K_1 matrix by equation (15-11) is

$$K^{-1} = \frac{L}{AE} \; \frac{2}{2+\sqrt{2}} \begin{bmatrix} \dfrac{1+2\sqrt{2}}{2\sqrt{2}} & \dfrac{1}{2\sqrt{2}} \\[3mm] \dfrac{1}{2\sqrt{2}} & \dfrac{1+2\sqrt{2}}{2\sqrt{2}} \end{bmatrix}$$

Using the second form of equation (16-7) for the external loads X_1 and Y_2 (where Y_2 is zero) gives

$$u_1 = \frac{1+2\sqrt{2}}{2+2\sqrt{2}} \; X_1 \; \frac{L}{AE}$$

$$v_2 = \frac{1}{2+2\sqrt{2}} \; X_1 \; \frac{L}{AE}$$

These values of deformation may be used with the K_3 submatrix in the equivalent of equation (16-9) to find the reactions at the pin joints. Thus,

$$Y_1 = \frac{AE}{L} \left(\frac{1}{2\sqrt{2}} \, u_1 + \frac{-1}{2\sqrt{2}} \, v_2 \right) = \frac{X_1}{2+2\sqrt{2}}$$

$$Y_3 = \frac{AE}{L} \; (-v_2) = \frac{-1}{2+2\sqrt{2}} \; X_1$$

$$X_2 = \frac{AE}{L} \left(\frac{-1}{2\sqrt{2}} \, u_1 + \frac{1}{2\sqrt{2}} \, v_2 \right) = \frac{-X_1}{2+2\sqrt{2}}$$

$$X_3 = \frac{AE}{L} \; (-u_1) = -X_1 \; \frac{1+2\sqrt{2}}{2+2\sqrt{2}}$$

Example 2

For the second example assume that an additional member is added to the system of Figure 16-5 to form the redundant structure of Figure 16-6. It may be noted in this figure that the joints 2, 3 and 4 are not allowed to deflect, so there is no need to determine the local or global matrices of members 2-3 and 3-4. It is necessary though to find the global matrix for member 1-4.

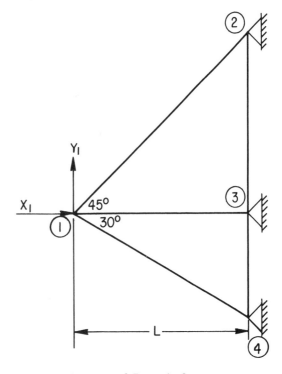

Figure 16-6. Truss of Example 2.

Member 1-4:

$$\theta_{14} = 330°; \; \text{length} = \frac{2}{\sqrt{3}} \, L; \; \cos\theta = \sqrt{3}/2;$$

$$\sin\theta = -1/2.$$

$$\begin{bmatrix} X_1 \\ Y_1 \\ X_4 \\ Y_4 \end{bmatrix}^L = \sqrt{3} \, AE/2L \begin{bmatrix} 1 & 0 & -1 & 0 \\ 0 & 0 & 0 & 0 \\ -1 & -0 & 1 & 0 \\ 0 & 0 & 0 & 0 \end{bmatrix} \begin{bmatrix} u_1 \\ v_1 \\ u_4 \\ v_4 \end{bmatrix}^L$$

$$
\begin{bmatrix} X_1 \\ Y_1 \\ X_4 \\ Y_4 \end{bmatrix}^G = \sqrt{3}\,A\,E/4L \begin{bmatrix} 1 & -1 & -1 & 1 \\ -1 & 1 & 1 & -1 \\ -1 & 1 & 1 & -1 \\ 1 & -1 & -1 & 1 \end{bmatrix} \begin{bmatrix} u_1 \\ v_1 \\ u_4 \\ v_4 \end{bmatrix}^G
$$

Adding corresponding terms of this matrix to the terms of K_{12}^G and K_{13}^G of Example 1 gives the global matrix K^G of the truss:

$$
\begin{bmatrix} X_1 \\ Y_1 \\ X_2 \\ Y_2 \\ X_3 \\ Y_3 \\ X_4 \\ Y_4 \end{bmatrix}^G = A\,E/L \begin{bmatrix} \dfrac{4\sqrt{2}+2+\sqrt{6}}{4\sqrt{2}} & \dfrac{2-\sqrt{6}}{4\sqrt{2}} & \dfrac{-1}{2\sqrt{2}} & \dfrac{-1}{2\sqrt{2}} & -1 & 0 & \dfrac{-\sqrt{3}}{4} & \dfrac{\sqrt{3}}{4} \\[2mm] \dfrac{2-\sqrt{6}}{4\sqrt{2}} & \dfrac{2+\sqrt{6}}{4\sqrt{2}} & \dfrac{-1}{2\sqrt{2}} & \dfrac{-1}{2\sqrt{2}} & 0 & 0 & \dfrac{\sqrt{3}}{4} & \dfrac{-\sqrt{3}}{4} \\[2mm] \dfrac{-1}{2\sqrt{2}} & \dfrac{-1}{2\sqrt{2}} & \dfrac{1}{2\sqrt{2}} & \dfrac{1}{2\sqrt{2}} & 0 & 0 & 0 & 0 \\[2mm] \dfrac{-1}{2\sqrt{2}} & \dfrac{-1}{2\sqrt{2}} & \dfrac{1}{2\sqrt{2}} & \dfrac{1}{2\sqrt{2}} & 0 & 0 & 0 & 0 \\[2mm] -1 & 0 & 0 & 0 & 1 & 0 & 0 & 0 \\[2mm] 0 & 0 & 0 & 0 & 0 & 0 & 0 & 0 \\[2mm] \dfrac{-\sqrt{3}}{4} & \dfrac{\sqrt{3}}{4} & 0 & 0 & 0 & 0 & \dfrac{\sqrt{3}}{4} & \dfrac{-\sqrt{3}}{4} \\[2mm] \dfrac{\sqrt{3}}{4} & \dfrac{-\sqrt{3}}{4} & 0 & 0 & 0 & 0 & \dfrac{-\sqrt{3}}{4} & \dfrac{\sqrt{3}}{4} \end{bmatrix} \begin{bmatrix} u_1 \\ v_1 \\ u_2 \\ v_2 \\ u_3 \\ v_3 \\ u_4 \\ v_4 \end{bmatrix}^G
$$

It is not necessary to reorder this global matrix since the external forces and unknown deflections are already on the top two rows. It is partitioned as shown by the dashed lines.

The inverse of the K_1 matrix is

$$
K_1^{-1} = L/A\,E(2+\sqrt{6}+2\sqrt{3})] \begin{bmatrix} 2+\sqrt{6} & \sqrt{6}-2 \\ \sqrt{6}-2 & 2+4\sqrt{2}+\sqrt{6} \end{bmatrix}
$$

Using this value in equation (16-7) gives

$$u_1^G = \frac{L}{A\,E}\; \frac{1}{2+\sqrt{6}+2\sqrt{3}}\; [(2+\sqrt{6})X_1 + (\sqrt{6}-2)Y_1]$$

$$v_1^G = \frac{L}{A\,E}\; \frac{1}{2+\sqrt{6}+2\sqrt{3}}\; [(\sqrt{6}-2)X_1 + (2+4\sqrt{2}+\sqrt{6})Y_1]\;.$$

These values of deformation can be used with K_3 to find the reactions at the pin joints, as was done in Example 1.

Internal Loads in Members

A complete analysis of a truss should include the determination of the axial loads in each member due to the applied external loading. Referring to Figure 16-4 and assuming that a positive internal load is tensile while a negative internal load is compressive, it is apparent that the internal load is given by $F_{ij} = k_{ij}\,[(u_j - u_i)\cos\theta_{ij} + (v_j - v_i)\sin\theta_{ij}]$, where k_{ij} is the local stiffness of the member $i\text{-}j$, or $A_{ij}E_{ij}/l_{ij}$.

This can be put into matrix form as

$$F_{ij} = k_{ij}\,[\cos\theta^{ij}\;\sin\theta^{ij}]\begin{bmatrix} u_j - u_i \\ v_j - v_i \end{bmatrix}. \tag{16-21}$$

In addition to the applied external loads it is possible that one member may have its length L changed by a small amount ΔL, and this will create an internal load P in the member. This change in length may be due to (1) local temperature changes, (2) a support motion, or (3) an incorrect length of member during fabrication. This load P is distributed among all the members of the truss and will affect the displacements of all joints.

Using the sign convention specified for equation (16-21), the value of P in the member becomes:

1. Member heated $(\Delta T)^\circ$ and having a linear coefficient of expansion α:

$$P = -A\,E\,\alpha\,(\Delta T) = -\frac{A\,E}{L}\,(\alpha\,L\,\Delta T). \tag{16-22}$$

If the member is cooled $(\Delta T)^\circ$:

$$P = \frac{A\,E}{L}\,(\alpha\,L\,\Delta T). \tag{16-23}$$

2. Support motion of (ΔL) tends to shorten the member:

$$P = -\frac{A\,E}{L}\,(\Delta L). \tag{16-24}$$

Support motion of (ΔL) tends to lengthen the member:

$$P = \frac{A\,E}{L}\,(\Delta L). \tag{16-25}$$

3. Member fabricated (ΔL) too long and forced into truss:

$$P = -\frac{A\,E}{L}\,(\Delta L). \tag{16-26}$$

Member fabricated (ΔL) too short and stretched to fit the truss:

$$P = \frac{A\,E}{L}\,(\Delta L). \tag{16-27}$$

The action of the internal load P is the same as an external load $-P$ acting along the member at the joints concerned. This equivalent external load $-P$ acting on the truss will produce a new set of displacements at the joints, or nodes. After they have been found by the procedure already discussed, they can be used to determine the internal loads on all members of the truss by the use of equation (16-21).

If a truss has external loads as well as members whose length is changed, the two cases can be dealt with separately and the net effects found by superposition.

The calculation of the internal loads induced in a truss due to external loading will be illustrated with reference to Example 1 as shown in Figure 16-5, where the external load is X_1.

$$F_{12} = \frac{AE}{\sqrt{2}\,L}\left[\frac{1}{\sqrt{2}} \quad \frac{1}{\sqrt{2}}\right]\begin{bmatrix} 0-u_1 \\ v_2-0 \end{bmatrix} = \frac{AE}{2L}\ \frac{X_1\,L}{AE}\ \frac{1-1+2\sqrt{2}}{2+2\sqrt{2}} = \frac{\sqrt{2}X_1}{2+2\sqrt{2}}$$

$$F_{13} = \frac{AE}{L}\,[1 \quad 0]\begin{bmatrix} 0-u_1 \\ 0-0 \end{bmatrix} = \frac{AE}{L}\ \frac{X_1\,L}{AE}\ \frac{-1-2\sqrt{2}}{2+2\sqrt{2}} = \frac{(-1-2\sqrt{2})X_1}{2+2\sqrt{2}}$$

$$F_{23} = \frac{AE}{L}\,[0 \ -1]\begin{bmatrix} 0-0 \\ 0-v_2 \end{bmatrix} = \frac{AE}{L}\ \frac{X_1\,L}{AE}\ \frac{1}{2+2\sqrt{2}} = \frac{X_1}{2+2\sqrt{2}}\ .$$

These values can be checked by summing the horizontal and vertical forces acting at any joint or mode by including the external loads or reactions along with the internal forces acting at that point.

If now one member has its length L changed an amount ΔL, the internal load P can be calculated by one of the equations (16-22) through (16-27); and then the negative of this load is assumed to act at the joints concerned. New values of the displacements can be then found due to $-P$ according to the procedure outlined in the examples. If the internal load is developed in member 1-2, the value of $-P$ would be broken into horizontal and vertical components X and Y at the joints. Having the displacements due to $-P$, the corresponding internal loads in all of the members can be found by the use of equation (16-21) as just shown.

The resultant displacements and internal loads can be found by superposition of the values found for the external loads and the values of $-P$.

References

1. Pestel, E.G. and F.A. Leckie. 1963. *Matrix Methods in Elastomechanics*. New York: McGraw-Hill.

17 elements of structural analysis

beam members

Introduction

Many structures are composed entirely or in part of beam members. In general the methods and principles involved in handling structures of this type apply equally well to those developed in Chapter 16 for axially loaded members. For example, the equations used must follow the basic principles of satisfying equilibrium and compatibility at any point; and, if the system is linear, the deformations must be proportional to the applied loads. The response of the structure may be based on stiffness or flexibility matrices or on the combination of the two (transfer matrices).

However, the matrices for a single beam member must be of the fourth order, since the applied loads could be a force and/or a bending moment, and the resulting deformations are generally a displacement v and a slope ϕ. In many beam structures it may be desirable to consider distributed loads acting along the beam as well as concentrated loads.

A convenient method of developing the required matrices is with reference to the response of a cantilever beam. Figure 17-1 gives the response in terms of the slope ϕ and deflection v at the tip of a cantilever due to a bending moment M or a force Y acting at the tip; and for a uniformly distributed load p. The only new symbol here is the moment of inertia I about the neutral axis. These equations are given in any elementary strength of materials text. The equations in the figure are lettered for easy reference.

In the case of axially loaded members the boundary conditions are generally either fixed or free. For beams there are four general types of supports: pinned, clamped, free or sliding. Figure 17-2 lists the boundary conditions that apply to each type of support.

Stiffness Matrix of Beam

Figure 17-3 shows a horizontal beam of length L, modulus of elasticity E and rectangular moment of inertia I which is subjected to concentrated forces Y_1 and Y_2 and bending moments M_1 and M_2. A positive moment is generally taken to be clockwise. These loads will produce vertical displacements v_1 and v_2 as well as slopes ϕ_1 and ϕ_2 at the ends of the beam. The method of finding the local stiffness matrix for this general case will now be outlined.

The relationship between the various loads can be found by writing the equilibrium equations:

$$Y_1 + Y_2 = 0$$

$$M_1 + M_2 + Y_1 L = 0.$$

These equations can be combined to give the matrices

201

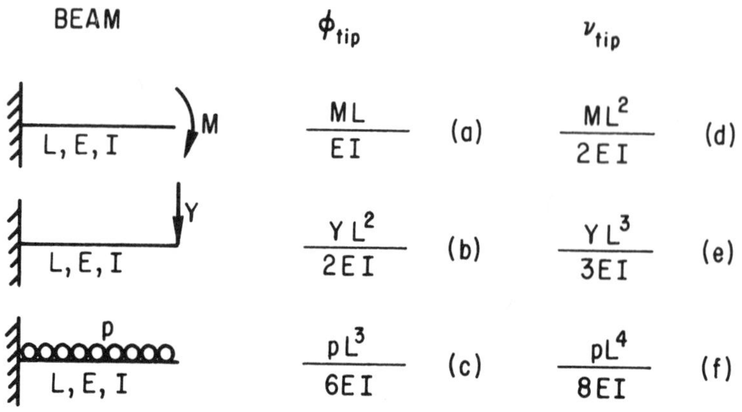

BEAM ϕ_{tip} v_{tip}

$\dfrac{ML}{EI}$ (a) $\dfrac{ML^2}{2EI}$ (d)

$\dfrac{YL^2}{2EI}$ (b) $\dfrac{YL^3}{3EI}$ (e)

$\dfrac{pL^3}{6EI}$ (c) $\dfrac{pL^4}{8EI}$ (f)

Figure 17-1. Deformation at tip of cantilever beam due to external loading.

$$\begin{bmatrix} Y_1 \\ M_1 \end{bmatrix} = \begin{bmatrix} -1 & 0 \\ L & -1 \end{bmatrix} \begin{bmatrix} Y_2 \\ M_2 \end{bmatrix} \qquad (17\text{-}1)$$

$$\begin{bmatrix} Y_2 \\ M_2 \end{bmatrix} = \begin{bmatrix} -1 & 0 \\ -L & -1 \end{bmatrix} \begin{bmatrix} Y_1 \\ M_1 \end{bmatrix}. \qquad (17\text{-}2)$$

Now assume that the right end of the beam is fixed or cantilevered at point 2, and that only the loads Y_1 and M_1 are acting on it. From Figure 17-1 two equations can be written for the deformations at end 1. These are

$$v_1 = Y_1 L^3/3 E I + M_1 L^2/2 E I$$

$$\phi_1 = Y_1 L^2/2 E I + M_1 L/E I .$$

These equations can be solved simultaneously to give

$$Y_1 = EI(12v_1/L^3 - 6\phi_1/L^2)$$

$$M_1 = EI(4\phi_1/L - 6v_1/L^2)$$

which can be put into matrix form as

$$\begin{bmatrix} Y_1 \\ M_1 \end{bmatrix} = E I \begin{bmatrix} 12/L^3 & -6/L^2 \\ -6/L^2 & 4/L \end{bmatrix} \begin{bmatrix} v_1 \\ \phi_1 \end{bmatrix}. \qquad (17\text{-}3)$$

If both sides of this equation are premultiplied by the matrix of equation (17-2) we obtain

$$\begin{bmatrix} Y_2 \\ M_2 \end{bmatrix} = E I \begin{bmatrix} -12/L^3 & 6/L^2 \\ -6/L^2 & 2/L \end{bmatrix} \begin{bmatrix} v_1 \\ \phi_1 \end{bmatrix}. \qquad (17\text{-}4)$$

When these steps are repeated for the case when the left end of the beam is fixed, the equations for the deformations from Figure 17-1 are

$$v_2 = Y_2 L^3/3 E I - M_2 L^2/2 E I$$

$$\phi_2 = M_2 L/E I - Y_2 L^2/2 E I$$

which can be solved simultaneously to give in matrix form

$$\begin{bmatrix} Y_2 \\ M_2 \end{bmatrix} = E I \begin{bmatrix} 12/L^3 & 6/L^2 \\ 6/L^2 & 4/L \end{bmatrix} \begin{bmatrix} v_2 \\ \phi_2 \end{bmatrix}. \qquad (17\text{-}5)$$

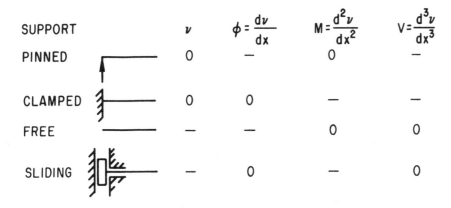

Figure 17-2. Beam boundary conditions.

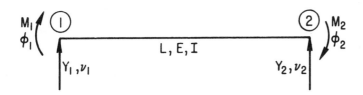

Figure 17-3. Beam loading and deformation in local coordinates.

Premultiplying both sides of equation (17-5) by the matrix of equation (17-1) yields

$$\begin{bmatrix} Y_1 \\ M_1 \end{bmatrix} = E\,I \begin{bmatrix} -12/L^3 & -6/L^2 \\ 6/L^2 & 2/L \end{bmatrix} \begin{bmatrix} v_2 \\ \phi_2 \end{bmatrix}. \tag{17-6}$$

Now equations (17-3), (17-4), (17-5) and (17-6) can be assembled into the final stiffness matrix of the beam:

$$\begin{bmatrix} Y_1 \\ M_1 \\ Y_2 \\ M_2 \end{bmatrix} = E\,I \begin{bmatrix} 12/L^3 & -6/L^2 & -12/L^3 & -6/L^2 \\ -6/L^2 & 4/L & 6/L^2 & 2/L \\ -12/L^3 & 6/L^2 & 12/L^3 & 6/L^2 \\ -6/L^2 & 2/L & 6/L^2 & 4/L \end{bmatrix} \begin{bmatrix} v_1 \\ \phi_1 \\ v_2 \\ \phi_2 \end{bmatrix}. \tag{17-7}$$

It may be observed that this matrix is symmetric. However, the sum of the elements of any row or column do not equal zero. This apparent discrepancy is due to the fact that the moment equilibrium equation involves forces Y as well as moments M as shown by equations (17-1) and (17-2).

The matrix in this equation can be put into dimensionless form to simplify its manipulation by incorporating the length term in the vectors:

$$\begin{bmatrix} Y_1 \\ M_1/L \\ Y_2 \\ M_2/L \end{bmatrix} = E\,I/L^3 \begin{bmatrix} 12 & -6 & -12 & -6 \\ -6 & 4 & 6 & 2 \\ -12 & 6 & 12 & 6 \\ -6 & 2 & 6 & 4 \end{bmatrix} \begin{bmatrix} v_1 \\ \phi_1 L \\ v_2 \\ \phi_2 L \end{bmatrix}. \tag{17-8}$$

Figure 17-4. First example of stepped beam.

It would be possible to obtain the flexibility matrix C of the beam in a similar manner; or it could be determined by finding the inverse of the stiffness matrix of equation (17-7) since $C = K^{-1}$, as shown in the "Formulation of Matrices" section of Chapter 16.

Rotation of Beam Members

The matrices of equations (17-7) and (17-8) are based on a local coordinate system of the member since the loads Y and the deformations are relative to the beam axis. If a beam is one part of a global system it may be necessary to rotate it to conform to a set of global coordinates x^G and y^G, as was done in Chapter 16 for a truss.

If the beam is rotated through an angle θ into a global position it is apparent that it will have components of the loading and the translational displacements in the horizontal as well as the vertical direction; whereas the slope term is independent of the coordinate system. The first step is to expand the local coordinate matrix to include the X and u terms. Doing this for equation (17-8) gives

$$\begin{bmatrix} X_1 \\ Y_1 \\ M_1/L \\ X_2 \\ Y_2 \\ M_2/L \end{bmatrix} = E\,I/L^3 \begin{bmatrix} 0 & 0 & 0 & 0 & 0 & 0 \\ 0 & 12 & -6 & 0 & -12 & -6 \\ 0 & -6 & 4 & 0 & 6 & 2 \\ 0 & 0 & 0 & 0 & 0 & 0 \\ 0 & -12 & 6 & 0 & 12 & 6 \\ 0 & -6 & 2 & 0 & 6 & 4 \end{bmatrix} \begin{bmatrix} u_1 \\ v_1 \\ \phi_1 L \\ u_2 \\ v_2 \\ \phi_2 L \end{bmatrix} \quad (1$$

while the transformation matrix T, corresponding to the matrix of equations (16-15) or (16-17), becomes

$$T = \begin{bmatrix} \cos\theta & \sin\theta & 0 & 0 & 0 & 0 \\ -\sin\theta & \cos\theta & 0 & 0 & 0 & 0 \\ 0 & 0 & 1 & 0 & 0 & 0 \\ 0 & 0 & 0 & \cos\theta & \sin\theta & 0 \\ 0 & 0 & 0 & -\sin\theta & \cos\theta & 0 \\ 0 & 0 & 0 & 0 & 0 & 1 \end{bmatrix} .$$

The rotated or global matrix is then found from equation (16-20), or $K^G = T^T K^L T$. If K^L is taken from equation (17-9), the corresponding value of K^G is

$$
\begin{bmatrix}
12\sin^2\theta & -12\cos\theta\sin\theta & 6\sin\theta & -12\sin^2\theta & 12\cos\theta\sin\theta & 6\sin\theta \\
-12\cos\theta\sin\theta & 12\cos^2\theta & -6\cos\theta & 12\cos\theta\sin\theta & -12\cos^2\theta & -6\cos\theta \\
6\sin\theta & -6\cos\theta & 4 & -6\sin\theta & 6\cos\theta & 2 \\
-12\sin^2\theta & 12\cos\theta\sin\theta & -6\sin\theta & 12\sin^2\theta & -12\cos\theta\sin\theta & -6\sin\theta \\
12\cos\theta\sin\theta & -12\cos^2\theta & 6\cos\theta & -12\cos\theta\sin\theta & 12\cos^2\theta & 6\cos\theta \\
6\sin\theta & -6\cos\theta & 2 & -6\sin\theta & 6\cos\theta & 4
\end{bmatrix}
$$

Stepped Beams

A beam having discontinuous changes in loading, moment of inertia or modulus of elasticity may be said to be "stepped" and can be considered as being made up of a number of sub-beams acting in series. Then an overall stiffness matrix can be found by the addition of corresponding elements of the individual stiffness matrices of each sub-beam. The procedure might best be developed with a simple example. The principles involved, of course, are applicable to more complex situations.

Figure 17-4 shows a beam of length 3L fixed at both ends and carrying a concentrated load Y_2 at the two-thirds point. The portion of the beam to the left of Y_2 is made of steel with a modulus of elasticity E, while the portion to the right of Y_2 is made of a brass with a modulus of E/2. The moment of inertia of the beam is constant over its entire length. It is desired to determine the reactions and deformations of the beam in terms of E, L and I.

This entire beam can be divided into two shorter sub-beams with nodes indicated by the circled numbers on the figure. Since there are three nodes the final stiffness matrix will be of the sixth order, so that it is convenient to expand the sub-matrices to this order.

For sub-beam 1-2 the expanded stiffness matrix equation based on equation (17-7) is

$$
\begin{bmatrix} Y_1 \\ M_1 \\ Y_2 \\ M_2 \\ Y_3 \\ M_3 \end{bmatrix}
= E\,I
\begin{bmatrix}
12/8L^3 & -6/4L^2 & -12/8L^3 & -6/4L^2 & 0 & 0 \\
-6/4L^2 & 4/2\,L & 6/4L^2 & 2/2\,L & 0 & 0 \\
-12/8L^3 & 6/4L^2 & 12/8L^3 & 6/4L^2 & 0 & 0 \\
-6/4L^2 & 2/2\,L & 6/4L^2 & 4/2\,L & 0 & 0 \\
0 & 0 & 0 & 0 & 0 & 0 \\
0 & 0 & 0 & 0 & 0 & 0
\end{bmatrix}
\begin{bmatrix} v_1 \\ \phi_1 \\ v_2 \\ \phi_2 \\ v_3 \\ \phi_3 \end{bmatrix} . \quad (17\text{-}10)
$$

In a similar manner the equation for sub-beam 2-3 is

$$
\begin{bmatrix} Y_1 \\ M_1 \\ Y_2 \\ M_2 \\ Y_3 \\ M_3 \end{bmatrix}
= E\,I/2
\begin{bmatrix}
0 & 0 & 0 & 0 & 0 & 0 \\
0 & 0 & 0 & 0 & 0 & 0 \\
0 & 0 & 12/L^3 & -6/L^2 & -12/L^3 & -6/L^2 \\
0 & 0 & -6/L^2 & 4/L & 6/L^2 & 2/L \\
0 & 0 & -12/L^3 & 6/L^2 & 12/L^3 & 6/L^2 \\
0 & 0 & -6/L^2 & 2/L & 6/L^2 & 4/L
\end{bmatrix}
\begin{bmatrix} v_1 \\ \phi_1 \\ v_2 \\ \phi_3 \\ v_3 \\ \phi_3 \end{bmatrix} . \quad (17\text{-}11)
$$

Adding corresponding terms of the matrices of equations (17-10) and (17-11) gives the overall stiffness matrix, K.

$$
K = E\,I
\begin{bmatrix}
3/2L^3 & -3/2L^2 & -3/2L^3 & -3/2L^2 & 0 & 0 \\
-3/2L^2 & 2/L & 3/2L^2 & 1/L & 0 & 0 \\
-3/2L^3 & 3/2L^2 & 15/2L^3 & -3/2L^2 & -6/L^3 & -3/L^2 \\
-3/2L^2 & 1/L & -3/2L^2 & 4/L & 3/L^2 & 1/L \\
0 & 0 & -6/L^3 & 3/L^2 & 6/L^3 & 3/L^2 \\
0 & 0 & -3/L^2 & 1/L & 3/L^2 & 2/L
\end{bmatrix} . \quad (17\text{-}12)
$$

An examination of Figure 17-4 shows that the boundary conditions for the entire beam are $v_1 = \phi_1 = v_3 = \phi_3 = 0$; and also that the external loads acting on it are $Y_1 = M_1 = M_2 = Y_3 = M_3 = 0$. Reordering the combined stiffness matrix, equation (17-12), gives the matrix equation

$$\begin{bmatrix} Y_2 \\ M_2=0 \\ Y_1=0 \\ M_1=0 \\ Y_3=0 \\ M_3=0 \end{bmatrix} = EI \begin{bmatrix} 15/2L^3 & -3/2L^2 & -3/2L^3 & 3/2L^2 & -6/L^2 & -3/L^2 \\ -3/2L^2 & 4/L & -3/2L^2 & 1/L & 3/L^2 & 1/L \\ -3/2L^3 & -3/2L^3 & 3/2L^3 & -3/2L^2 & 0 & 0 \\ 3/2L^2 & 1/L & -3/2L^2 & 2/L & 0 & 0 \\ -6/L^3 & 3/L^2 & 0 & 0 & 6/L^3 & 3/L^2 \\ -3/L^2 & 1/L & 0 & 0 & 3/L^2 & 2/L \end{bmatrix} \begin{bmatrix} v_2 \\ \phi_2 \\ v_1 \\ \phi_1 \\ v_3 \\ \phi_3 \end{bmatrix}.$$

The matrix of this equation can be partitioned as shown, and the inverse of the K_1 matrix is

$$K_1^{-1} = \frac{2L}{111\,EI} \begin{bmatrix} 8L^2 & 3L \\ 3L & 15 \end{bmatrix}.$$

Using this in the deformation equation gives

$$v_2 = \frac{2L}{111\,EI}\, 8L^2\, Y_2 = \frac{16\,L^3}{111\,EI}\, Y_2$$

$$\phi_2 = \frac{2L}{111\,EI}\, 3L\, Y_2 = \frac{6\,L^2}{111\,EI}\, Y_2 .$$

The values of v_2 and ϕ_2 just found can be used to determine the reactions Y_1, M_1, Y_3, and M_3 with the aid of submatrix K_3.

$$Y_1 = \frac{-3EI}{2L^3} \left(\frac{16\,L^3}{111\,EI}\right) Y_2 - \frac{3\,EI}{2\,L^2} \left(\frac{6\,L^2}{111\,EI}\right) Y_2 = \frac{-11}{37}\, Y_2$$

$$M_1 = \frac{3\,EI}{2\,L^2} \left(\frac{16\,L^3}{111\,EI}\right) Y_2 + \frac{EI}{L} \left(\frac{6\,L^2}{111\,EI}\right) Y_2 = \frac{10}{37}\, Y_2 L$$

$$Y_3 = \frac{-6EI}{L^3} \left(\frac{16\,L^3}{111\,EI}\right) Y_2 + \frac{3EI}{L^2} \left(\frac{6\,L^2}{111\,EI}\right) Y_2 = \frac{-26}{37}\, Y_2$$

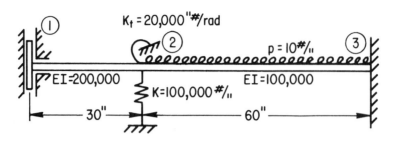

Figure 17-5. Second example of stepped beam.

$$M_3 = \frac{-3EI}{L^2}\left(\frac{16\,L^3}{111\,EI}\right)Y_2 + \frac{EI}{L}\left(\frac{6\,L^2}{111\,EI}\right)Y_2 \; = \; \frac{-14}{37}\,Y_2\,L$$

Transfer Matrices

It has been mentioned that transfer matrices (1) can be used to analyze a structure. They will be used here to determine points on the elastic curve of a beam and also to find key points on the shear and moment diagrams.

Frequently, a beam has discontinuities in its properties such as sudden changes in E or I, and also in the loading, due to several concentrated loads along its length; that is, it is "stepped." To analyze a beam of this type, it can be considered to be composed of a number of short sub-beams acting in series. These sub-beams are selected so that any discontinuities occur at their ends. Then, with the aid of Figure 17-1 and the basic principles of strength of materials, it is possible to write simple equations relating the displacement v, slope ϕ, moment M, and shear V at one end of a section in terms of corresponding values at the other end. These equations can be put in matrix form to be used with column vectors having the terms $\{v\;\phi\;M\;V\}$. This type of matrix is known as a *field matrix* F_{ij}, where the subscripts refer to the numbers of the nodes at the end of the section.

At the points or nodes where the sections are joined there may be a sudden change in the shear due to an external force or a flexible support; or there may be a sudden change in the moment M

due to an applied external moment at this point. Simple equations can be written to include these changes and put into matrix form. These matrices are known as *point matrices* P_i, where i is the number of the node where the discontinuity occurs.

After all of the matrices have been formulated they can be placed in the proper order and multiplied together to obtain the column vectors $\{v\;\phi\;M\;V\}$ at points along the beam. Some of the values in these vectors will be zero due to the boundary conditions of the beam. When these zeros are inserted in the vectors it becomes possible to find the values at points along the beam.

This general discussion will be illustrated by an example of the weightless beam shown in Figure 17-5, which has a "sliding" bearing (see last case of Figure 17-2) at the left end, while the right end is "fixed." Between these two points is a flexible support consisting of a rectilinear spring whose stiffness k is 100,000 lb./in., and a torsional spring whose stiffness k_t is 20,000 in.-lb./radian. The left portion of the weightless beam is unloaded and its rigidity is EI = 200,000 lb.-in.^{2}; while the right portion carries a distributed load p of 10 lb./in. and its rigidity is EI = 100,000 lb.-in.^{2}. It is desired to determine the shape of the elastic curve of this beam and also find key points on the shear and moment diagram.

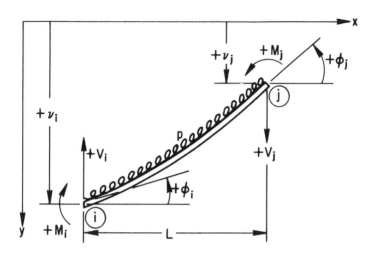

Figure 17-6. Section of beam of length L carrying distributed load of p lb/in.

It may be observed that the entire beam can be divided into two sections so there are three nodes which are labeled 1, 2 and 3 on the figure. Hence, two field matrices F_{12} and F_{23} and one point matrix P_2 can be formulated.

It is convenient to change the sign conventions in setting up these matrices. Figure 17-6 shows a typical section of a beam of length L carrying a distributed load p. Values of positive terms are taken as shown on this figure. This sketch can be used directly to find the field matrix F_{23}. It can be used to find F_{12} by letting p = 0. The equations and matrices will be written using symbols and the subscripts i and j employed to make the derivations more general. Later, when the beam of Figure 17-5 is analyzed, numerical values will be substituted for the symbols.

From strength of materials and an inspection of Figure 17-6:

$$V_j = V_i - p L \qquad (17\text{-}13)$$
and
$$M_j = M_i + V_i L - p L^2/2 . \qquad (17\text{-}14)$$

If the beam of Figure 17-6 is considered to be cantilevered at node i, the equation for the relative slope and deflection at node j can be written using the equations of Figure 17-1:

$$\phi_j = M_j L/E I - V_j L^2/2 EI - p L^3/6 EI \quad (17\text{-}15)$$
and
$$v_j = -M_j L^2/2 E I + V_j L^3/3 EI \qquad (17\text{-}16)$$
$$+ p L^4/8 EI .$$

It is necessary to express ϕ_j and v_j in terms of M_i, V_i, ϕ_i and v_i. By inspection of Figure 17-6 and substituting equations (17-13) and (17-14) in equations (17-15) and (17-16) we obtain

$$\phi_j = \phi_i + (L/EI)M_i + (L^2/2EI)V_i - p L^3/6EI \quad (17\text{-}17)$$

and

$$v_j = v_i - L \phi_1 - (L^2/2EI)M_i - V_i L^3/6 EI \quad (17\text{-}18)$$
$$+ p L^4/24 EI .$$

Now the field matrix F_{23} can be written using equations (17-13), (17-14), (17-17) and (17-18). If there are no external loads on a beam the field and point matrices are of the fourth order. However, in sub-beam 2-3 each of the final equations contains a constant term which must be included. This can be done by augmenting the matrix and adding a fifth term of unity to the column vectors. The matrix equation for section 2-3 then becomes

$$
\begin{bmatrix} v \\ \phi \\ M \\ V \\ 1 \end{bmatrix}_j =
\begin{bmatrix}
1 & -L & -L^2/2EI & -L^3/6EI & p\,L^4/24\,EI \\
0 & 1 & L/EI & L^2/2EI & -p\,L^3/6\,EI \\
0 & 0 & 1 & L & -p\,L^2/2 \\
0 & 0 & 0 & 1 & -pL \\
0 & 0 & 0 & 0 & 1
\end{bmatrix}
\begin{bmatrix} v \\ \phi \\ M \\ V \\ 1 \end{bmatrix}_i
$$

$$(17\text{-}19)$$

The field matrix F_{12} can be obtained in a similar manner. However, the only difference between sub-beams 1-2 and 2-3 is that sub-beam 2-3 carries the distributed load p, while sub-beam 1-2 does not. Letting $p = 0$ in the field matrix F_{23} of equation (17-19) gives the F_{12} field matrix:

$$
\begin{bmatrix} v \\ \phi \\ M \\ V \\ 1 \end{bmatrix}_j =
\begin{bmatrix}
1 & -L & -L^2/2EI & -L^3/6EI & 0 \\
0 & 1 & L/EI & L^2/2EI & 0 \\
0 & 0 & 1 & L & 0 \\
0 & 0 & 0 & 1 & 0 \\
0 & 0 & 0 & 0 & 1
\end{bmatrix}
\begin{bmatrix} v \\ \phi \\ M \\ V \\ 1 \end{bmatrix}_i
$$

It is not necessary to augment the F_{12} matrix since no constants are involved. However, it is done here to make it conformable for multiplication with the other matrices in a later step.

The point matrix at node 2, or P_2, is also augmented for the reason just given. The deflection and slope must be the same to the right and left of node 2, hence $v_2^R = v_2^L$, and $\phi_2^R = \phi_2^L$. A vertical displacement v_2 would produce a shear force on the beam due to the movement of the rectilinear spring of scale k. Hence, there is a step change in the shear of $k\,v_2$ at this point, to give the relationship $V_2^R = V_2^L + k\,v_2$. In a similar

manner the torsional spring stiffness of k_t induces a step change in the bending moment at point 2, or $M_2^R = M_2^L + k_t\phi_2$. Putting these relationships in the augmented point matrix P_2 gives

$$
\begin{bmatrix} v \\ \phi \\ M \\ V \\ 1 \end{bmatrix}_2^R =
\begin{bmatrix}
1 & 0 & 0 & 0 & 0 \\
0 & 1 & 0 & 0 & 0 \\
0 & k_t & 1 & 0 & 0 \\
k & 0 & 0 & 1 & 0 \\
0 & 0 & 0 & 0 & 1
\end{bmatrix}
\begin{bmatrix} v \\ \phi \\ M \\ V \\ 1 \end{bmatrix}_2^L
$$

Having formulated the necessary matrices and designating a vector as a boldface **V**, it may be seen that the equations can be written by starting at either end of the beam. The following equations are based on working from left to right.

$$\mathbf{V}_2^L = \mathbf{F}_{12}\,\mathbf{V}_1 \tag{17-20}$$

$$\mathbf{V}_2^R = \mathbf{P}_2\,\mathbf{V}_2^L = \mathbf{P}_2\,\mathbf{F}_{12}\,\mathbf{V}_1 \tag{17-21}$$

$$\mathbf{V}_3 = \mathbf{F}_{23}\,\mathbf{V}_2^R = \mathbf{F}_{23}\,\mathbf{P}_2\,\mathbf{F}_{12}\,\mathbf{V}_1. \tag{17-22}$$

From an inspection of Figure 17-5 it may be noted that the boundary conditions as listed in Figure 17-2 are that $\phi_1 = V_1 = \phi_3 = v_3 = 0$. These could be inserted in the appropriate vectors after all of the multiplications indicated in equation (17-22) are completed, but the amount of work and chance of possible errors are reduced by inserting them before the multiplication is done.

In the multiplications that follow, numerical values based on Figure 17-5 will replace the symbols used previously.

$$
\begin{bmatrix} v \\ \phi \\ M \\ V \\ 1 \end{bmatrix}^L_2 =
\begin{bmatrix}
1 & -30 & \dfrac{-9}{4000} & \dfrac{-27}{1200} & 0 \\
0 & 1 & \dfrac{3}{20000} & \dfrac{9}{4000} & 0 \\
0 & 0 & 1 & 30 & 0 \\
0 & 0 & 0 & 1 & 0 \\
0 & 0 & 0 & 0 & 1
\end{bmatrix}
\begin{bmatrix} v \\ \phi=0 \\ M \\ V=0 \\ 1 \end{bmatrix}_1 =
\begin{bmatrix} v_1 - \dfrac{9M_1}{4000} \\ \dfrac{3}{20000}\,M_1 \\ M_1 \\ 0 \\ 1 \end{bmatrix}^L_2
\qquad (17\text{-}23)
$$

$$
\begin{bmatrix} v \\ \phi \\ M \\ V \\ 1 \end{bmatrix}^R_2 =
\begin{bmatrix}
1 & 0 & 0 & 0 & 0 \\
0 & 1 & 0 & 0 & 0 \\
0 & 20000 & 1 & 0 & 0 \\
100000 & 0 & 0 & 1 & 0 \\
0 & 0 & 0 & 0 & 1
\end{bmatrix}
\begin{bmatrix} v_1 - \dfrac{9M_1}{4000} \\ \dfrac{3}{20000}\,M_1 \\ M_1 \\ 0 \\ 1 \end{bmatrix}_2 =
\begin{bmatrix} v_1 - \dfrac{9M_1}{4000} \\ \dfrac{3}{20000}\,M_1 \\ 4\,M_1 \\ 100000v_1 - 225M_1 \\ 1 \end{bmatrix}^R_2
\qquad (17\text{-}24)
$$

$$
\begin{bmatrix} v \\ \phi \\ M \\ V \\ 1 \end{bmatrix}_3 =
\begin{bmatrix}
1 & -60 & \dfrac{-18}{1000} & \dfrac{-36}{100} & 54 \\
0 & 1 & \dfrac{6}{10000} & \dfrac{18}{1000} & -3.6 \\
0 & 0 & 1 & 60 & -18000 \\
0 & 0 & 0 & 1 & -600 \\
0 & 0 & 0 & 0 & 1
\end{bmatrix}
\begin{bmatrix} v_1 - \dfrac{9}{4000}M_1 \\ \dfrac{3}{20000}\,M_1 \\ 4\,M_1 \\ 100000v_1 - 225M_1 \\ 1 \end{bmatrix}^R_2
\qquad (17\text{-}25)
$$

$$
\begin{bmatrix} v=0 \\ \phi=0 \\ M \\ V \end{bmatrix}_3 =
\begin{bmatrix}
-35999\ v_1 + 80.91675\ M_1 + 54 \\
1800\ v_1 - 4.04745\ M_1 - 3.6 \\
-6000000\ v_1 - 13496\ M_1 - 18000 \\
100000\ v_1 - 225\ M_1 - 600
\end{bmatrix}
\qquad (17\text{-}26)
$$

The two top terms of the vector of equation (17-26) form a pair of simultaneous equations:

$$-35999 \, v_1 + 80.91675 \, M_1 + 54 = 0$$

$$1800 \, v_1 - 4.04745 \, M_1 - 3.6 = 0.$$

Solving these equations gives $v_1 = -1.34899$ in. and $M_1 = -599.93$ in.-lb. If these values are back substituted into the vectors of equations (17-23), (17-24), (17-25) and (17-26) the values of the variables at key points along the beam are found to give the vectors

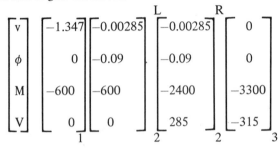

If additional values are desired along the beam they could be obtained by using more sub-beams that terminate at the points where the values are desired. The corresponding point matrices at these points would be unit matrices, I.

References

1. Pestel, E.G. and F.A. Leckie. 1963. *Matrix Methods in Elastomechanics.* New York: McGraw-Hill.

18 elements of structural analysis

finite element method

Introduction

The structures considered thus far can be idealized as an assembly of structural shapes carrying direct, bending and torsion loads either singly or in combination. Another type of structure consists of continuous members such as plates, shells or pressure vessels. The deformations and stresses in structures of this type can be approximated by the use of the "finite element" method, which follows the principles and general procedures already outlined. It was originally developed about 15 years ago to determine stresses and deformations in aeronautical structures, but since then has been extended to cover such diverse fields as heat conduction, seepage flow, vibration, creep and plasticity problems (1).

In the finite element method the surface is divided into a set of imaginary triangular, trapezoidal or rectangular elements joined together at their corners, or "nodes." It is then possible, as described later, to determine the displacements of these nodes when the surface is loaded externally. It is also possible to find the average normal and shear stresses existing in each element due to these loads. More elements, or a finer mesh, give greater accuracy, especially near points where the stress level changes rapidly. However, the size of the resulting matrices increases rapidly as more elements and nodal points are used.

This discussion is limited to finding the response of a loaded plane surface using triangular elements, although it is possible to solve three-dimensional problems; and rectangular or trapezoidal elements can be used. The emphasis here is to outline the procedure of formulating the matrix solution from the viewpoint of strength of materials, rather than dealing with the programming difficulties of handling large matrices in the computer.

General Procedure

The first step is to make a sketch of the surface to be studied, with the axes for the global coordinates located. Then this surface should be divided into elements and the nodes specified. The subdivision of the surface requires good engineering judgment and experience. It is desirable that the triangles be made as equilateral as possible (1, p. 230).

The next step is to determine the stiffness properties of each element in terms of its local (element) coordinates. These properties can then be formed into an element matrix, which can then be transformed from local to global coordinates.

Superposition of the individual global matrices for each element to form a general global matrix of the entire surface requires only simple addition of the corresponding terms of the global matrices of the various elements.

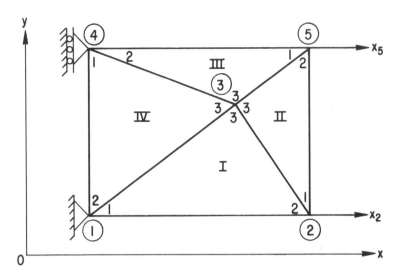

Figure 18-1. Bar with four elements.

If the general global matrix is designated as [**K**], the equilibrium equation relating the nodal force vector $\{F\}$ and the resulting deformation vector $\{\delta\}$ is $\{F\} = [K] \{\delta\}$ or $\{\delta\} = [K^{-1}] \{F\}$.

Since the general matrix [K] does not include the boundary conditions at the supports of the surface, it is necessary to reorder it by partitioning in accordance with equation (16-6). Then the procedure outlined in the "Trusses" section of Chapter 16 can be followed to solve for the nodal displacements throughout the surface; the external forces and reactions; and, finally, it is possible to determine the forces acting at each nodal point in the directions of the assumed global coordinates as well as the average normal and shear stresses in each element.

Each of these steps will be discussed in detail in the following sections. To be more specific, this will be done with reference to a bar having a uniform thickness that is divided into four triangular elements as shown in Figure 18-1. These elements can be numbered in any sequence; and, for clarity, Roman numerals are used for this in the figure. A pair of global perpendicular coordinate axes, x and y, is set up with an origin 0 located at their inter-

section. The global nodal points are also numbered in any sequence, and these numbers are enclosed in circles on the figure. The position of any element is specified by the x and y coordinates of its global nodal points.

When the element is loaded the nodal points will be deformed. A deformation in the x direction is specified as u, while a deformation in the y direction is designated as v. This is in accord with the nomenclature used previously.

Evaluation of Element Stiffness Properties

Local coordinates must be used in determining the local stiffness matrix of an element. The nodes are numbered 1, 2, 3 in a counterclockwise direction as shown in Figure 18-1.

It is assumed that the edges of each element remain straight when the element is loaded; i.e., that the displacements of the nodes are a linear function of x and y. Hence:

$$u = a_1 + a_2 x + a_3 y \qquad (18\text{-}1)$$

and

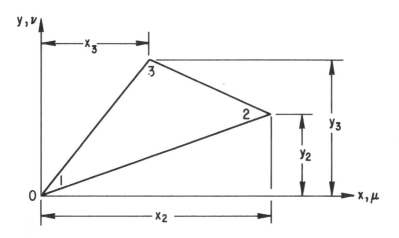

Figure 18-2. Coordinates of an element.

$$v = a_4 + a_5 \ x + a_6 \ y. \qquad (18\text{-}2)$$

Equations (18-1) and (18-2) can be expressed in matrix form as $\{\delta\} = [M] \ \{a\}$. Applying equations (18-1) and (18-2) to a particular element gives

$$u_1 = a_1 + a_2 \ x_1 + a_3 \ y_1$$

$$u_2 = a_1 + a_2 \ x_2 + a_3 \ y_2$$

$$u_3 = a_1 + a_2 \ x_3 + a_3 \ y_3$$

$$\qquad\qquad\qquad\qquad (18\text{-}3)$$

$$v_1 = a_4 + a_5 \ x_1 + a_6 \ y_1$$

$$v_2 = a_4 + a_5 \ x_2 + a_6 \ y_2$$

$$v_3 = a_4 + a_5 \ x_3 + a_6 \ y_3 \ .$$

Equation (18-3) can be written in matrix form as

$$\{\delta\} \ = \ [A] \ \{a\} \qquad (18\text{-}4)$$

or

$$
\begin{bmatrix} u_1 \\ u_2 \\ u_3 \\ v_1 \\ v_2 \\ v_3 \end{bmatrix}
=
\begin{bmatrix}
1 & x_1 & y_1 & 0 & 0 & 0 \\
1 & x_2 & y_2 & 0 & 0 & 0 \\
1 & x_3 & y_3 & 0 & 0 & 0 \\
0 & 0 & 0 & 1 & x_1 & y_1 \\
0 & 0 & 0 & 1 & x_2 & y_2 \\
0 & 0 & 0 & 1 & x_3 & y_3
\end{bmatrix}
\begin{bmatrix} a_1 \\ a_2 \\ a_3 \\ a_4 \\ a_5 \\ a_6 \end{bmatrix}
\qquad (18\text{-}4')
$$

It is frequently desirable to reorder this matrix equation to

$$\{\delta\} = [C] \ \{a\} \qquad (18\text{-}5)$$

or

$$\begin{bmatrix} u_1 \\ v_1 \\ u_2 \\ v_2 \\ u_3 \\ v_3 \end{bmatrix} = \begin{bmatrix} 1 & x_1 & y_1 & 0 & 0 & 0 \\ 0 & 0 & 0 & 1 & x_1 & y_1 \\ 1 & x_2 & y_2 & 0 & 0 & 0 \\ 0 & 0 & 0 & 1 & x_2 & y_2 \\ 1 & x_3 & y_3 & 0 & 0 & 0 \\ 0 & 0 & 0 & 1 & x_3 & y_3 \end{bmatrix} \begin{bmatrix} a_1 \\ a_2 \\ a_3 \\ a_4 \\ a_5 \\ a_6 \end{bmatrix} \qquad (18\text{-}6)$$

Several comments can be made regarding equations (18-4') and (18-6).

First, it may be observed that the a's are generalized coordinates. If the values of δ are known for a given element, the corresponding values of the a's could be obtained from the matrix equations $\{a\} = A^{-1} \{\delta\}$, or $\{a\} = [C^{-1}] \{\delta\}$. However it will be shown that there is no need to determine these values.

Second, in applying these equations to a particular element, it is convenient to set the local origin at point 1 of the element while keeping the local axes parallel to the global axes as shown in Figure 18-2. Then the values of x_1 and y_1 in the equations are zero, and there is no need to rotate the element.

The unit strains can be related to the generalized coordinates by taking the derivatives of the displacement functions as given by equations (18-3). Thus,

$$\epsilon_x = \partial u/\partial x = a_2 \qquad \epsilon_y = \partial v/\partial y = a_6$$

$$\gamma_{xy} = \partial u/\partial y + \partial v/\partial x = a_3 + a_5.$$

This gives the matrix equation

$$\begin{bmatrix} \epsilon_x \\ \epsilon_y \\ \gamma_{xy} \end{bmatrix} = \begin{bmatrix} 0 & 1 & 0 & 0 & 0 & 0 \\ 0 & 0 & 0 & 0 & 0 & 1 \\ 0 & 0 & 1 & 0 & 1 & 0 \end{bmatrix} \begin{bmatrix} a_1 \\ a_2 \\ a_3 \\ a_4 \\ a_5 \\ a_6 \end{bmatrix}$$

or

$$\{\epsilon\} = [B] \{a\} \qquad (18\text{-}7)$$

Hence, by equations (18-4) and (18-5) $\{\epsilon\} = [B] [A^{-1}] \{\delta\}$ and $\{\epsilon\} = [B] [C^{-1}] \{\delta\}$.

Next, it is necessary to relate the unit stress σ to the unit strain ϵ in each element. From the theory of elasticity this relationship for an isotropic material, expressed in matrix form, is

$$\begin{bmatrix} \sigma_x \\ \sigma_y \\ \tau_{xy} \end{bmatrix} = \frac{E}{1 - \mu^2} \begin{bmatrix} 1 & \mu & 0 \\ \mu & 1 & 0 \\ 0 & 0 & \frac{1-\mu}{2} \end{bmatrix} \begin{bmatrix} \epsilon_x \\ \epsilon_y \\ \gamma_{xy} \end{bmatrix}$$

where μ is Poisson's ratio of the element material. Then,

$$\{\sigma\} = [D] \{\epsilon\} = [D] [B] \{a\}, \qquad (18\text{-}8)$$

or

$$\{\sigma\} = [D] [B] [A^{-1}] \{\delta\} \qquad (18\text{-}9)$$

$$\{\sigma\} = [D] [B] [C^{-1}] \{\delta\} \qquad (18\text{-}10)$$

With this background it is now possible to derive equations, based on the principle of strain energy, to determine the local stiffness matrix K^L for the nodes or points of each element.

For static loading of a conservative element the kinetic energy is zero; and the work W done by the applied loads on the corresponding displacements is stored as strain energy U. It is important to realize in this connection that U depends only on the final deformation of the structure, and hence is independent of the loading history. Strain energy can be developed by 1) axial loads, 2) shear forces, 3) bending moments and 4) torques. These can exist in any combination and amount; and these various types will be designated by the subscript i.

Figure 18-3a shows a general (nonlinear) force-displacement curve for any type of loading i. The area under this curve is $\int_0 F_i(u_i)\, du_i$, or the work done by the force F_i. The corresponding

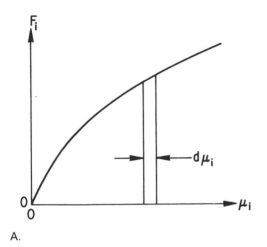

A.

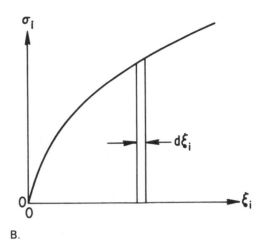

B.

Figure 18-3. Nonlinear force-displacement and corresponding stress-strain plots for any type of element loading.

stress-strain plot is shown as Figure 18-3b. The area under this curve is $\int_0 \sigma_i \, \partial\epsilon_i$, or the strain energy per unit volume. Note that this area has the units of psi, and when multiplied by the volume the units are lb.-in. If all four types of loading exist, the total energy stored is

$$U = \sum_{i=1}^{4} \int_v \left(\int_0 \sigma_i \, d\epsilon_i \right) dV. \quad (18\text{-}11)$$

If the structure is linear, i.e., has a constant E, the curves of Figures 18-3a and 18-3b are straight lines, and $F_i = k_{ii}^L \, u$. Then,

$$W = \int_0 k_{ii}^L \, u \, du = 1/2 \, k_{ii}^L \, u_i^2 = 1/2 \, F_i \, u_i \quad (18\text{-}12)$$

Also the unit stress and strain are linear so that $\int_0^{\epsilon i} \sigma_i d_{\epsilon i} = 1/2 \, \sigma_i \epsilon_i$ and equation (18-11) can be written as

$$U = 1/2 \sum_{i=1}^{4} \int_v \sigma_i \epsilon_i \, dV. \quad (18\text{-}13)$$

Thus, if the structure is linear and the stresses are linearly distributed over the cross sections, equation (18-13) can be written as

$$U = \frac{1}{2} \int \frac{P^2(x)}{E \, A(x)} \, dx + \frac{1}{2} \int \frac{M^2(x)}{E \, I(x)} \, dx$$
(normal)   (bending)

$$+ \frac{1}{2} \int \frac{V^2(x)}{G \, A(x)} \, dx + \frac{1}{2} \int \frac{T^2(x)}{G \, J(x)} \, dx \, .$$
(shear)   (torsion)

The right side of equation (18-12) can be put into matrix form as $W = U = 1/2 \, \{u\}^T \, \{F\}$; or $U = 1/2 \, \{F\}^T \, \{u\}$. Substituting $[k^L] \, \{u\}$ for $\{F\}$ gives $U = 1/2 \, \{u\}^T \, [k^L] \, \{u\}$.

The internal virtual work dW of any differential volume dV of a member is the product of the actual stress σ and the virtual strains through which they move, or $dW = \epsilon^T \sigma dV$ or by equations (18-7) and (18-8): $dW = a^T B^T D a \, dV$, or $W = a^T [\int_v B^T D B \, dV] \, a$. Hence,

$$k^L = \int_v B^T \, D \, B \, dV \quad (18\text{-}14)$$

where equation (18-14) is the generalized stiffness matrix of the element. The volume of the element $= \int_v dV =$ (thickness) (area).

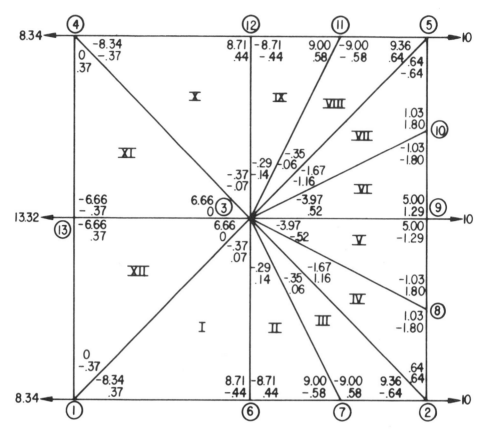

Figure 18-4*a*. External loads and reactions on a 1 inch cube, and internal loads acting in each of 12 elements of cube. All loads given in kips.

Applying equations (18-3) or (18-4) to equation (18-14) relates the generalized coordinates a to the normal coordinates δ. $[k^L] = [A^{-1}]^T [k] [A^{-1}]$ or

$$[k^L] = [C^{-1}]^T [k] [C^{-1}] . \qquad (18-15)$$

Once the local stiffness matrix has been found for each element, it is necessary to formulate them into global coordinates. If the coordinates of the points were taken as shown in Figure 18-2, it is not necessary to "rotate" the element into position. Some analysts formulate the element stiffness matrix with local nodes 1 and 2 horizontal regardless of its position on the surface. Then it is necessary to use equation (16-20) to place it at the correct angle in the surface.

In either case the subscripts of the F_i, u_i, v_i, and k^L_{ij} terms must be changed from the local values of 1, 2, and 3 to the corresponding global values. Then superposition of the element matrices to formulate the global matrix becomes one of simple addition of corresponding elements. The resulting global matrix $[K^G]$ is then inserted into the equation $\{F^G\} = [K^G] \{\delta^G\}$.

The next step is to determine the boundary conditions in terms of global notation. For the case shown in Figure 18-1, these would be that $u_1^G = v_1^G = u_4^G = 0$. The external loads must also be specified, and these must be applied at the nodes. In Figure 18-1 all of the external loads are zero except for X_2^G and X_5^G.

After these values are specified it is necessary

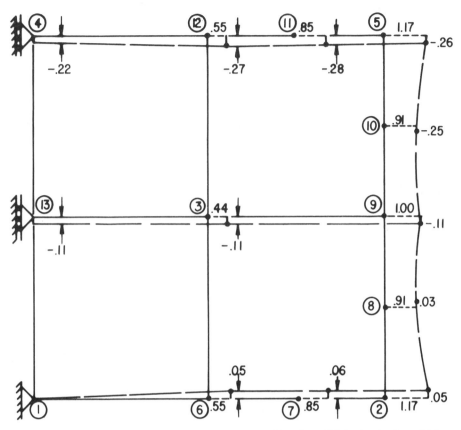

Figure 18-4*b*. Deformations of cube in mils due to external loading. Original cube in solid lines; deformed cube in dashed lines.

to reorder the global matrix as was done in Chapters 16 and 17, putting the rows containing the boundary conditions and the corresponding reaction forces on the bottom rows. This gives a global matrix that can be partitioned as in equation (16-6), and the equations involving the submatrices K_1 and K_3 solved to find first the unknown deformations and then the reactions at the supports.

After the global deformations have been found, the average normal stresses in the x and y direction and the shear stress in each element can be found with the aid of equations (18-9) or (18-10). The δ vector in these equations refers to the deformations within the element. These can be found by using one node as a datum point and subtracting its global u and v values from the global values of the other nodes to obtain the local element deformations. If desired, these stresses can be used to find the resultant maximum normal and shear stresses in the element by the use of the equations

$$\tau_{max} = \sqrt{(\sigma_x - \sigma_y)^2/4 + \tau^2_{xy}}$$

$$\sigma_{max} = (\sigma_x + \sigma_y)/2 + \tau_{max}.$$

In some analyses it may be desirable to determine the internal horizontal and vertical loads acting at the nodes of each element. This can be done with the equation $\{F_{int}\} = [K^L] \{\delta^L\}$, where the δ^L vector is the relative deformation used to determine the average stresses in the element, and K^L is the local stiffness matrix of the element.

Example

To clarify the use of the finite element method consider the bar shown in Figures 18-4a and b. It is made of steel ($\mu = 0.3$ and $E = 30(10)^6$) and is a 1 in. cube (length = depth = thickness = 1 in.). The surface is approximated by 12 triangular elements which are designated by the Roman numerals on Figure 18-4a. The global numbers of the nodes are enclosed in circles on both parts of the figure. The external loads on the bar are each 10 kips in a horizontal direction at nodes 2, 9 and 5 as shown in part a. The supports are as shown in part b, hence the boundary conditions are $u_1 = v_1 = u_4 = u_{13} = 0$. It is desired to find the deformations (u and v) at each node as well as the average stresses in each element..

Since there are 13 nodes, the size of the global matrix is 26x26. When the four boundary conditions are inserted the size of the reordered matrix is reduced to 22x22. It is not practical to include the complete calculations for this example here, but the results as given by the computer are summarized in the figures.

In Figure 18-4a the reactions at nodes 1 and 4 are given as 8.34 kips, while the reaction at node 13 is 13.32 kips. The internal loads in each element at the nodes are given in kips, rounded off to two decimal places, by two numbers. The upper number represents the horizontal force, and the lower number the vertical force. It should be observed that these forces are in balance at each node and also for each element.

In Figure 18-4b the unloaded bar is shown by solid lines along with the supports, but the elements have been omitted for clarity. The deformations at the nodes in a horizontal and vertical direction have been magnified and are shown as short dashes. The numbers given on these lines are the deformations in mils. To aid in the visualization of the bar deformation, long dashed lines are drawn connecting nodes that lay on a straight line when the bar was unloaded.

It is not feasible to include the average stress values for each element as found by the computer on these figures. However, they are tabulated in

Table 18-1
Average Stress Values

Element	σ_x	σ_y	τ_{xy}	τ_{max}	σ_{max}
I	33363	290	-1488	16603	33430
II	36012	1085	-2321	17617	36166
III	38848	515	-2836	19375	39057
IV	13403	2057	-9271	10869	18599
V	31737	-7214	4131	19909	32171
VI	31737	-7214	-4131	19909	32171
VII	13403	2057	9271	10869	18599
VIII	38848	515	2836	19375	39057
IX	36012	1085	2321	17617	36166
X	33363	290	1488	16603	33430
XI	26637	1488	0	12574	26637
XII	26637	1488	0	12574	26637

Table 18-1, and it may be observed that they are symmetrical about the horizontal centerline of the bar.

General Comments

If the procedure is formulated as a computer program it is a simple matter to make variations in the loading, the boundary conditions or the surface properties by merely changing the appropriate data cards. Large computer programs have been developed and they are continually being improved (2).

While the method is basically an approximate one, it yields satisfactory results from an engineering viewpoint (3; 4, pp 100-103; 1).

It is possible to incorporate into the analysis the effect of internal loads in the members due to internal strains or local temperature variations by incorporating the principles outlined at the end of Chapter 16.

The amount of work involved (or computer time, hence cost) in solving problems can frequently be drastically reduced by taking full advantage of any symmetry existing in the surface being studied. For example, if it is desired to find the stresses and deformation in a ring that is to carry a concentrated load acting on its diameter, only one quarter of the ring need be analyzed by selecting proper boundary conditions. Hence, before plunging into any analysis, it would be desirable to devote time to considering the possibilities of symmetry.

It should be obvious that this discussion is but an introduction to a technique that is still being developed. It is presented with the hope that the interested reader will be stimulated to delve further into its possibilities. For these purposes a good starting point would be references 1, 2, 4, 5, 6 and 7 plus the innumerable technical papers that are being issued. As noted previously, the application of the method has now gone far beyond the calculation of linear stress analysis of members(1).

References

1. Zienkiewicz, O.C. 1967. *The finite element method*. New York: McGraw-Hill.
2. Khol, R. 1968. Computer stress analysis. *Machine Design* (Nov. 21) pp. 136-145.
3. DeHart, R.C. and L.F. Greimann. 1969. *Penetrations in shells under external pressure*. A.S.M.E. Paper No. 69-WA/Unt-4.
4. Zienkiewicz, O.C. and G.S. Holister. 1965. *Stress analysis*. New York: Wiley.
5. Rubinstein, M.F. 1966. Matrix computer analysis of structures. Englewood Cliffs: Prentice-Hall.
6. Argyris, J.H. and S. Kelsey. 1960. Energy theorems and structural analysis. London: Butterworth.
7. Turner, M.J., R.W. Clough, H.C. Martin and L.J. Topp. 1956. Stiffness and deflection analysis of complex structures. *J. Aero. Sciences* (Sept.), pp. 805-854.

part 7: frans gerritsen

Ice Problems in Coastal Waters

The Effect of Ice on Hydraulic Structures

Frans Gerritsen is Chairman of the Department of Ocean Engineering of the University of Hawaii.

After receiving a B.S. degree with honors (1943) and an M.S. (1950) in civil engineering from the Technical University, Delft, Holland, he continued his research (begun in 1948 in hydraulics engineering)—at Delft Hydraulics Laboratory in harbor and coastal engineering until 1951.

Gerritsen served as a coastal engineer for the Dutch Board of Roads, Waterways and Harbors until 1956. He was a professor of coastal engineering and then coastal and harbor engineering at the University of Florida at Gainesville from 1956-1962. Returning to the Dutch Board of Roads, he served as Chief Engineer of the Northern Delta Division until 1969.

Among his projects have been tidal land reclamation in the Republic of Korea, the damming of a North Sea inlet in West Germany, coastal development in Kuwait, closing of tidal gaps and design and planning of the DELTA project in the Netherlands, and improvement of the San Nicolas Harbor in Aruba.

He has published over 25 papers, articles, and books including contributions to Volumes 1, 3, and 4 of this series; a Dutch civil engineering text, *Coastal Protection, Dike Construction and Water Management* (1963) as co-author and *Stability of Coastal Inlets* with Dr. Per Bruun (1960).

Professional affiliations include: the Royal Dutch Society of Graduate Engineers, the American Society of Civil Engineers, the International Association of Hydraulic Research, the Permanent International Association of Navigational Congresses, the Marine Technology Society, the Engineering Association of Hawaii and the Society of the Sigma Xi.

19

ice problems
in coastal waters

Introduction

In the temperate and arctic zones, coastal waters are affected by ice conditions during a greater or smaller part of the year. The problems that arise in these areas are

1. Difficulties to navigation;
2. Difficulties in discharge of high runoff in rivers after periods of severe winter conditions;
3. Problems connected with the design, construction and operation of hydraulic structures.

In this chapter we will direct our attention to the first two problems mentioned; the last subject will be discussed in the next chapter on ice. Examples will be taken from conditions on the North American continent and from western Europe.

Ice Problems in Navigation

Navigation in temperate and arctic zones has always been affected by problems associated with ice conditions. In the coastal waters of the North American continent above the American-Canadian borderline, ice conditions hamper navigation during a considerable part of the year.

The first example of an important navigation route affected by ice conditions is the St. Lawrence Seaway (Figure 19-1). The St. Lawrence Sea-

way was discovered and explored by European adventurers looking for a northwest passage. In 1535, Jacques Cartier sailed about 1000 miles into the interior until his ship was halted by rapids. The history of this waterway's development is a very long one—the last sections were constructed only in recent years.

One of the seaway's most essential problems is ice formation which makes the route unnavigable during 3-4 months of the year. Needless to say, this is quite an economic drawback to a sound seaway project development. In January, temperatures over the estuary average below 10°F.

Icebreakers have been used extensively on the seaway to break up ice jams to prevent serious flooding (1) and to assist ships trapped in the ice in early winter. More and bigger icebreakers, aerial ice surveys and improved aid to navigation are easing the situation.

The purpose of this chapter is not to give a complete account of all problems in this field but rather to indicate the existing problems for which solutions need to be found. We will now take a closer look at the problems associated with the attempts to find a navigable northwest passage as mentioned before and will discuss the recent attempt to find a modern solution.

The discovery of new and presumably rich oil fields along the north shores of Alaska (Prudhoe Bay) (Figure 19-2) created the problem of trans-

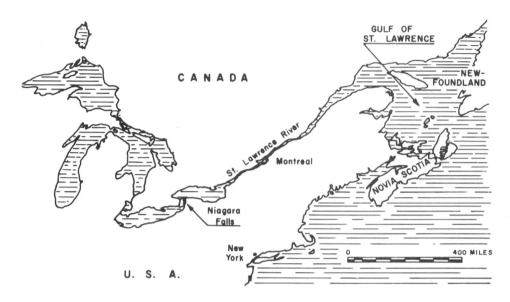

Figure 19-1. St. Lawrence seaway.

porting the oil to the industrial sites of the U.S. eastern seaboard. After carefully evaluating a number of alternatives, the oil companies involved (Humble Oil and Atlantic Richfield) decided that the problem could possibly be solved by using supertanker-icebreakers through the Northwest Passage. They decided to refit one of the largest tankers in existence and to make a voyage from the east coast of North America to Alaska and return; this experimental program cost approximately 40 million dollars.

The *Manhattan*, the largest tanker flying the U.S. flag, was chosen for this adventure. The *Manhattan* had a 108,400 dwt capacity, measured 940 1/2 feet long with maximum draft of 52 feet, 9 3/8 inches and had twin screws of 43,000 shp in total.

Before we go into the technical details of this operation, it is interesting to remember some of the historical attempts at finding the Northwest Passage. The search for such a passage from western Europe to the Far East encompasses a dramatic story that began almost five centuries ago (see Reference 2).

In 1497, a few years after Columbus came back to Europe from his trip to the West, the English sent a Genovese navigator named Giovanni Cabato, better known as John Cabot, to seek the northwest connection between the Atlantic and Pacific oceans. This unsuccessful attempt was repeated several times in later years.

In history books we read the names of Martin Frobisher (1576), John Davis (three attempts between 1585 and 1587) and Henry Hudson (1610); Davis reached Lancaster Sound, the entrance of the true passage, but his discovery was not recognized at that time. Henry Hudson, with his ship, the *Discovery*, sailed into the bay named after him; he found death when his crew put him adrift in a boat to perish in the icy winter.

In 1616 the *Discovery*, under the command of Robert Bylot, a rehabilitated mutineer, reached Smith Sound at a latitude of $78°$, which was not reached again for 237 years. In 1819 Parry sailed through Lancaster Sound and reached $113°$ west longitude. The two ships of Sir John Franklin disappeared forever in 1845 after his expedition had been sighted entering Lancaster Sound. Although

Figure 19-2. Northwest passage to Prudhoe Bay.

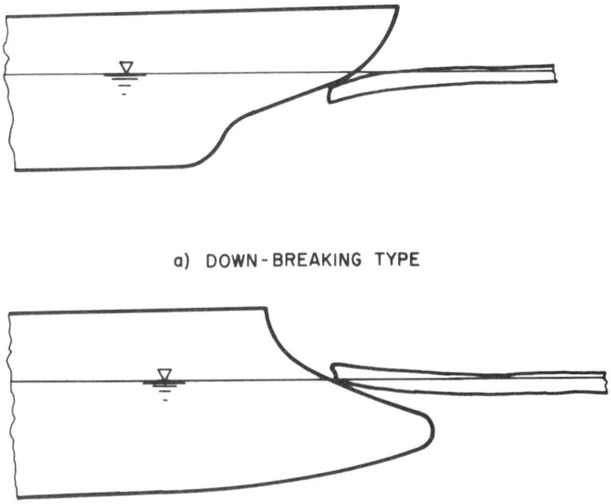

a) DOWN-BREAKING TYPE

b) UP-BREAKING TYPE

Figure 19-3. Two types of icebreaking bows.

McClure rounded Banks Island from the west and demonstrated the possibility of a northwest passage, it was Roald Amundsen who made the first complete crossing from east to west, which took him three years. His voyage began in 1903.

The *Manhattan*'s adventurous voyage into the icy arctic sea may mark the beginning of a new phase in the search for the Northwest Passage. The future will show the importance of this venture. Various aspects of the *Manhattan's* voyage are discussed in references 3 through 7.

Let us now look at the technical details that made the *Manhattan* suitable for this undertaking. It was clear from the beginning that the ship as it was would not be able to fight the arctic ice, and that important changes had to be made. The most essential change was her new icebreaking bow, which used the down-breaking principle. The bow inclines initially at 18° and bends to the conventional angle of 30° further down; in this way, ice-breaking efficiency is increased by possibly 60%. In principle, two types of icebreaker bows can be used: the down-breaking bow and the up-breaking bow (Figure 19-3). Experience has shown that the down-breaking type is more effective. To reduce

ice friction, the bow of the *Manhattan* was also widened to extend 8 feet beyond the main hull.

Other modifications included welding a 9-foot, 1 1/4-inch thick ice belt around the hull to strengthen it against ice forces, internal reinforcement with steel girders, strengthening of propeller shafts and screws, installing machinery collision shocks and constructing a helicopter flight check. The ship was also provided with a heeling system: the two bow thrusters serve as high-capacity ballast pumps, pumping 2000 tons of water from one side of the ship to the other in a maximum of 75 seconds, producing a rocking motion of 3° to both sides. It also included the installation of all types of modern equipment such as a satellite navigational system and computers that stored all data collected during the voyage.

Gruber (2) gives the following description of the voyage:

The Manhattan departed Philadelphia on August 28, 1969, and proceeded by way of

Figure 19-4. Air bubbling system is effective in keeping Stockholm Harbor open.

Halifax, and Thule, Greenland. Almost immediately, she encountered extensive pack ice in Baffin Bay, and had a chance to shake down her equipment. At various times during the voyage she was assisted by the U.S. Coast Guard icebreakers North-wind and Staten Island and the Canadian icebreakers Sir John A. McDonald and Louis St. Laurent. On September 5 she entered Parry Channel and headed westward through the Canadian Archipelago in the 'northwest passage.'

The first severe test came as the Manhattan probed the heavy ice in McClure Strait, north of Banks Island. Winds sweeping across the open Arctic Ocean often jam the dense polar pack into the mouth of the strait, with heavy pressure and ridging, making it potentially the most difficult part of a transit. Faced with just this situation, the tanker was stuck in ice three times, and the decision taken to proceed via the more ice-free Prince of Wales Strait between Banks and Victoria Islands.

Proceeding via this alternate route she soon passed the Mackenzie delta and steamed up the Alaskan coast, arriving in sight of the

drilling rigs at Prudhoe Bay on September 19. After onloading a symbolic barrel of Alaskan crude oil, she sailed on to Point Barrow, her successful westward transit completed.

On her return voyage the Manhattan spent almost a month in Melville Sound, deliberately seeking out different ice conditions. Maneuvering tests were run to see how large ships would perform under difficult circumstances comparable to those which would be encountered during much of the year. These tests concluded, she continued her eastward transit and arrived at New York City on November 12, having covered over 11,000 miles on the round trip.

Despite the reinforcements of the icebreaker, the *Manhattan* suffered a hull rupture due to ice impact, so that redesign of some hull characteristics will be required if additional experimental voyages are carried out. Because of the difficulties involved and the damage incurred, future explorations along this route are not at all certain. However, a vast amount of information was obtained on this voyage, which will be of great benefit to further studies.

The spectacular voyage of the *Manhattan* stresses the great value of icebreakers in keeping coastal sea lanes open. We should realize that using icebreakers in the conventional way is still a rather crude approach and that refinements will be necessary to further improve the system's efficiency such as possibly the additional use of explosives or the development of special tools such as ice-saws. Other methods of preventing navigation problems may usually be applied on a smaller scale. Among these are the use of air bubbling systems, special methods of cutting and removing ice, etc.

As an example, harbor basins in the port of Stockholm are kept open with an air bubbling system on the harbor bottom along the sides of the piers (Figure 19-4). It should be realized, however, that air bubbling systems are effective only when the flow generated by the system is water of a somewhat higher temperature from the basin depth. In salt water areas, the temperature gradient in vertical direction is usually very small and the air bubbling system consequently is less effective.

A large portion of northwestern Europe is favorably affected by the relatively warm water of the Gulf Stream, which runs along the east coast of the United States from Florida to Europe. It crosses the Atlantic Ocean at about 40° latitude and helps countries in northwest Europe keep their harbors and sea lanes open during the winter season. Even harbors at high latitudes benefit from it, such as the port of Hammerfest in northern Norway, which is open all year. Most Norwegian fiords which form many natural harbors are ice free.

Opposite the Scandinavian peninsula, the northern part of the Baltic is under less favorable conditions. This sea has a low salinity and is frozen during most of the year. The southern part of the Baltic Sea is in open connection with the North Sea through the Skagerak. This part is usually navigable, except in severe winters when ice conditions can seriously hamper navigation.

A ferry boat ride from the northern part of Germany to the southern tip of Sweden during a severe winter is a rewarding experience. During this trip the big ferry boat plows through the heavy ice most of its way.

The economic significance of the problem of keeping sea lanes open has created a great interest in methods and techniques for solving this problem. The ice breaking techniques used in the United States and Canada have also been widely used by western Europe countries and Russia. Ice engineering plays a major role in Russia, as may be easily understood. The Russians are studying ice problems in both the laboratory and in the field and have developed powerful icebreakers.

In the Netherlands, ice breaking on rivers began in 1861 (8). The icebreakers were primarily used to break up ice fields and ice jams in securing the discharge of river flow toward the sea after severe winter conditions. The first ice breakers used on the Dutch rivers were steamboats; paddle wheel boats did a better job than boats with screw propulsion. Further experiments with icebreakers were carried out in the winters of 1875, 1879, 1880 and 1890. In the winter of 1890-91 the scale of operations was increased; of nine steamboats put into action, three were paddle wheel boats and six had screw propulsion. In the official report, however, the judgment on performance was still rather negative and the resultant effects of this operation seemed to have been rather disappointing.

In the severe winter of 1928-29 more powerful boats were put into operation and the results were indeed more successful. A similar experience during the winter of 1939-40 confirmed that for adequate ice breaking, under certain conditions, a minimum amount of horsepower is required. A relationship exists between ice thickness, the speed of the icebreaker crushing the ice and the amount of horsepower needed, as is demonstrated by Figure 19-5. The dotted lines are extrapolations based on observed data.

Only after 1940 were special requirements for ice breaking performance added to the ship. They consisted mainly of a uniquely designed bow, a stronger hull and adapted machinery and propulsion.

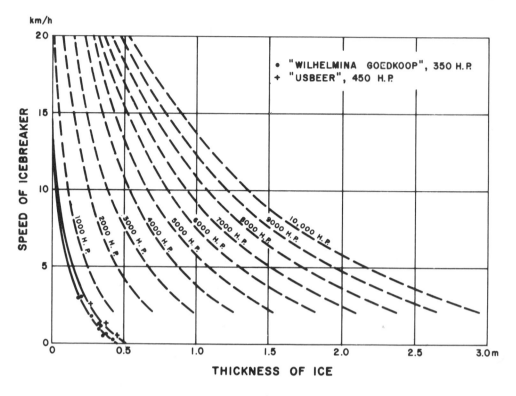

Figure 19-5. Relationship between speed of crushing ice, ice thickness and horsepower.

The breaking of ice for maintaining navigable conditions was essential for the ports of Amsterdam and Rotterdam. In 1929 several ice committees were organized to consider the possibilities of keeping navigation going as long as possible.

The water in Rotterdam Harbor is brackish and has a varying salinity with the tide phase. Therefore, chances for the formation of a solid ice deck are very slim, but the possibility of ice floes freezing together into large and heavy units does occur during severe winter conditions, and at that time the use of icebreakers has proven useful and economical. In the winter of 1962-63 a total of 1500 ice break hours were utilized to aid navigation in Rotterdam Harbor. The more powerful boats in the inland connection with Germany and also the powerful and adequately shaped push bows have been a great help in stimulating navigation traffic on the inland shipping routes.

Figures 19-6 and 19-7 show ice breaking action on Dutch rivers.

Effects of Ice Conditions on River Discharge

Ice conditions on rivers can seriously hamper the discharge of runoff to sea, particularly after a severe winter when thawing sets in abruptly. This type of condition has given rise to serious floodings in many cases and should be approached with great care. As an example, the conditions that prevail on the Dutch rivers and estuaries will be discussed here, but it should be noted that this type of problem is of a general nature and occurs in many other areas as well.

Climatic Conditions in Holland

Before entering into a discussion of the ice problems on the Dutch rivers, it will be necessary

Figure 19-6. Ice breaking on Dutch rivers.

Figure 19-7. Ice breaking on Dutch rivers.

to say a few words about the climatic conditions in the winter.

Holland, as one of the West European countries, is situated on the borders of the North-Sea and is therefore generally under the influence of the Gulf stream. This makes most winters in Holland rather mild, as far as freezing is concerned. At the same time, however, the country borders to the large continental mass of Germany and Russia to the East. During certain winters the high pressure area above the continental mass extends itself beyond the border of the North-Sea, taking Holland in the grip of severe winter conditions.

The above described situation gives the Netherlands a continental winter every now and then with prevailing easterly winds and temperatures far below the freezing level. On the average, only during one out of five winters does the ice cover in the rivers become landfast (8). The average monthly temperatures as well as the maximum and minimum values are presented in Figure 19-8 (9). The lowest temperature recorded in the past was −27.4°C in 1942.

The Dutch River and Estuary System

Three rivers discharge their runoff towards the sea through the Dutch estuarine system via the Scheldt, the Meuss (Maas) and the Rhine (Figure 19-9). Of these three rivers, the Rhine is by far the most important. The source can be found in the mountains of Switzerland; its average discharge is about 2000 meters^3/second, but under certain conditions this value can rise sharply (up to 13,000 meters^3/second).

Not far from the eastern border the river Waal branches from the Rhine in westerly direction; near Arnhem the river Yssel carries part of the Rhine water to the Yssel lake (formerly Zuiderzee).

The Rhine, Waal and Maas rivers are each discharging through the delta area in the southwest part of the country. During severe winters the main concern is eliminating restrictions for the discharge of river water due to ice formation in the Rhine and Waal rivers. Many times in the past, ice conditions on these rivers have created hazardous high water conditions often leading to a break of

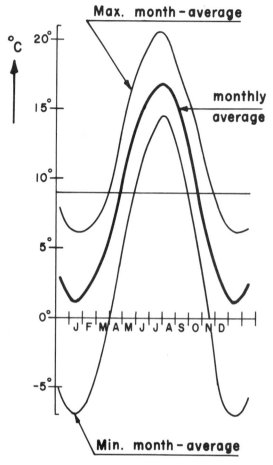

Figure 19-8. Average, maximum and minimum temperatures in Holland.

the river dikes and the flooding of large areas, causing a great deal of damage.

Ice Formation on Rivers

Before discussing the conditions that prevail on the Dutch rivers, let us first look at the way ice is formed. Ice formation on rivers is distinctly different from lakes or canals, where currents are very weak or totally absent. In the latter case, when air temperature drops below the freezing point, the body of water in the lake or canal loses heat to the air due to convection, evaporation and radiation. This leads to lowering of the water tem-

perature. In this process the water near the surface may get considerably below the freezing point before an ice sheet near the surface actually starts forming. Heat losses are compensated by sun warmth heat supply through the bottom and by water turbulence. Heat loss from the water to the air is considerably reduced by the formation of an ice sheet, which works as a protective shield against heat losses.

The process of fresh water freezing is also affected by the fact that water of $4°C$ has its greatest density, so the water cooled at the surface, e.g., to $-0°C$, does not have a tendency to sink to the bottom. On the contrary, because of the lesser density of the upper layers the system of water layers is a stable one.

If the water velocity exceeds a certain value, the process of ice formation will be different from the formation of a continuous ice sheet. Due to water turbulence, the vertical temperature gradient will tend to disappear; although some supercooling may occur (in the case of very clean water), ice crystals will soon be formed throughout the fluid, where solid particles, like suspended matter, will function as crystallization centers.

Two mechanisms of the forming of a continuous ice cover on a river can be distinguished: *the frazil ice evolution process* and the *border ice process* (10). In the first process, frazil particles are formed, agglomerate into flocks and come to the surface to grow into floes; these floes grow together to form an ice deck in cases of low-velocity flow.

Above a critical velocity, an ice cover fed by ice pieces will stop to progress upstream and a hanging ice dam will be formed. A criterion for this condition as given by Kivisild will be discussed later in this chapter.

A study on border ice growth is given by Devik (11). This process includes the formation of an ice cover from the borders of the river, developing towards the center. After an ice cover forms on a river, its continued growth depends on the thermal exchange with the atmosphere. The process of ice formation with or without the formation of an ice cover is demonstrated by Figure 19-10 (12).

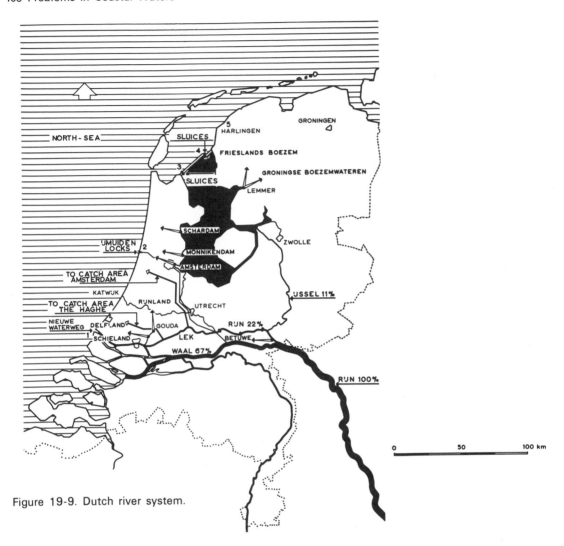

Figure 19-9. Dutch river system.

Whether or not an ice-coat can be formed depends on the water surface velocity and the air temperature below the freezing point, as shown in this figure.

Below a maximum velocity $V_{A\ max}$ an ice-coat will be formed that will stay. Between $V_{A\ max}$ and $V_{B\ max}$ frazil ice can be formed, but no ice-coat; however, an earlier formed ice-coat can stay. In zone C frazil ice can be formed but no ice-coat can remain. In zone D no ice can be formed at all.

On a river an ice cover usually starts when the velocity is too low to enable the formation of an ice bridge (13). The cover then progresses by the packing of ice floes against the ice bridge.

Although in theory the progression of an ice cover consisting of a thin sheet is feasible, in practice the progression mechanism of an ice cover takes place rather by the piling up of ice floes. Ice floes may shift underneath the previous ones and pile up under the cover until it is thick enough to be stable and progress upstream. The cover then progresses with a well-determined minimum thickness. If the river is neither shallow or rough, the thickness of the upstream edge t follows from the expression (13):

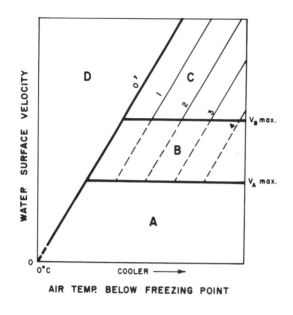

Zone A   Ice - coat can be formed and stay.

Zone B   Frazil ice can be formed but no ice-coat
          an earlier ice-coat can stay.

Zone C   As B, but earlier ice - coat cannot stay
          for longer time.

Zone D   No ice can be formed or stay.

y
0, 1, 2, 3, 4 indicate quantity of frazil ice
production per unit of time.

Figure 19-10. Schematic diagram of ice formation (from Swedish State Power Board Paper 25-5A-8).

$$\frac{V}{2gh} = \frac{\rho - \rho^1}{\rho^1} \quad \frac{t}{h} \quad (1 - \frac{t}{h})$$

$$\surd / \surd \ (2gh) = \surd \ [(\rho - \rho')\ (t/h)\ (1 - t/h)]$$

in which λ is the mean water depth and $\surd$ the corresponding water velocity. ρ and ρ^1 are the densities for water, respectively ice.

For each value of the average velocity, V, two solutions to the equation exist, but only the one corresponding to the lowest value of t is physically possible. According to this process, rivers can be covered with ice due to stream-upward progression of the ice cover. On the Dutch rivers, the growth rate of an ice cover ranges from about 9-36 kilometers per 24 hours. An average figure is about 1 kilometer per hour.

The process of ice cover formation as described above has an upper boundary. If the upstream velocity V exceeds a critical value V_c, then the ice floes carried with the river no longer stay on the surface but are washed under the ice cover

to form an "ice jam" or, as some authors call it, a "hanging ice dam" (14).

Investigations by Hans R. Kivisild (14) on many Canadian rivers have shown that the Froude number at the upstream end is a likely criterion for determining whether ice running down the river will be attached to the upstream end or will be carried underneath the ice cover that is already formed.

Kivisild's observations show that an average critical Froude number is about $F_c = V_c/\surd(gh) = 0.08$. As the Froude number approaches this value, the pack of ice tends to become more massive.

Returning to Dutch river conditions we are able to observe several of the phenomena mentioned above with respect to ice formation on the upper parts of the Rhine and Wall rivers. It can be observed that in certain areas like sharp bends and near bridges an ice bridge is likely to start forming.

In order to check Kivisild's criterion with respect to a critical Froude number the Dutch rivers have indeed confirmed the value of F_c suggested by Kivisild, as demonstrated in Figure 19-11. The

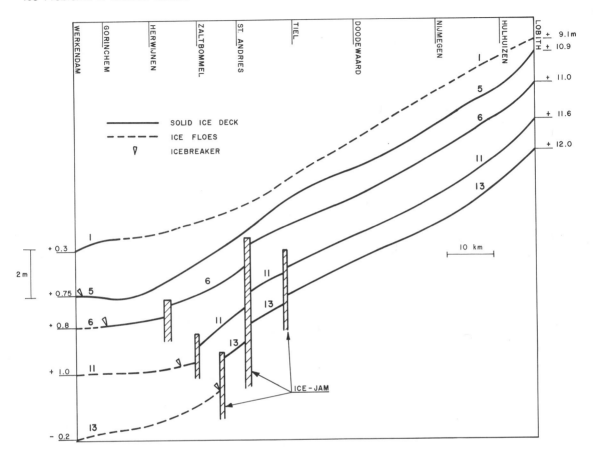

Figure 19-11. Ice jams on Waal River as demonstrated by discontinuities in slope of water level.

criterion does not hold for the lower portion of the river where water motion is tidal and velocities fluctuate between a positive and negative value.

A careful analysis of observed water levels along the river made it possible to determine water surface lines with reference to distance. When the Froude number exceeded the value of 0.08 the water level was likely to make a jump demonstrating the existence of locally increased velocities caused by a decrease in available flow area, in this way confirming Kivisild's criterion.

Both a continuous ice sheet on the river and ice jams will jeopardize the discharge of runoff, in case thawing conditions are setting in, in the river's upper regions. In both cases the hydraulic resist-

ance of the river will be considerably increased so that the discharge-stage relationship is adversely affected and water rises to dangerous levels. It is evident that all available technical means will be required to prevent the occurrence of such conditions. Possibilities for preventing such undesired conditions are as follows:

1. The use of icebreakers;
2. The use of explosives, particularly for blowing up ice dams (15);
3. Taking advantage of heat supply to river sections by using warmed cooling water from power plants.

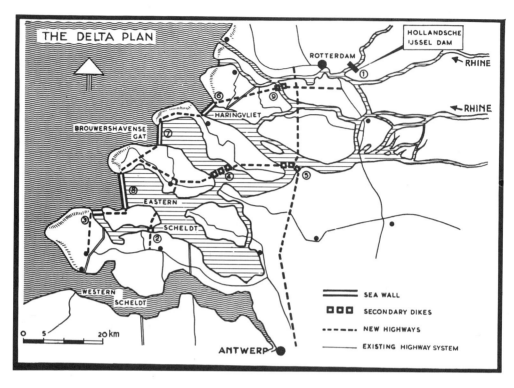

Figure 19-12. Delta Project.

Breaking the ice pack with icebreakers is done from the lower part of the river upwards so that ice floes, being separated from the pack, can be discharged toward the sea.

The increased use of river water for cooling in thermal power plants has increased the water temperature in winter by 1-2°. This has contributed considerably to lesser ice conditions during recent severe winters.

Influence of Delta Project

To protect the low-lying parts of southwest Holland against high storm tides, the Delta Project is being carried out (Figure 19-12). It encompasses the closure of a number of large estuaries by building a dam from one bank to the other. In the Haringvliet Dam a large sluice is being built to enable the discharge of part of the Rhine runoff in cases of high river flow (Figure 19-13). An important consequence of the project has been the for-

mation of a large fresh water lake (Zeeland Lake) south of the secondary dam through the Volkerak.

It is evident that such a major change in hydraulic conditions is likely to affect the discharge conditions of ice (16). In the past the Volkerak channel was a major route for the discharge of ice toward the sea due to prevailing easterly and northeasterly winds, but it is now no longer possible to use this discharge route. Consequently, ice has to be discharged through the 17 openings of the Haringvliet Sluice, about 1 kilometer long (Figures 19-14 and 19-15), or through the estuary sections near Rotterdam. Water circulation studies on the Haringvliet have made clear that the wide Haringvliet Estuary will not be the best way to discharge ice because of high ice productions on the Haringvliet itself, which has fresh water and long periods of low velocities.

Measures to improve discharge of ice through the Haringvliet could be the following:

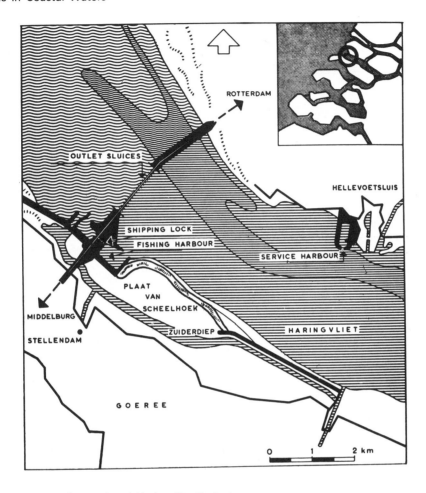

Figure 19-13. Layout of completed Haringvliet Project.

1. Opening the sluices during ice conditions and reestablishing a tidal motion (sea water) on the Haringvliet;
2. Opening the sluices during one or more ebb periods, thereby increasing the drift velocity in seaward direction.

The first possibility did not create sufficient relief to justify the temporary reestablishment of a salt water estuary in the Haringvliet (with all its disadvantages). The second solution was not accepted because manipulating the sluices in the suggested way would intolerably increase salt wa-ter conditions on the Rotterdam waterway, there-by spoiling fresh water intake conditions in the larger part of the estuary system. The best method was to discharge ice through the northern branches of the system, for instance, Nieuwe Waterway and Oude Maas (see Figure 19-16).

Calculations have been carried out by means of an electric analog computer (representing the entire estuary system) to determine whether the discharge of ice along the northern branches is a hydraulic feasibility or whether other ways should be found to cope with the current problems. The computations have shown that ice discharge along

Figure 19-14. Dam under construction in Haringvliet Estuary.

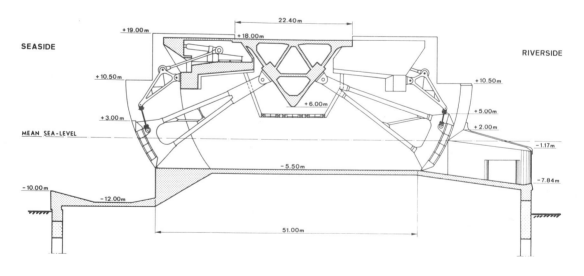

Figure 19-15. Cross-section of Haringvliet Sluice.

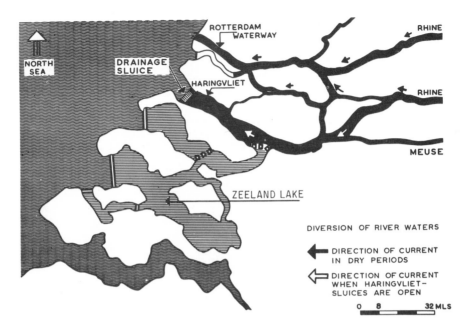

Figure 19-16. Discharge of ice after completion of Delta Project.

the New Waterway seems to be a feasible solution, although less desirable from the standpoint of navigation. Discharge along the routes of Lower Merwede, Oude Mass and New Waterway is less promising because of small resultant ebb drift velocities in the neighborhood of Dordrecht. Further studies and full-scale experimentation will be required to arrive at a satisfactory solution of the problem.

References

1. Land, H.L. 1959. Ice problems in the St. Lawrence River. *Proceedings 8th Congress of the IAHR*, Montreal, Seminar No. 1 (Paper 21-SI).
2. Gruber, Michael. 1970. The new Northwest Passage, *Sea Frontiers* 10(1). (January-February).
3. Anonymous. 1969. Northwest Passage, what could it mean. *Ocean Industry* 4(7) (July).
4. ____. 1969. *Manhattan* begins battering voyage up Northwest Passage. *Ocean Industry*, 4(9) (September).
5. ____. 1969. Retrofitting the *Manhattan* for the Arctic. *Ocean Industry* 4(11) (November).
6. Baker, Charles H. 1969. Voyage of the *Manhattan*. *Ocean Industry* 4(8) (August).
7. Chabot, Paul L. and Peterson, Robert. 1969. Northwest Passage, choosing the route through the ice. *Ocean Industry* 4(8) (August).
8. Oudshoorn, H.M. 1970. *Ice cover formation and associated hydrodynamic effects in the lower part of the river Rhine.* Ice symposium, IAHR. Reykjavik.
9. Wemelsfelder, P.J. 1956. Research concerning the calorie balance in a river during a frost period. *De Ingenieur* 8. Bouw—en Waterbouwkunde 4.
10. Michel, B. 1966. *Ice covers in rivers.* Proceedings of a conference held at Laval University, Quebec, Nov. 10-11. National Research Council of Canada, Tech. Mem. 92, compiled by L.W. Gold and G.P. Williams, Appendix IV (C).
11. Devik, O. 1964. Present experience on ice problems connected with the utilization of

water power in Norway. *Meddelelsenr* 12, fra, Hydrologisk Avdeling.

12. Scherman, Karl Arthur. 1959. On ice difficulties of open water courses of hydro-electric plants. *Proceedings 8th Congress of the IAHR*, Montreal, Seminar 1 (Paper 25-SI).

13. Pariset, E. and Hausser, R. 1959. Formation of ice covers on rivers. *Proceedings 8th Congress of the IAHR*, Montreal, Seminar 1 (Paper 3-SI).

14. Kivisild, Hans R. 1959. Hanging ice dams. *Proceedings 8th Congress of the IAHR*, Montreal, Seminar 1 (Paper 1-SI).

15. Kley, J. Van der. 1965. The use of explosives for clearing ice. *Communications, Rijkswaterstaat* 7.

16. Santema, P. and Svasek, J.N. *The effect of the damming-up of the tidal estuaries on the ice in the rivers of Zeeland and Zuid Holland*. Report of the Delta Committee, Part V, Contribution IV-5.

20

the effects of ice on hydraulic structures

Introduction

The forces exerted on structures by ice are very much dependent on the pertinent characteristics of the ice. First, we will briefly review the structural properties of ice and then discuss the effect on hydraulic structures. The problems are very much complicated and we are only in the first stage of comprehension. Much more research is necessary to clear up a number of uncertainties.

The engineering problems concerning the effect of ice on structures will be dealt with in a few examples, such as:

1. Forces on bridge piers;
2. Forces on piles and pile structures;
3. Design of a small harbor in a sea ice area;
4. Measures to keep hydraulic structures operating;
5. Design of the Haringvliet Sluice in the Delta Project with reference to ice.

Ice Formation

The behavior of sea ice is different from fresh water ice and needs special consideration. In Chapter 19 we briefly touched on ice formation in fresh water. We distinguished between lakes and rivers (in relation to the water velocity) and found that surface ice can be formed by cooling of the surface layers and by the formation of ice crystals in the somewhat supercooled water of a running river.

The presence of frazil ice in flowing water may have important consequences for hydraulic structures such as power plants because it may create serious obstructions and endanger machinery operation. One will always attempt to create a solid ice cover upstream of the power plant to avoid the effect of large amounts of frazil ice dams at the plant. A solid ice sheet, however, creates problems of a different nature because of the forces that may be exerted on the structure.

We mentioned in Chapter 19 that fresh water has its highest density at $4°C$, so surface layers being cooled by the colder air tend to remain on the surface if they are not mixed with deeper water due to turbulence or circulation. In sea water the process of ice formation is different. For a salinity above 24.7%, the density of sea water increases with lower temperatures, whereby surface water sinks down and warmer water from deeper parts rises. This circulation prevents surface layer freezing when the deeper water has a higher temperature. The temperature at which saline water transforms into ice depends on the salinity content. A higher salinity gives a lower freezing point, as shown in Table 20-1.

When freezing begins, very small, colloidal, dish-shaped ice particles appear that develop into hexagonal bars (ice crystals) 1-2 centimeters long,

Table 20-1
Effect of Salinity on Freezing Point

Salinity	$0°/^\infty$	$10°/^\infty$	$30°/^\infty$	$35°/^\infty$
Freezing point ($°C$)	0	−0.53	−1.08	−1.91

which give the water surface a dull look. These ice crystals do not contain salt, but the water density in the immediate vicinity increases and sinks. Finally, a network of ice crystals encloses a certain amount of salt water. A rapid decrease in temperature gives rise to a greater salt content than a slower process.

A representative horizontal cross section of ice crystals is shown in Figure 20-1 (1). Between the ice crystals we find wedge-shaped regions of high salinity content called liquid brine. The relative amount of entrapped brine is consequently dependent on the ice sheet growth rate. The mechanism of brine migration, although not completely understood, significantly affects the characteristics of the ice being formed. The amount of salt in sea ice decreases with time. (Some sea ice even becomes sufficiently salt free after the first winter to produce potable water.)

The salinity of the ice is then dependent on both the depth in the ice sheet and on time. The inclusion of brine diminishes the mechanical strength of sea ice.

Strength of Ice

Ice is crystalline material. When loads are applied to it, the material shows both elastic and plastic behavior. The plastic behavior is directly related to the crystal structure. It is usually built up of long, pencil-like grains and is then called columnar ice. In snow ice the crystals have more of a granular structure and this is called granular ice. Crystalline materials like ice deform plastically by slip on preferred crystallographic planes. The plane perpendicular to the axis of hexagonal symmetry of the ice crystal (the so-called C axis) is

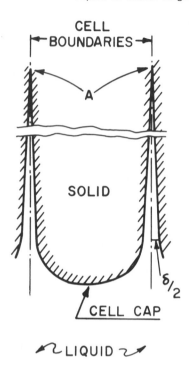

Figure 20-1. Ice crystal in salt water (after Tiller, 1963).

called the basal plane; it plays a very important role in the plastic deformation (2). Slip in ice occurs most easily on the basal plane. In other planes much higher shear stresses are required to cause deformation, depending on the value of the shear stress.

Because of the combination of plastic and elastic properties, the stresses induced in ice masses are dependent on time-load relationships. This has a considerable influence on the test results that have been applied and accounts for a great deal of wide scatter in values given by different authors for the elastic constants of ice. Furthermore, both the elastic and plastic properties of ice are time dependent. Because the formation of ice is also dependent on weather conditions, we may expect to trace some ice properties back to changes in these conditions.

Elastic properties are determined by the value of the elastic modulus E. As observed before, the

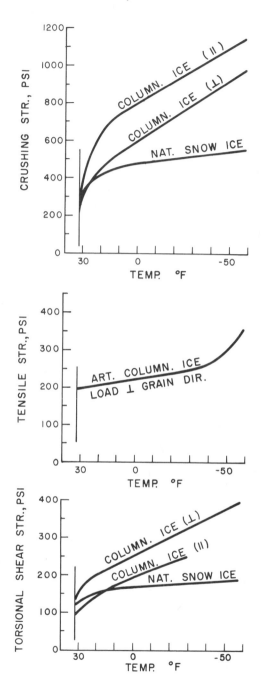

Figure 20-2. Elastic properties of ice as a function of temperature (after Butkovich, 1954).

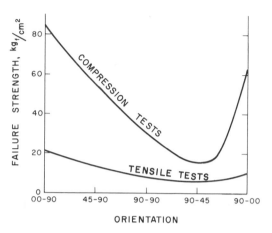

Figure 20-3. Dependency of strength on crystal orientation (after Peyton, 1966).

value of E will be dependent on the loading characteristics. The elastic properties prevail when no time is given to plastic deformation. For periods of loading less than about 10 seconds, and depending on the temperature, the deformation of ice is almost entirely recoverable and can be considered elastic. For periods greater than 10 seconds the ice will undergo plastic deformation (3). The value of the "dynamic" E can be obtained by studying dynamic effects of small deflection such as the propagation of a shock wave through an ice cover.

Butkovich (1958, as reported by Weeks and Assur in ref. 5) recommended that when the strength characteristics of ice are determined by a loading experiment, the rate of applying the load should be greater than 7 psi/second. This has been adopted as a standard for most tests since that time. There is evidence, however, that for uniaxial compression tests the rate of applying the load should be even higher (e.g., 20 psi/second) to be independent of loading rate.

Butkovich (4) observed earlier (1954) that the compressive, shear and tensile strength increases with decreasing temperatures, as shown in Figure 20-2.

Peyton, as reported by Weeks and Assur (5), has shown that the orientation of crystals in the ice has a very significant influence on both compression and tensile strength (Figure 20-3). Such

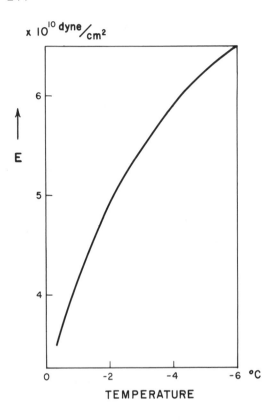

Figure 20-4. Relation between elastic modulus E and temperature (after Peschanskii, 1960).

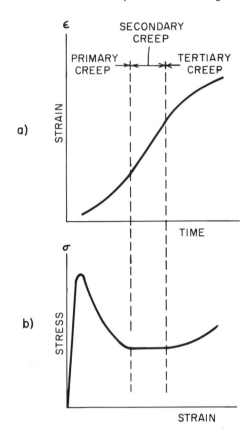

Figure 20-5. Creep test and constant strain rate test (after Krausz, 1966).

could be expected from the previously mentioned characteristics that the direction of the basal plane in the ice crystals with respect to force direction is a vital parameter. If the force acts perpendicular or parallel to the basal plane, shear stresses in this plane tend to disappear giving a high resistance to crushing. Figure 20-4 shows the dependency of the elastic modulus E on temperature.

Most studies on the plastic behavior of ice have been made under conditions of constant load and for strain exceeding 1-2%. Laboratory experiments on the plastic behavior of ice show that the dependence of strain rate on stress can be represented by the expression (2): $\dot{\epsilon} = A\sigma^n$ in which $\dot{\epsilon}$ is strain rate and σ, n, are time- and temperature-dependent constants.

Figure 20-5 shows relationships between creep

test (*a*) and constant strain rate test (*b*) for granular-grained ice.

An understanding of the plastic behavior of ice is vitally important in designing field measurements and to the engineering evaluation of these data. When ice has been subject to deformation some of its strength characteristics may decrease.

Strength of Sea Ice

Strength of sea ice can be expressed in terms of the brine content of the ice. It has been shown that $\sigma = \sigma_0 [1 - \sqrt{(v/c)}]$ in which σ is strength of sea ice, σ_0 is basic strength at zero brine volume, v is relative brine volume and c is a constant involving geometric parameters of brine channels.

The crystal characteristics and orientation give different types of ice with different characteristics, also depending on salinity content. Similar to fresh water ice, sea ice is a vico-elastic material—sustained loading produces time-dependent deformation.

Forces Exerted by Ice

The forces exerted by ice on structures depend on a great many variables, among which the most important are (6):

1. Movement of ice cover relative to the structure;
2. Flexibility of the structure;
3. Characteristics of the ice cover and
4. Properties of the ice.

The first factor plays an essential role in large ice fields on lakes and reservoirs, where movements of large areas of ice may take place, often due to unpredictable circumstances. When a structure is in the middle of such an unbroken ice area, ice field movements may cause great forces on it. Maximum forces will probably be found by multiplying the maximum crushing pressure of ice by the area of contact between the ice sheet and the structural elements upon which the ice force is working. The total forces are also very much influenced by the flexibility characteristics of the structure on which the ice force is working. This is similar to the problem of forces on structures in general.

Analysis shows that ice cut by a structure produces a certain frequency of the pressure fluctuations. A hazardous situation occurs when the frequency of the pressure fluctuations corresponds to the natural frequency of the structural member so that high stresses in these members can be generated.

Ice cover characteristics are related to the way the ice cover was formed; it makes a difference whether a structure is affected by a solid ice sheet of homogeneous properties or whether a cover of irregular ice floes has frozen together in an irregu-

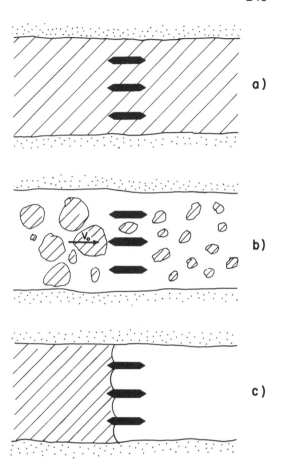

Figure 20-6. Different load situation for bridge piers.

lar way. The thickness, extent and movement of solid ice covers and the size, thickness and speed of ice floes are predominant parameters with reference to forces.

Ice Forces Exerted on Bridge Piers

Forces working on bridge piers caused by the presence of ice on rivers can be distinguished as follows:

1. Forces caused by thermal expansion of an ice field; these forces not only work in the direction of the river flow but perpendicular to the main direction of the piers as well;

2. Forces exerted by floating ice fields hitting the pier;
3. Forces exerted by a closed ice deck, bridged upstream of piers; and
4. Forces due to ice jam formed against bridge piers.

Figure 20-6 presents different load situations.

Forces Due to Thermal Expansion

An ice sample, being warmed, exerts pressure against its confined ends. Similarly, ice sheets in nature are subject to the same process if a temperature rise penetrates into the ice as a temperature wave from the surface. The maximum force exerted by the ice sheet may be limited due to failure by buckling.

Applying the results of end-constrained ice samples to practical problems we may use the relation (7): $\sigma = \alpha E \theta' t$ in which σ is compressive stress, E is elastic modulus, α is the linear thermal expansion coefficient, $\theta' = (d\theta/dt)$ is rate of warming and t is time. The viscous characteristics of the ice limit the maximum stresses, however, under prolonged loading.

The thermal expansion can create a thrust on structures like bridge piers. Values for the maximum amount of thrust to be reckoned with vary considerably. Assur has developed a mathematical theory which seems to agree with experimental results.

Values used by Canadian and Russian ice engineers vary considerably, but forces of 60-80 ton force/meter² may be the right order of magnitude.

Considering bridge piers the thermal force may do most harm by working against the long side of the pier (in a direction perpendicular to the flow).

Forces Due to Moving Ice Floes

With respect to the action of moving ice flows on piers, much field research was done by Russian scientists between 1934 and 1958, as reported by Korzhavin (8). It was found that ice floes carried

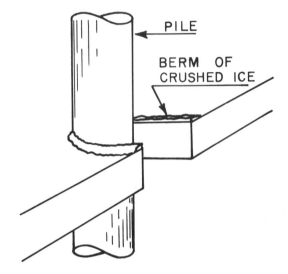

Figure 20-7. Cutting of ice by pile (after Peyton).

by the flow were cut by the piers (Figure 20-7) and that an inclined nose facilitates the breaking up. The cut had the same width as the pier.

As the rate of deformation increases, the compressive strength decreases, according to the expression

$$R_{compr.} = \frac{R_1}{\sqrt[3]{V_O}}$$

V_O being the ice floe velocity and R_1 the average ice pressure at $V_O = 1$ meter/second. This formula is valid for values of $V_O > 0.1$ meter/second. For R_1, a value of 50 ton force/meter² is indicated, which seems rather low compared to an average compressive strength of ice of 280 ton force/meter².

However, local compression due to pier penetration in the ice floe was found to increase interaction forces two to two and a half times. A form factor (m) for a pier with a triangular form was suggested, as given by the expression $m = 0.39 (1 + \alpha)$, α being one-half of the pier nose angle.

In Canada, design engineers have been using the average crushing strength (280 ton force/

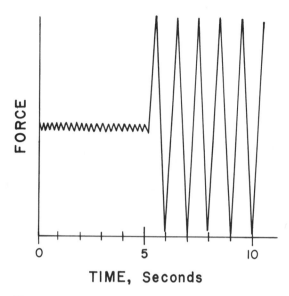

Figure 20-8. Oscillating ice forces on pile (after Peyton).

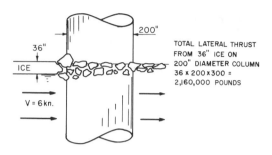

Figure 20-9. Ice crushed against pile (after Peyton).

meter^2) as a basis for their design of highway bridges across rivers (9). Field measurements, obtained by pressure devices into two bridge piers over a four-year period have yielded unit pressures from moving ice-sheets in the range of 70-113 ton force/meter^2 (7 to 11.3 kgf/cm^2) for ordinary to moderate conditions (10). The value of 280 ton force/meter^2 specified in Canadian codes may therefore be too conservative.

This criterion presumably gives a safe upper limit to the load, but the amount of safety is unknown and probably unnecessarily large compared with the Russian figures.

Some Canadian designers have accepted the values used in dam design, ton force/meter. With an average thickness of 2 feet (0.6 meter), this corresponds to 50 ton force/meter^2, which comes close to the figures mentioned in the Russian publication and also corresponds to figures based on thermal expansion.

Forces Due to an Ice Deck Formed Upwards of the Piers

The forces exerted on piers when ice floes

have formed a continuous ice sheet upstream from the bridge are dependent on:

1. The size of the continuous ice sheet;
2. Roughness of underside ice deck and
3. Hydraulic conditions on the river (bottom roughness, depth, slope).

If the ice deck connects the banks, the latter may exert shearing forces on the ice deck and reduce the forces on the piers.

Forces on the piers become maximum when the ice is loose from the banks. They can be computed based on the friction that the water flow underneath the ice deck exerts on the ice deck.

Local Ice Jams

The presence of bridge piers in a river may give rise to the formation of a local ice jam (as explained in Chapter 19). In that case a head difference may be formed, exerting an additional horizontal force on the hanging ice dam and consequently on the piers.

In a given situation the conditions should be analyzed to determine whether or not such an ice dam can be formed or whether its formation can be corrected by carrying out preventive measures.

Forces on Piles or Pile Structures

As an example let us consider the problems associated with the erection of oil platforms in

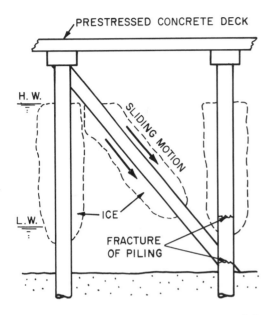

Figure 20-10. Failure in dock construction (after Peyton).

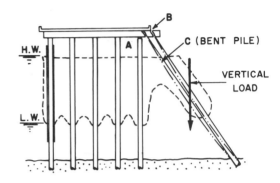

Figure 20-11. Failure in dock construction (after Peyton).

Alaskan coastal waters in connection with the recent discovery of oil in this area. Problems of this type have been described by Peyton (11).

Recently, oil platforms were constructed in Cook Inlet, near Anchorage, Alaska. Tidal range in this area is 35 feet or more and tidal velocities are in the order of 6 knots (3 meter/second). The predominant forces are generated by large ice floes that hit the platform legs. Since these platforms are in salt water, we will have to consider the properties of sea ice.

Peak forces from slowly moving ice fields on structures such as piles have their greatest value at low velocities of the ice floes, as discussed before. The peak strength (and consequently force) occurs at speeds which are not even visually perceptible.

Observations of loadings on test piles have shown that slow speeds not only cause greater forces but also generate oscillations in the forces so that maximum forces are considerably increased. It was found that the oscillating ice failing frequency was independent of the resonant frequency of the structure. In Cook Inlet, the oscilla-

tion frequency was about one per second (Figure 20-8).

Ice floats affecting oil platforms in Cook Inlet can be over 1 mile in diameter and over 3 feet thick. When such a floe hits a pile, the ice is crushed (Figure 20-9) and a large force is exerted on the pile. A rough approximation of the crushing strength is 300 psi for sea ice in this area. A 36-inch ice sheet acting on a 200-inch diameter column gives an ice force of 2,160,000 pounds per pile. On top of everything, this force may oscillate from zero to full load at one cycle per second.

Types of pile structure failures under ice conditions in coastal waters are depicted in Figures 20-10 and 20-11. In both cases, the use of batter piling was responsible for the failures that occurred.

Uplift Forces

Piles and pile structures (such as docks) are not only subjected to the effect of horizontal forces, they are also affected by vertical forces due to the freezing of surface ice onto the structure in tidal zones. Because of rising of the water level, vertical uplift forces are acting on the piles; these forces are counteracted by the weight of the structure and the friction in the bottom.

In cases of single piles and large tidal range the vertical uplift force may tend to lift the structure

Above. Figure 20-12. Ice grown on concrete piling.

out of the soil. Usually the ice breaks, however, in the vicinity of the pile. After a certain time, a heavy ice layer may cover the piles, as shown in Figure 20-12.

The previously mentioned examples show that cloggings of ice may create additional forces when ice gets loose from the pile after thawing has set in.

Small Harbor Design (Layout) in Coastal Areas with Sea Ice

An example of harbor design in an arctic zone is presented (Figure 20-13), from which the following steps can be distinguished:

1. A closed harbor (*a*) will not work because a solid ice cover will form inside;

Below. Figure 20-13. Designing small boat harbor under ice condition.

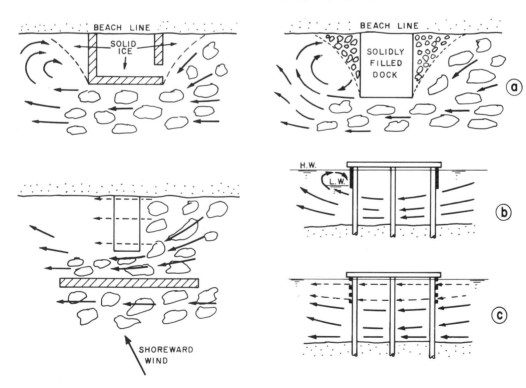

Figure 20-14. Heated side seal in Haringvliet Sluice.

Figure 20-15. Measures for keeping gates in operating condition during severe winters: electrically heated steel plates in Sweden.

2. A solidly filled dock (*b*) will be somewhat better, but boats will not be very protected;
3. An open dock with solid skirting (*c*) gives additional leeway, although the vertical eddy due to the tidal current will tend to keep ice near the dock;
4. An open dock with permeable skirting (*d*) and a protective ice barrier (*e*) may give a solution for both ebb and flood flow.

Ice Problems in the Haringvliet Sluice Design (Delta Plan in the Netherlands)

As mentioned in Chapter 19, discharge of ice from the Dutch rivers toward the North Sea is an important problem which must be given serious consideration.

The Haringvliet Sluice (Figure 20-14) performs an important function in the discharge of water (runoff) and ice after the spring break-up and was specially designed to accommodate its purpose. The following problems had to be considered:

1. Determination of an acceptable width between piers so as to stay within economic limits and still perform its function properly;
2. Sufficient vertical clearance to operate icebreakers in the sluice;
3. The design of an icebreaker nose on the river piers;
4. The necessity of designing protective bottom aprons on both sides of the sluice to protect the bottom both during ebb and flood flow. During ice winters it was considered possible to reopen the sluice gates to reestablish tidal flow on the estuary;
5. To prevent ice formation on the gates, inside the gates and on the side seals;
6. To keep at least one set of gates in operating condition during a severe winter.

As a result of a program of studies and investigation, which also included the use of ice laboratory experiments, construction of an electric

Figure 20-16. Measures for keeping gates in operating condition during severe winters: heating side seals in Sweden by means of infared lamps.

analogous model and mathematical computations, the Haringvliet Sluice was provided with a heating system for the side seals of the river side gates. Additional measures such as heating the inside of the river side gate and air bubbling systems were studied but were not found necessary because of the great structural strength of the gate elements and of the suspension rods. To study solutions to these types of problems, the author and two colleagues of the Dutch Ministry of Roads and Waterways made a trip to Sweden, where severe freezing conditions are experienced every winter.

Figure 20-15 through 20-17 show some of the solutions that are used in Sweden to keep gates of power plants in operating condition. They include the use of heated side seals and heated strips on pier structures, heating the inner sides of gates, use of circulating pumps and air bubble systems. They served as important guides for designing the final solution for the Haringvliet Sluice.

Figure 20-17. Measures for keeping gates in operating condition during severe winters: circulation of water by means of flygt pumps (Sweden).

References

1. Tiller, W.A. 1963. *Principles of solidification in the art and science of growing crystals.* New York: Wiley, p. 276-312.
2. Krausz, A.S. 1966. *Plastic deformation of fresh-water ice.* Proceedings of a conference held at Laval University, Quebec, 10-11, November. Tech. Mem. 92, National Research Council of Canada.
3. Gold, L.W. 1966. *Elastic and strength properties of fresh water ice.* Proceedings of a conference held at Laval University, Quebec, 10-11 November. Tech. Mem. 92, National Research Council of Canada.
4. Butkovitch, T.R. 1954. *Ultimate strength of ice.* U.S.A. C.R.R.E.L. Research Report 11, Hannover, New Hampshire.
5. Weeks, W.F. and Assur, A. 1966. *The mechanical properties of sea ice.* Proceedings of a conference held at Laval University, Quebec, 10-11 November. Tech. Mem. 92, National Research Council of Canada.
6. Gold, L.W. 1966. *Deformation and strength properties of ice.* Proceedings of a conference held at Laval University, Quebec, 10-11 November. Tech. Mem. 92, National Research Council of Canada. Appendix III (D).
7. Assur, A. 1959. Maximum lateral pressure exerted by ice sheets. *Proceedings 8th Congress of the IAHR*, Montreal, Seminar 1 (Papers 22-IS).
8. Korzhavin, K.N. 1959. Ice affecting engineering structures on the Siberian rivers during ice-run. *Proceedings 8th Congress of the IAHR*, Montreal (Paper 16-SI).
9. Neill, C.R. 1966. *Bridge piers and similar isolated structures,* Appendix III (A). Proceedings of a conference held at Laval University, Quebec, 10-11 November. Tech. Mem. 92, National Research Council of Canada.
10. _____. 1970. *Ice pressure on bridge piers in Alberta, Canada.* IAHR. Ice Symposium, Reykjavik.
11. Peyton, Harold. 1968. Ice and marine structures. Part 1: The magnitude of ice forces involved in design. *Ocean Industry* 3(3) (March); Part 2: Sea ice properties. *Ocean Industry* 3(9) (September). Part 3: The importance of design alternatives. *Ocean Industry* 3(12) (December).

part 8: li-san hwang

**Behavior of Explosion-Generated Waves
Numerical Modeling of Tsunamis—A Step Toward Forecasting Tsunami Heights in a Warning System**

Dr. Li-San Hwang is Director of the
Engineering Division at Tetra Tech, Inc.,
Pasadena, California. He received his B.S.
degree at the National Taiwan University, his
M.S. at Michigan State University, and his
Ph.D. in Civil Engineering at the California
Institute of Technology in 1965. After gradua-
tion, he joined National Engineering Science
Company in Pasadena, where he became the
Manager of the Hydrodynamics Department.

Dr. Hwang has made significant contributions
in the fields of harbor agitation, tsunamis, and
explosion generated waves. He developed a
numerical model of tsunami generation and
propagation which is capable of deterministic
calculation of the behavior of a major tsunami
over the entire Pacific Basin. Currently, Dr.
Hwang is directing the development of coastal
and harbor pollution-transport models including
the fate of oil spills in an arbitrary environment.
He is the co-author of a specialized handbook of
explosion wave phenomena as well as numerous
technical papers and lectures.

21

behavior of explosion-generated waves

Introduction

This chapter presents a general survey of the surface wave phenomena associated with explosion sources. The treatment will be, of necessity, much abbreviated in details and will attempt, instead, to summarize the diverse features from generation to shoreline interaction in a unified manner; sources listed in the references will provide more extensive treatment of individual aspects. The primary impetus to such study has been, of course, the obvious one of military weapons effects. Beyond this, however, are both the intrinsic interest in explosion wave properties from a hydrodynamic viewpoint and the concern for safety in potential engineering usage of large-yield devices in the marine environment (such as for harbor or canal excavation). These considerations have led to the development of an extensive body of both theoretical and experimental results which will be of ever-increasing value to the coastal engineering community.

The first topic to be covered here will be the generation process. This may be visualized as simply the displacement of a considerable volume of water by an expanding "bubble" of explosion products and water vapor. The energy released by the explosion is partially lost by venting to the atmosphere and by radiation and heating; of the remaining fraction imparted to the fluid, a further portion is lost in the violent turbulent flows associated with the plume and base surge. Not surprisingly then, the energy radiated as surface waves represents a minor fraction of the whole. To develop a deterministic model accounting for the details of generation would clearly be a monumental task. In practice, the complexities of the initial near-source events are ignored and a much simplified model involving only a surface displacement is adopted. This fictitious source is then calibrated with experimental data so that the wave history at a distance is correctly predicted.

The second regime of interest is that of propagation in nonuniform depth. Determination of refraction and shoaling is not straightforward since the wave system is an organized train of groups of dispersive waves. Special procedures are required to follow the system and reconstruct the wave history properly at an arbitrary point.

The third and final regime is that of nearshore transformation and effects. A number of topics are subsumed here including breaking, wave set-up, shelf oscillations, harbor agitation and shoreline run-up. For each, explosion waves demand a somewhat specialized treatment and may exhibit features not normally encountered in coastal engineering applications.

An indication of the relation of explosion waves to other wave classes of coastal engineering interest is given by consideration of periods, heights and organization. The frequencies associated with a large explosion are typically lower

Figure 21-1. Plume and base surge resulting from 125 pounds. TNT exploded near the surface at Lake Ouachita, Arkansas (courtesy of Waterways Experiment Station).

than those of significant wind waves (perhaps overlapping the region of long swell) but are much greater than those of tsunamis. In height, explosion waves may be much greater than either wind waves or large tsunamis but are less formidable than the tsunami in shoreline effects. The final distinction is degree of organization. Wind waves are irregular in both amplitude and frequency; a tsunami shows a degree of organization in that a train of waves is present although the character of the generating disturbance introduces a large degree of erratic complexity. Explosion waves, on the other hand, are highly organized, showing predictable patterns of both amplitude and frequency variation.

It can be seen, then, that explosion waves are a distinct class, intermediate in many ways between wind waves and tsunamis but unlike either

in essential features. They are, therefore, of independent interest and, although governed by the same underlying laws, require separate treatment. The succeeding sections will delineate the phenomena and methods encountered in the three regimes of generation, propagation and nearshore effects. Useful general sources to supplement the discussions to follow are Van Dorn, Le Méhauté and Hwang (1968) and LeMéhauté (1971); original sources are noted in the text.

Wave Generation

Observed Features

An explosion wave system is produced by the tendency of the free surface to return to the mean level under gravitational forces following the de-

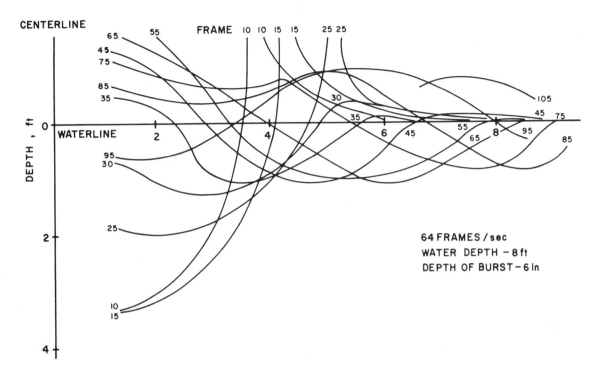

Figure 21-2. Cavity shape versus time (obtained in a wedge tank). The size of the yield is equivalent to 1 pound TNT sphere in open water (Kiebel, 1969).

formations and velocities imparted to it by the explosion. While the wave system, once formed, can be considered to conserve energy during its subsequent development, the generation process appears to be extremely inefficient in that only a small fraction of the total available energy is radiated as organized wave motion. As much as 50% of the thermal energy available in a submerged (nuclear) explosion is lost as irreversible heating or shock in the water during the first bubble expansion. Much of the remainder (excepting that small fraction appearing as wave motion) generates disorganized turbulent motion and then dissipates to heat. This phenomenon is readily apparent for an explosion at shallow depth by the successive appearances (Figure 21-1) of the massive spray dome, column, plume and base surge as precursors to the issuance of waves from the central region. For detonations at great depth or above the free surface, the losses increase with the result that

significant waves occur for only a restricted range of depth of burst. In the case of deep submergence (see Cole, 1948; Kot, 1964, among others) the explosion bubble undergoes repeated oscillations which radiate successive shock waves carrying away energy. (The bubble may, in fact, collapse entirely, leaving a mass of warm fluid and essentially no surface disturbance.) A detonation above the surface is preferentially dissipated into the atmosphere; the waves generated by the surface overpressure are consequently small.

It may be envisioned, then, that the significant generation mechanism is the venting of the bubble at which instant the free surface is effectively "cratered." This initial crater, with dimensions determined by yield and depth of burst, collapses under gravity and produces the outward radiating wave train. The details of this process have been recently investigated experimentally (Kriebel, 1969) with small charges detonated at

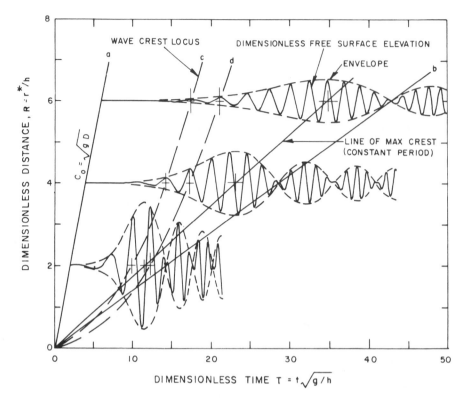

Figure 21-3. Theoretically calculated wave trains. The three wave trains show amplitude-time recorded at nondimensional distance R = 2, 4 and 6 (Van Dorn, 1965).

the apex of a wedge tank. High-speed photography permits observation of bubble expansion, venting and subsequent wave formation; a typical record of the surface profile at successive times is shown in Figure 21-2.

The subsequent propagation (in constant depth) is indicated schematically in Figure 21-3, which shows three successive stages in the development of the wave train. The vertical and horizontal axes in this figure are dimensionless distance $R = r/h$ and dimensionless time $T = t\sqrt{g/h}$, respectively. The three curves show the amplitude-time histories of a wave train generated at the origin as they would be recorded at nondimensional distances, R, of 2, 4 and 6. The symmetrical dashed curves bounding the wave train comprise the wave envelope and serve to define the distribution of energy within the train. The precise shape of the

envelope depends on the initial source conditions, whereas the space-time coordinates of the individual waves are independent of the source and depend only on the water depth. A characteristic of the wave envelope is that any identifiable portion of it—say, a node or antinode—propagates at uniform velocity, as shown by the straight lines a and b. On the other hand, the trajectory of a particular wave is a curve, concave upwards, because the waves are continuously accelerating towards the limiting phase velocity $C = \sqrt{gh}$ of the wave front. Thus, the waves travel faster and pass through the successive nodes of the wave envelopes, and therefore progressively more waves are in each envelope segment with increasing time or distance.

A second important feature of the wave envelope is that its amplitude, as measured along any

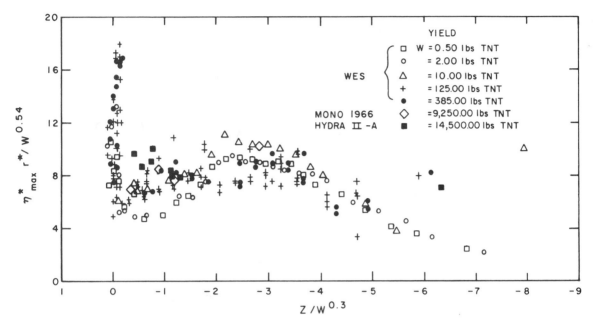

Figure 21-4. Empirical scaling fit relating the maximum wave height η^*_{max} with distance from explosion r*, yield and depth of explosion (data provided by Waterways Experiment Station).

straight line through the origin of the $R - T$ diagram, is inversely proportional to its distance from the origin, owing to the combined effects of circumferential spreading and the stretching of individual groups. Thus, the height of the highest wave in the upper wave train of Figure 21-3 is about 1/3 that of the corresponding wave in the lower train, the former having traveled three times farther.

Experiments of the kind shown in Figure 21-2 could, in principle, be adopted to determine the early cavity and flow field from which the later trains, as shown in Figure 21-3, would follow directly. The bulk of available experimental data (obtained primarily at Waterways Experiment Station, Vicksburg, Miss.) is of a different sort, however, and the existing prediction method was developed accordingly. This quantity of data covering a large range of yields, water depths and depths of burst consists primarily of observations of length (frequency) and height of the maximum wave at specified distances from the source. Fortunately, this data has been found to be accepta-

bly scaleable in terms of yield and detonation depth (Whalin, 1965). The wave number, k_{max}, associated with the maximum wave of the first envelope is given approximately by:

$$k_{max} \approx 0.41W^{-0.288}[1.185 + 0.37\,(Z/W^{0.3})],$$
$$0 > (Z/W^{0.3}) \geqslant -0.5$$
$$k_{max} \approx 0.41W^{-0.288},\ -0.5 > (Z/W^{0.3}) > -7.5. \quad (21\text{-}1)$$

In these expressions, W must be expressed in pounds of TNT and Z (negative) in feet; k_{max} is then the dimensional wave number in feet^{-1}. For depths of burst greater than $7.5W^{0.3}$ data is lacking.

The second item for which data exists in sufficient quantity is the amplitude of the first maximum wave, η_{max}. Figure 21-4 summarizes existing observations for explosions in deep water. Again, the required dimensions are feet and pounds of TNT equivalent (LeMéhauté, 1971).

Equation (21-1) and Figure 21-4 comprise the entire foundation of the prediction scheme to be described later. It is to be noted that the scaling groups found by Whalin (1965) to give the best compression of data points are not precisely those to be expected from dimensional analysis, which gives exponents of 1/2, 1/3 and 1/4, for example, instead of 0.54, 0.3 and 0.288. This indicates a lack of similitude which is not unexpected. Nevertheless, the range of yields incuded in the data is extremely great and the scatter is acceptably small, indicating that satisfactory predictions will be possible; of course, extrapolation to extremely large yields will be of lesser reliability.

A discussion of Figure 21-4 is in order. As the charge position is moved downwards from the free surface the curve exhibits a very high, narrow maximum followed by a minimum and a second—but lower—maximum, after which the more scattered results suggest that further maxima and minima may occur. The first two maxima have been named the "upper and lower critical depths" (UCD and LCD, respectively) because the largest waves (statistically) occur at these charge depths. No sound physical explanation has yet been advanced for the upper critical maximum despite persistent efforts. Since the entire region interior to this maximum is filled with data points, it would appear to be a precarious stability condition that results in maximum effects, and one that is not readily reproducible. Nevertheless, the possibility that a near-surface explosion might produce waves of the peak magnitude must be assumed when making wave predictions.

It was noted that Figure 21-4 pertains to explosions in "deep" water, which was not defined. The relative water depth is clearly dependent on the explosion yield; even the deep ocean must be considered "shallow" for some (quite large) yield, for example. It is observed that the wave-making efficiency of a given yield is relatively independent of depth if the depth is greater than about $6W^{0.3}$ feet (Whalin and Divoky, 1966). For lesser depth, however, the generated heights are less and appear to be roughly halved at $h = W^{0.3}$. LeMéhauté et al. (1967) suggest modifying the ordinate of Figure 21-4 by the factor

$$\left\{ \frac{1}{2} + \frac{1}{10} \left(\frac{h}{W^{0.3}} - 1 \right) \right\} \quad \text{if } 1 \leq \frac{h}{W^{0.3}} \leq 6$$

This should be considered as only a rough guide pending further investigation.

Generation Model

A suitable framework in which to incorporate the information of equation (21-1) and Figure 21-4 is needed. A very general treatment of the wave produced by an arbitrary surface or bottom disturbance has been given by Kajiura (1963); in the case of an axisymmetric surface disturbance in constant depth, the Kajiura formulation reduces to that of Kranzer and Keller (1958) which has been generally adopted as the generation model and which has been extended by Van Dorn (1964), Whalin (1965), Whalin and Divoky (1966) and Hwang and Divoky (1967) in a variety of applications. The model will not be developed here but will be simply described and the pertinent relations stated in the form currently thought most reliable.

Briefly stated, the model assumes the source to be an initially motionless axisymmetric deformation of the free surface in constant, finite depth. The governing equations are then formulated as an initial value problem which may be solved with the appropriate boundary conditions by an integral transform technique. The resulting analytical solution is then approximated by the method of stationary phase to yield a manageable explicit expression. The result is

$$\eta(r, t) = \frac{\bar{\eta}(\lambda)}{rh} \sqrt{\frac{khV(kh)}{-V'(kh)}} \bigg|_{kh}$$

$$= \lambda \quad \cos\left(\frac{\lambda}{h} r - t \sqrt{\frac{g\lambda}{h} \tanh \lambda} \right) \quad (21\text{-}2)$$

In this expression $V(kh) = \frac{1}{2}\sqrt{gh}\,(\tanh\,kh/kh)^{\frac{1}{2}}$ $(1 + 2kh/\sinh\,2kh)$ which is a group velocity, and λ is that value of kh found from $V = r/t$.

The factor $\bar{\eta}$ is the Hankel transform of the initial deformation, $\xi(r)$:

$$\bar{\eta}(k) = \int_0^\infty \xi(r)J_0(kr)r\,dr.$$

The most satisfactory form of $\xi(r)$ known is the parabolic crater and lip (of zero net volume) described by

$$\xi(r) = \eta_0\,[2(r/R_0)^2 - 1],\ r \leqslant R_0$$

$$= 0 \qquad\qquad r > R_0.$$

For this choice, the Hankel transform is found to be

$$\bar{\eta}\,(k) = (\eta_0 R_0\,/\,k)\,J_3\,(kR_0)$$

which, inserted into equation (21-2), specifies the wave history at any point and time.

The two constants, η_0 and R_0, are the depth and radius of the initial deformation and are to be correlated with the two measured wave features, k_{max} and η_{max}. Because of our simplifying assumptions, η_0 and R_0 may be quite unlike true values such as might be inferred from results like Figure 21-2. They are fictitious dimensions of the source to be chosen in such a way that k_{max} and η_{max} are properly reproduced.

The correlation is simply achieved by equating the model values at the first envelope maximum to the observed values. Under the assumption of deep water the following very simple formulas obtain:

$$R_0 = 4.20/k_{max} \tag{21-3}$$

and

$$\eta_0 = 1.63\,\eta_{max}r/R_0. \tag{21-4}.$$

The procedure to predict the explosion wave history (in constant depth) is, finally: 1) with specified values of yield, W, and depth of burst, Z

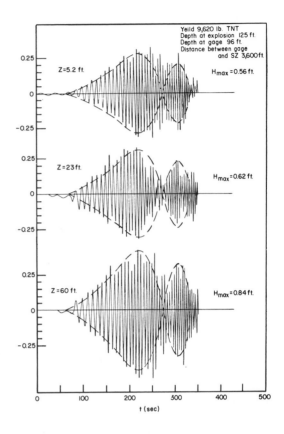

Figure 21-5. Comparison of 1966 Mono Lake Experiments with theory. Broken lines are obtained by theory (courtesy of Van Dorn).

the corresponding k_{max} and $\eta_{max}r$ are found from equation (21-1) and Figure 21-4, respectively; 2) the fictitious cavity parameters η_0 and R_0 are computed from equations (21-3) and (21-4); 3) with these values, equation (21-2) specifies the complete space and time history. Figure 21-5 shows an example obtained in this manner for the 1966 Mono Lake test series involving a detonation of 9620 pounds in a depth of 125 feet (Wallace and Baird, 1968). In the figure, the envelopes are the computed values, while the wave traces are as recorded. For convenience, the wave by wave comparison is not shown (the cosine term in equation 21-2), although it is quite accurate in successive arrival times.

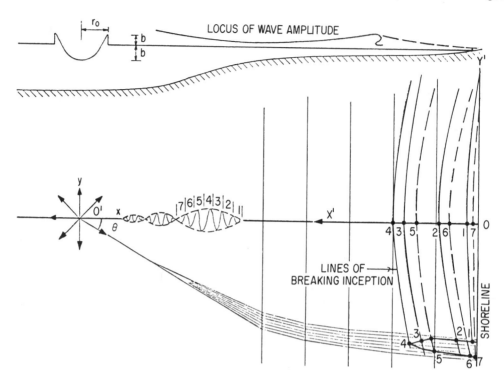

Figure 21-6. Schematic illustration of different wave rays as function of frequencies. The straight lines are contour lines. The first wave train is subdivided into seven components of frequencies, and each one follows paths determined by Snell's law.

Propagation in Nonuniform Depth

As the expanding circular wave front encounters depth variations it suffers the usual effects of refraction and shoaling. Accounting for these is complicated by the circular and dispersive character of the wave system. For example, wave elements of different frequencies initially traveling in the same direction will gradually diverge owing to their differing rates of refraction. The wave history at any point is then the superposition of components coming from different directions. The shoaling problem is correspondingly difficult.

A useful approximate procedure has been developed by Van Dorn (1964). (See also Van Dorn, LeMéhauté and Hwang, 1968, or LeMéhauté, 1971. An equivalent development has been given by Van Mater and Neal, 1970.) This method re-

quires that bottom contours be approximated by straight lines (not necessarily parallel) and that the laws of geometric optics apply. The latter requires: 1) that the wave amplitude be small; 2) that the water depth vary slowly over a wave length; 3) that the phase velocity at any point be that of the corresponding constant depth.

Under these assumptions, the computation proceeds in the following steps:

1. The wave envelope spectrum obtained from equation (21-2) is subdivided into a number of component frequencies, ω_i, each identified by an initial amplitude η_1 (ω_i) in water depth h_1 at the explosion site, which is taken as the origin of Cartesian coordinates (see Figure 21-6).

2. The region shoreward of the explosion

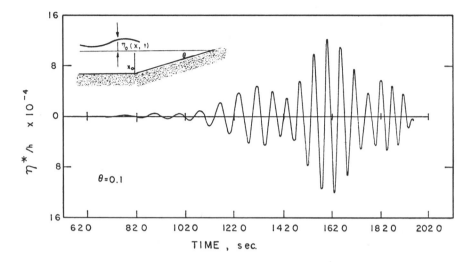

Figure 21-7. Theoretical wave profile showing the interference between incident and reflected wave trains at $x_0/4$ from the shore. The input wave train is shown in Figure 21-8.

point, assumed to be the area of interest, is subdivided into zones radiating from the source, and a suitable number of depth contours are approximated by drawing straight lines across each zone.

3. A family of wave rays for each above frequency is then computed to further subdivide each zone. Such subsets of contours intersected by refracted rays comprise elemental regions within which energy is to be conserved for each frequency.

4. The wave amplitude at every contour is then computed from the conservation of energy flux; also computed are arrival time, wavelength, period and order number by which the local wave phase can be identified. Test criteria should also be applied to determine whether the local wave phase is stable or whether it breaks and must therefore be treated differently (as discussed later).

5. Step 4 is then repeated contour by contour until some reference contour near shore is reached.

6. Steps 4 and 5 are then repeated for each ray and each frequency to determine the time history of the wave field as a function of distance along shore within each zone.

The foregoing scheme is practical only for digital computer application; the appropriate equations, therefore, need not be reproduced here but can be found in the cited sources. For many conditions of practical interest, the wave front propagates a sufficient distance before entering shoaling water that it may be considered effectively one-dimensional. In such a case, and if the angle of incidence is not too oblique, a crude estimate of the wave characteristics can be obtained by simply applying refraction and shoaling coefficients to the amplitude envelope. The justification for this is two fold. First, after a long propagation distance the wave envelope tends to contain a great number of waves with very slow variation in height and frequency; in a sense, they are so well dispersed as to be locally periodic. Second, the rate of frequency dispersion lessens in shoaling water as the phase and group velocities approach equality. Under

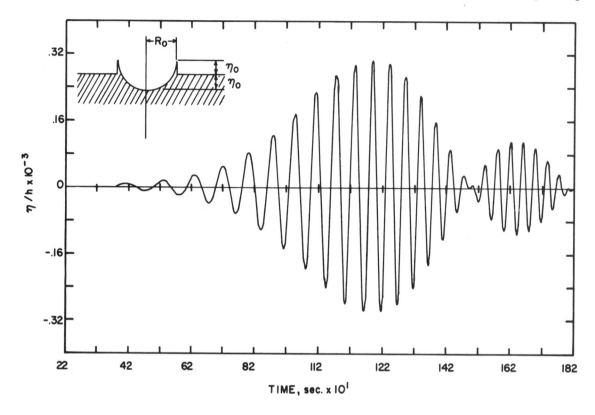

Figure 21-8. The wave train at beach toe, x_0. This wave train was generated at a distance 62 miles away and by an initial cavity, η_0 = 300 feet deep and R_0 = 3300 feet in radius. The water depth is 16,400 feet. (Hwang et al., 1969).

these conditions the system is adequately treated as a succession of periodic waves with the attendant simplifications. Since fixed points on the amplitude envelope are associated with fixed values of frequency, shoaling and refraction corrections are straightforward.

An additional possibility is the occurrence of reflection from depth variations (excluding at this point shreline reflection). The many theoretical attempts to derive a suitable formulation for the reflection coefficient for periodic progressive waves in water of nonuniform depth have been reviewed by LeMéhauté (1967). Of these, the works of Miche (1944) and Roseau (1952) have the most general application to the problem of explosion waves advancing shoreward from deep water. The Roseau theory applies to a particular family of

bottom profiles, some of which quite closely resemble the continental shelves and their terminal escarpments. Application of this theory to the explosion wave problem (LeMéhauté, Whalin et al., 1966) indicates that in practical cases correction for wave reflection is insignificant for all frequencies of interest, even the long leading waves. In the limiting case of small reflection, the Roseau theory tends to the linear conservation of energy flux (in one dimension). This conclusion is further supported by wave tank experiments (LeMéhauté, Snow and Webb, 1966).

Nearshore Behavior

As the wave train arrives to very shallow water at a coastline it undergoes a series of transforma-

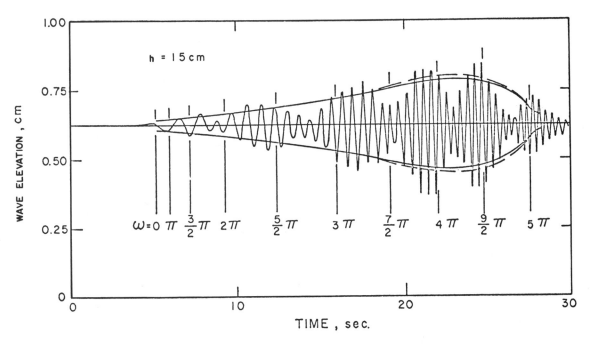

Figure 21-9. Experimental wave profile showing the interference between incident and reflected waves. The waves were measured at water depth of 15 cm.

tions and produces a variety of effects which are of primary concern to coastal engineers. In this section, we consider first nonbreaking wave effects including beating, run-up, nonlinear phenomena and harbor oscillations. Subsequently, attention is given to breaking waves and the development of wave set-up and shelf oscillations.

Amplitude Beats

Upon reaching the shoreline, a dispersive wave train is, to some extent, reflected seaward, the leading portions arriving back at any intermediate relevant point before later portions have yet passed that point on the way to the beach. Thus, the observed offshore surface motion consists partly of incident waves and partly of reflected waves, the exact motion depending on their relative heights and phases, so that a beating effect is seen at such points. It is to be noted that the envelope of the run-up history does not evidence such beat-

ing. This is because there is no phase difference between the incident and reflected waves right at the shoreline, since, in fact, they are the same wave.

The beating phenomena associated with reflection of dispersive wave trains has been theoretically investigated (Hwang, Fersht and Le-Méhauté, 1969) for the case of total reflection, corresponding to nonbreaking waves. The wave pattern is calculated at various distances from the shore and it is found that the irregularity due to superposition of the reflected wave increases as the distance from the shore increases. Figure 21-7 is an example of a computed time history for an offshore point and shows a beat superimposed on the normally smooth modulation envelope. The incoming wave train recorded at the beach toe x_0 is shown in Figure 21-8. This result is confirmed by the experimental observations of Van Dorn (1966). Figure 21-9 shows a dispersive train in a

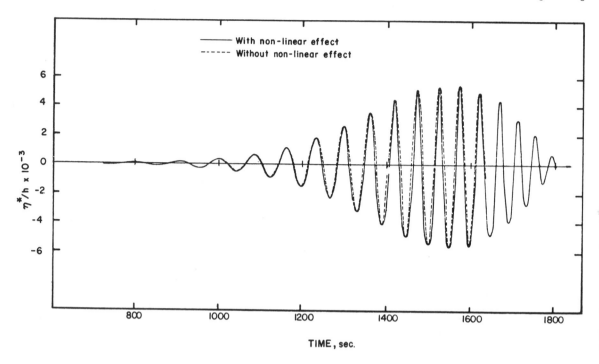

Figure 21-10. Wave run-up and the influence of nonlinear effects showing sharpening of crests and flattening of troughs. The history of run-up calculated at the shoreline on the basis of incoming wave specified in Figure 21-8.

tank with a sloping beach. It can be seen that the beats are quite prominent.

If the waves break, this beating effect will be attenuated, but some residual discrepancy will still remain between the unreflected theoretical model for the incoming wave and the measured one, since beaches never totally absorb wave energy.

The importance of beating lies in the problem of data interpretation. In Figure 21-9 the incident wave amplitude is actually given by the envelope lines; if this were not known a priori the correlation of run-up with measured offshore height, for example, would produce meaningless results.

Nonlinear Effects and Stability

As the wave train enters very shallow water, the deep water equations become progressively inadequate. A more suitable formulation has been

developed by Hwang, Fersht and LeMéhauté (1969) which is specifically adaptable to explosion wave trains. The primary feature not seen from a linear analysis is a steepening of crests and a flattening of troughs in accordance with finite amplitude periodic wave theories.

An example of this nonlinear effect for an explosion wave train is shown in Figure 21-10. The dashed and solid curves represent the linear and nonlinear formulations, respectively. The crest steepening is apparent. There is, however, no difference in peak displacements so that, for most applications, the linear development of Hwang et al. will suffice. It may be noted here that even in the limit of zero depth (i.e., the run-up history) the peak excursions are as given by the linear analysis, only the intermediate time history being influenced by the higher order terms.

The second item to be described here is the problem of wave stability. As a single long wave is followed into shoaling water it may, under certain conditions, split into two or more shorter undulations. Madsen and Mei (1969) and Mei (1970) have investigated this transformation and indicate that if a dispersion parameter, D, given by

$$D = \eta \, (L^2 / h^3) \qquad (21\text{-}5)$$

where L is the wave length, is of order unity or greater, such wavesplitting is to be expected given sufficient time. They show, in particular, that a single wave climbing over a slope and onto a shelf will, for large D, quickly degenerate into three distinct crests followed by a series of small undulations.

This effect is of particular importance for explosion waves, although no study has as yet been performed to delineate the topography and wave conditions under which it will occur. There is some indication that splitting may have been important in certain of the shoreline records obtained during the Mono Lake explosion wave tests of 1965 and 1966, although this has not been fully investigated.

Nonbreaking Wave Run-up

The climb of a wave on an impermeable beach has received extensive investigation which will not be detailed here. Many of the relations determined for periodic waves may be applied to the explosion wave system if a judicious choice of parameters is made; the run-up summary of LeMéhauté, Koh and Hwang (1968) will be most useful in this regard.

A method was developed specifically for dispersive wave trains by Hwang, Fersht and Le-Méhauté (1969) and has been mentioned regarding nonlinear effects. Basically, their analysis proceeds from a transformation of variables (Carrier and Greenspan, 1958) which maps the moving shoreline into a fixed line, thereby making run-up determination quite simple. An example of this technique, which requires computer evaluation, is also given in Figure 21-10.

Application of any run-up prediction method to a real environment is complicated by irregular topography, frictional damping and the like. An instructive picture of the vagaries of explosion wave run-up in an irregular environment is contained in data reported by Wallace and Baird (1968).

Harbor Oscillations

We turn now to a particularly important aspect of coastal explosion wave effects. An explosion train contains significant waves with a broad range of frequencies, the early arrivals being very long and the later being progressively shorter. Therefore, if a train is incident at a given harbor, it will at some time during its history supply an excitation at the fundamental harbor mode and at all harmonics. If the exciting waves are of even relatively small height, the harbor agitation may grow to significant levels. The same argument applies to ship mooring characteristics with a consequent threat of broken lines, damaged fenders and so on.

The problem is aggravated by the fact that explosion waves from a sizeable source tend to show the greatest amplitude for frequencies lower than ordinary sea or swell. The harbor, designed for sea and swell protection, may be particularly sensitive to this lower frequency excitation.

In order to estimate harbor effects one must first determine the harbor's response function. This can be done by using a method recently developed by Hwang and Tuck (1970). Their procedure gives the amplification factor versus incident wave number and was specifically developed for harbors of arbitrary shape. Furthermore, no entrance approximation is involved (such as the arbitrary imposition of a nodal line); instead, the harbor and the sea are considered as a whole, not separately. The response curve assumes periodic excitation, but it can be applied to an explosion wave train if there are a sufficient number of waves in each envelope that the wave to wave frequency variation is small.

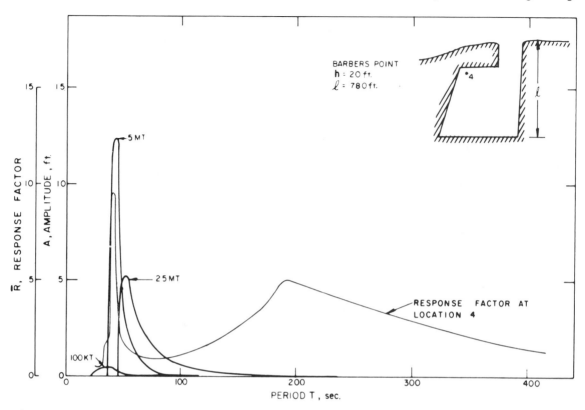

Figure 21-11. Wave amplitude inside Barbers Point Harbor. The wave heights were calculated on the basis that the explosions occurred at Johnston Island.

A sample calculation for Hawaiian harbors to assess the potential hazard associated with large-yield tests at Johnston Island has been made. The result of the calculation is shown in Figure 21-11. The curve labeled $\overline{R}$ is the amplification factor which reaches a peak of about 9 at a wave period of about 40 seconds. The remaining three curves are the explosion wave amplitudes for the specified yields detonated at UCD about 1000 miles to the southwest. These amplitude curves (plotted versus period for convenience) contain the factor $\overline{R}$ as well as smaller corrections for refraction and shoaling. Only the first envelopes are plotted; succeeding envelopes lie progressively nearer the origin.

This particular example shows a peculiar feature in that the largest yield (and therefore the largest incident waves) does not produce the maximum harbor agitation. The reason is simply that the resonant peak occurs at a frequency corresponding to a null in the envelope, whereas for the somewhat smaller source the resonant and envelope peaks are well matched. While this hypothetical example is somewhat extreme for practical concern, it does indicate quite well the paramount importance of resonant amplification in the assessment of explosion wave effects.

Breaking Waves

Breaking Criteria

As an explosion wave shoals on a slope, a point is reached at which the wave height exceeds

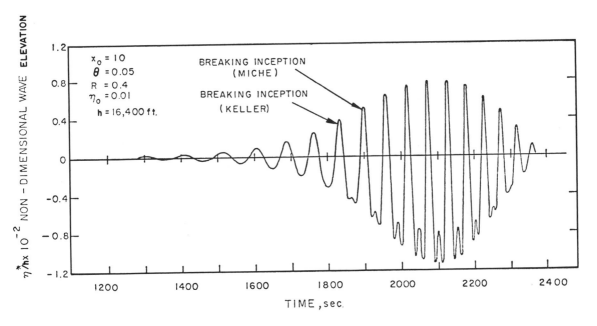

Figure 21-12. Wave run-up on a sloping beach of a nonperiodic wave train in the case where the breaking limit is reached.

a certain limiting value and the wave breaks. No fully satisfactory breaking criterion exists. The most prevalent assumption is that the McCowan solitary wave breaking criterion (Munk, 1949) applies: $H_b/h_b = 0.78$. The subscript b denotes conditions at the breaking point. This simple criterion is derived from the assumption that the particle velocity at the crest equals the wave celerity; that this is the actual breaking mechanism has not been demonstrated. Many other criteria are found in the literature. For example, Street and Camfield (1966) give an empirical relation: $H_b/h_b = 0.75 + 25\alpha$ where α is the beach slope in radians. Le-Méhauté and Koh (1967) examine a quantity of experimental data and suggest

$$H_b = 0.76 \, H_o \tan^{1/7} \alpha \, (H_o/L_o)^{-1/4}$$

The numerical techniques for dispersive wave trains developed by Hwang, Fersht and LeMéhauté (1969) show the occurrence of breaking but, unfortunately, require computer evaluation. As shown in Figure 21-12, breaking is evidenced nu-merically by the appearance of a hump within the wave trough (which has no physical reality but instead indicates that either the leading or trailing crest has broken). Interestingly, the example indicates breaking between the criteria of Miche (1951) and Keller (1961). These criteria, for steep and gentle slopes respectively, define a maximum deep water wave steepness such that breaking will not occur on a given slope; waves of greater steepness will break but the corresponding height or depth is not specified.

Breaking Wave Transformation and Set-up

Once broken, the wave dissipates energy and diminishes in height as it proceeds shoreward. The processes of shoaling, breaking dissipation, bottom friction dissipation and convergence or divergence of orthogonals through refraction are all operative in a real environment and contribute to a great complexity of surf-zone behavior. We shall consid-

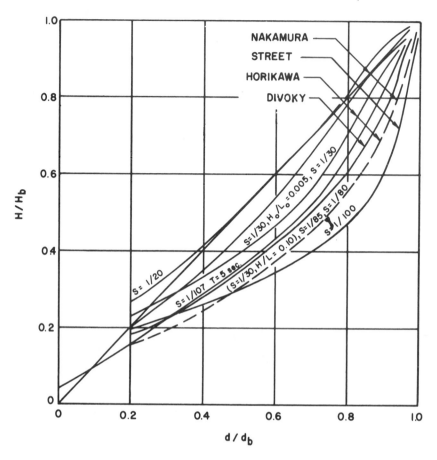

Figure 21-13. Summary of wave transformation data (shoreward of the breaking point).

er the idealized case of normal incidence at a plane beach.

Experiments on wave height transformation have been reported by Street and Camfield (1966), Horikawa and Kuo (1966), Nakamura et al. (1966), and Divoky et al. (1970) among others. A summary of experiments is shown in Figure 21-13. Typically there is a rapid decline in height just after breaking followed by a more gentle decline near shore.

Phenomenological energy-balance models of the behavior illustrated in Figure 21-13 have been developed by LeMéhauté (1962), Horikawa and Kuo (1966) and Divoky et al. (1970) with moderate success. However, there is another important pro-

cess operative in the surf zone which must be considered. This is the phenomenon of wave set-up, a rise in the mean water level caused by the passage of the breakers. This effect is caused by the steady decrease of momentum flux which is balanced by an increase of hydrostatic pressure in the form of increased depth. Set-up is particularly important for explosion waves, since at the shoreline it may be as great as the offshore wave height. Figure 21-14 is a nearshore record obtained during the 1965 Mono Lake explosion wave tests; the set-up is clearly a dominant feature. Since the explosion waves arrive in groups, one may expect that the set-up history should show a corresponding modulation. Experiments reported by Hwang (1970)

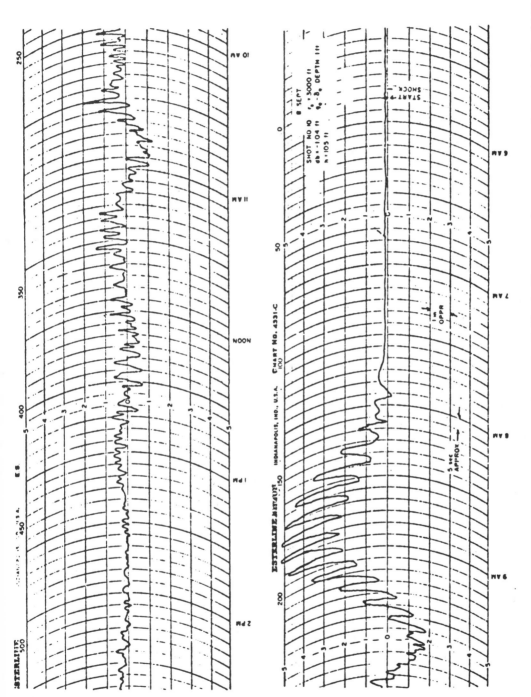

Figure 21-14. Wave record obtained by Van Dorn at Mono Lake, California. The wave train wave generated at water depth of 104 feet by an explosion of 9260 pounds of TNT. The wave trains were measured at water depth of 1 foot, 3000 feet away from explosion (courtesy of Van Dorn, Scripps Institute of Oceanography).

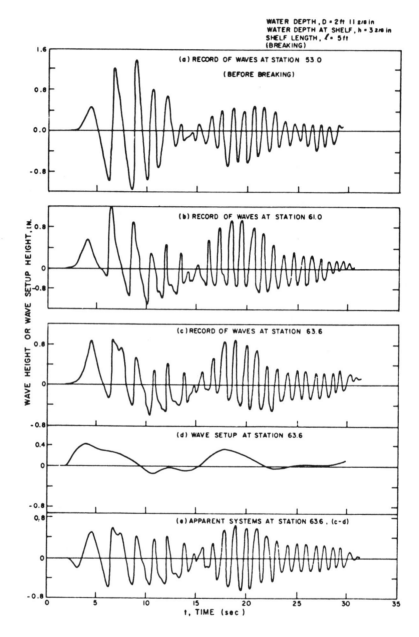

Figure 21-15. Wave profiles and wave set-up.

confirm this expectation as shown in Figure 21-15. Set-up is large under the wave train maxima and is negative at the nodes (as recorded under the experimental set-up shown in Figure 21-16).

It is readily appreciated that wave set-up on a shelf under the transient conditions of explosion waves could excite a shelf oscillation. This is illustrated by the Mono Lake observation shown in

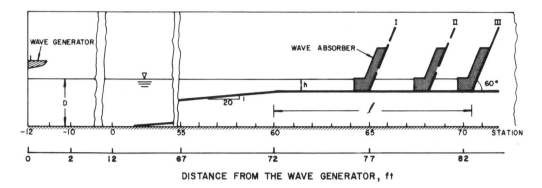

Figure 21-16. Experimental arrangement for performing wave set-up and shelf oscillation measurements of Figure 21-15.

Figure 21-14; the large peak of wave set-up is associated with the first envelope maximum while the later fluctuations are shelf oscillations. It is conceivable that for certain conditions of bottom topography and frequency of incoming wave train maxima a forced resonance could develop, although this has never been observed. Note that successive envelopes are of diminishing amplitude and therefore may produce much less set-up than the first, largest envelope (Hwang, 1970).

Returning to the problem of breaking wave transformation, one realizes that set-up should be considered. The theoretical framework has been developed in a series of papers by Longuet-Higgins and Stewart (1960, '61, '62 and '64). Adaptations of their work to the problem of breaking wave transformation have been made by Bowen, Inman and Simmons (1968) and by Hwang and Divoky (1970). Bowen et al. assumed that the breaking wave height remains a constant fraction of the total mean water depth and from this deduce a linear wave set-up variation and a height decay of the general sort shown in Figure 21-13. Hwang and Divoky (1970) solved simultaneously for set-up and wave height after breaking using conoidal wave theory and a simplified energy dissipation model. The results showed good agreement with experiment for both set-up and height decay. Figures 21-17 and 21-18 show typical results obtained by Hwang and Divoky. In Figure 21-17 computa-

tions are compared with height decay data of Horikawa and Kuo (1966) for a slope of 1/65. In Figure 21-18 a comparison is shown with set-up data of Saville (1961). It is seen that in general the agreement with observation is quite satisfactory in both figures.

A Summary for Making Rough Estimates of Wave Characteristics

The previous sections summarize the description of the behavior of explosion-generated waves. Calculations are sometimes quite time consuming but may be necessary when detailed information is required. In some cases, a quick estimation is needed for operational purposes. For this reason a simplified method which provides only the properties of the maximum waves is given.

As shown in Figure 21-19, a nuclear explosion of yield W is assumed to occur in deep water off the continental slope. The explosion generates a wave train propagating in all directions. The maximum wave amplitude η_{max} of the wave train in deep water is related to the distance from the explosion, r, and the yield, W, as follows:

$$\eta_{max} = (18\,W^{0.54}/r)\text{feet}, \quad [r] = \text{feet}, \quad [W]$$

$$= \text{pounds TNT} \qquad (21\text{-}6)$$

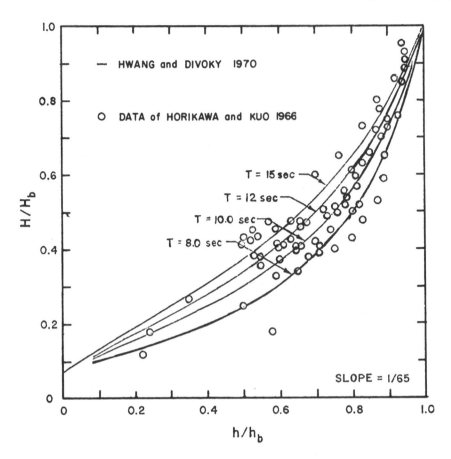

Figure 21-17. Comparison of the model of Hwang and Divoky with data of Horikawa and Kuo for wave height decay in the breaking zone.

assuming that the detonation occurs at the upper critical depth. The wave period, τ, of this maximum wave is

$$\tau = 1.63 \, W^{0.15} \text{ seconds, [W] - pounds TNT.}$$

As the maximum wave propagates towards the continental shelf, its amplitude decreases as a result of radial spreading until shoaling and refraction effects become important. The calculation of wave amplitude including these effects is rather complicated as discussed in the preceding section. Here, for simplicity, we assume that the explosion is rather far away from shallow water so that the waves are almost two-dimensional when they arrive. Thus, we may calculate the maximum wave amplitude η_{max} by using equation (21-6) until the water depth, h is equal to one-quarter of the wave length, $\frac{1}{4}L_{max}$. (We assume that shoaling becomes important when $h = \frac{1}{4}L_{max}$.) From there on, the maximum wave height may be calculated by simply multiplying by the shoaling coefficient. Refraction must also be accounted for and may be determined from linear theory to a good approximation.

As the wave increases its amplitude by shoaling to the point that the water depth is insufficient to transmit the wave energy, the wave will break;

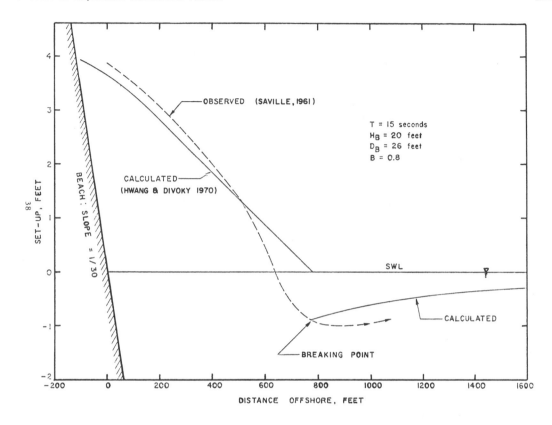

Figure 21-18. Comparison of the Hwang and Divoky model of breaking wave set-up with data of Saville.

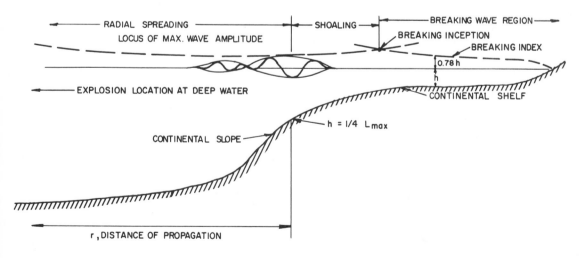

Figure 21-19. Schematic drawing of maximum wave transformation as it propagates towards the shore.

this occurs at the intersection of the wave height curve and the breaking index line, as shown in Figure 21-19, and as a first approximation is at H = 0.78h. After the wave breaks, it propagates towards the shore as a spilling breaker if the bottom slope is gentle.

The estimation of wave run-up is extremely difficult because of the usually complicated shore geometry. However, if the shoreline is assumed to be locally straight, one may estimate the run-up according to standard procedures such as the curves presented in Beach Erosion Board TR 4. However, if the shore includes possibly resonant bays or harbors, or if the incoming waves are not well dispersed and vary rapidly in height or period, it will be necessary to resort to more accurate computations.

Acknowledgments

The authors are indebted to a number of individuals who have contributed both directly and indirectly to this work. Bernard LeMéhauté and William Van Dorn have provided the benefit of their expertise through numerous discussions. A. Cox, K. Olsen, J. Warner and R. Ballard, among others, have encouraged our continued interest in explosion wave properties. Thanks also to R. Whalin and D. Bucci of the Waterways Experiment Station who provided the excellent photo reproduced in Figure 21-1.

References

1. Cole, R.H. 1948. *Underwater explosions*. Trenton: Princeton University Press.
2. Divoky, D., LeMëhautë, B. and Lin, A. 1970. Breaking waves on gentle slopes. *J. Geophys. Res.* vol. 75, no. 9.
3. Horikawa, K. and Kuo, C.T. 1966. A study on wave transformation inside surf zone. *Proc. Am. Soc. Civil Engrs.* 10th Conf. on Coastal Engineering, 1, 217-233.
4. Hwang, Li-San. 1970. Wave set-up of non-periodic wave train and its associated shelf oscillation. *J. Geophys. Res.* vol. 75, no. 21, 4121-4130.
5. ____ and Divoky, D. 1967. *Summary of techniques of explosion wave predictions*. **Tetra Tech** Rept. TC-103, vol. 2, Pasadena, California.
6. ____. 1970. *Breaking wave set-up and decay on gentle slopes*. 12th Coastal Engineering Conference, Washington, D.C.
7. ____, Fersht, S. and LeMéhauté. 1969. Transformation and run-up of tsunami-type wave trains on a sloping beach. 13th Convention of I.A.H.R., Japan.
8. ____ and Tuck, E.O. 1970. On the oscillations of harbors of arbitrary shape. *J. Fluid Mechanics.* vol. 42, part 3, 447-464.
9. Kajiura, K. 1963. The leading waves of tsunamis. *Bull. Earthquake Res. Inst.* 41, 535-571.
10. Keller, J.B. 1961. *Tsunami's water waves produced by earthquakes.* Tsunami Hydrodynamics Conf., Honolulu, Hawaii.
11. Kot, C.A. 1964. *Theoretical study of bubble behavior in underwater explosions.* U.S.N.-R.D.L. Tech. Rept. 747.
12. Kranzer, H.C. and Keller, J.D. 1959. Water waves produced by explosions. *J. Applied Physics* vol. 30, no. 3.
13. Kriebel, A.R. 1969. *URS 679-5.* Burlingame, Calif.: URS Research Company.
14. Laitone, E.V. 1961. *Higher approximation to nonlinear water and the limiting heights of cnoidal, solitary and Stoke's waves.* Inst. Eng. Res. Tech. Rept. Series 89, issue 6, University of California.
15. LeMéhauté, B. 1962. On the non-saturated breaker theory and the wave run-up. *Proc. 8th Conf. on Coastal Eng.* Mexico City, 77-92.
16. ____. 1966. *Hydrodynamic effects of nuclear explosions,* vol. I: state of the art. NESCO Rept. S 244-1.
17. ____. 1970a. Theory of explosion-generated waves. In *Advances in Hydroscience,* vol. 7, Ven Te Chow, ed. New York: Academic Press.
18. ____. 1970b. Explosion-generated water waves. *8th Symposium on Naval Hydrodynamics.* Pasadena, California.
19. ____, Snow, G. and Webb, L. 1966. *Gravity waves on bottom slopes.* NESCO Rept. S245A; Defense Atomic Support Agency Rept. DASA-167-1
20. ____, Whalin, R.W. and Divoky, D. 1966. *Explosion waves and run-up.* vol. 3, Mono Lake Experiments. NESCO Rept. SN300.
21. ____, Hwang, Li-San, and Butler, L. 1967. *Explosion-generated wave environment in shallow water.* Defense Atomic Support Agency, Rept. DASA 1963.
22. ____ and Koh, R.C.Y. 1967. On the breaking waves arriving at an angle to the shore. *J. Hydr.*

Res. vol. 5, no. 1.

23. ____, Koh, R.C.Y. and Hwang, Li-San. 1968. A synthesis on wave run-up. *J. of Waterways and Harbors Division, Proc. Am. Soc. Civil Engrs.* 94, no. WW1, 77-92.

24. Longuet-Higgins, M.S. and Stewart, R.W. 1960. Changes in the form of short-gravity waves on long waves and tidal currents. *J. Fluid Mechanics* vol. 8 part 4, 565-583.

25. ____. 1961. The changes in amplitude of short-gravity waves on steady non-uniform currents. *J. Fluid Mechanics* vol. 10, 529-549.

26. ____. 1962. Radiation stress and mass transport in gravity waves. *J. Fluid Mechanics,* 13, 481-509.

27. ____. 1964. Radiation stress in water waves, a physical discussion with applications. *Deep Sea Res.* vol. 2, 529-563.

28. Madsen, O.S. and Mei, C.C. 1969. The transformation of a solitary wave over an uneven bottom. *J. Fluid Mechanics,* 39, 781.

29. Mei, C.C. 1970. Evolution of solitary and periodic long waves with dispersion. *Symposium on Long Waves.* University of Delaware, Newark, Delaware.

30. Miche, R. 1944. Mouvements ondulatoires de la mer en profondeur constante au decroissante. *Annals des Ponts et Chaussees.*

31. ____. 1951. Le pouvoi reflechissant des ouvrages maritimes. *Annales des Ponts et Chaussees.* Paris: Ministere des Traveus Publics et des Transports.

32. Munk, W.H. 1949. The solitary wave theory and its application to surf problems. *Annals of the N.Y. Academy of Sciences,* vol. 51, art. 3, 343-572.

33. Nakamura, M., Shiraishi, H. and Saski, Y. 1966. Wave decaying due to breaking. *Proc. 10th Conf. on Coastal Eng.* ASCE.

34. Roseau, M. 1952. Contributions a la theorie des ondes liquides de gravite en profoundeur variable. *Publications Scientifiques et Techniques du Ministere de l'air* 275.

35. Saville, T. 1961. Experimental determination of wave set-up. *Proc. Second Tech. Conf. on Hurricanes,* 242-252.

36. Street, R.L. and Camfield, F.E. 1966. Observations and experiments on solitary wave deformation. *Proc. Am. Soc. Civil Engrs.* Conf. on Coastal Engineering, 1, 289-301.

37. Van Dorn, W.G. 1964. Explosion-generated waves in water of variable depth. *J. Mar. Res.* 22, no. 2.

38. ____. 1965. Tsunami. In *Advances in Hydroscience.* Ven Te Chow, ed. New York: Academic Press.

39. ____1966. Run-up recipe for periodic waves on uniformly sloping beaches. *Proc. Am. Soc. Civil Engrs.,* 10th Conf. on Coastal Engineering, 1, 349-363.

40. ____, LeMéhauté B. and Hwang, Li-San. 1968. *Handbook of explosion-generated water waves.* vol. 1: state of the art. ONR Washington, D.C., Tetra Tech Rept. TC130.

41. Van Mater, R.R., Jr. and Neal, E. 1970. *On the prediction of impulsively-generated waves propagating into shallow water.* 8th Symposium on Naval Hydrodynamics, Pasadena, California.

42. Wallace, N.R. and Baird, C.W. 1968. *Run-up on an irregular shoreline—Mono Lake tests.* Oceanographic Services, Inc., U.S. AEC Rept. NVO-297-2.

43. Whalin, R.W. 1965. Water waves generated by explosions (progation theory for the area near the explosion). *J. Geophys. Res.* 70, no. 22.

44. ____ and Divoky, D. 1966. *Water waves generated by shallow water explosions.* NESCO Rept. S-359.

22

numerical modeling of tsunamis—forecasting heights in a warning system

Introduction

Nothing can be done to prevent the occurrence of tsunamis. Thus, coastal civilizations will always be vulnerable to these cataclysmic events. The only defensive strategy for the vulnerable regions is to understand the behavior of tsunamis. Then an adequate warning system can be devised and structures can be built which will minimize the damaging effects of tsunamis.

Tsunamis are caused by large shallow undersea earthquakes which involve tectonic displacements of the sea bed. Part of the energy of the seismic disturbance is taken up by very low amplitude long water waves which travel from the generation area throughout the ocean basin. As the waves approach coastal areas, the amplitudes increase to form huge masses of water which strike coastlines with devastating results. The passage of these waves near the bays and harbors also can excite the natural frequency of these basins so that large waves are produced even in supposedly protected areas. Obviously, loss of life and considerable property damage result.

Certain areas of the Pacific basin are known to be breeding grounds for earthquakes, and certain coastal areas are repeatedly damaged by tsunamis. With these facts in mind a tsunami warning system

has been established. Earthquakes which occur in a tsunami generation area are immediately reported to the warning center. At the warning center the earthquake location is used to predict arrival times for the tsunami at vulnerable coastal stations. This information is then relayed to the areas which might be affected. The system as it operates today only provides the information of arrival time. Much work remains to be done before the system can give accurate information on waves along the coast where damage is likely to occur. The major work for establishing such a system should fall into three categories:

1. *Ground Motion:* Estimating the bottom displacement with sufficient accuracy from seismic records and extensive foreknowledge of the type of motion likely to be experienced by a given epicentral region and the magnitude of the earthquake. Though considerable study has been made of this facet of the tsunami problem, it is felt that present-day knowledge is only a first approximation to the nature of the source. Extensive research is needed to establish reasonably accurate information.
2. *Tsunami Generation and Propagation:* Once the ground displacement is specified it is rel-

atively easy to compute the surface wave disturbances and their propagation across the Pacific.

3. *Coastal Response:* The impact of a given tsunami at a given place is just as dependent on the particular nature of that place as it is on the tsunami itself. Thus, a warning system must incorporate the evaluation of local behavior.

We have limited ourselves to a discussion on the latter two subjects, which are of interest to coastal engineers.

In regard to tsunami generation, Ichiye (1958), Honda and Nakamura (1951), Webb (1962), Momoi (1964) and Kajiura (1963) have investigated the surface disturbance resulting from a sea-bed motion under restrictions of constant depth and simplified bottom displacement. Van Dorn (1965) has presented an analytical formulation based on Kajiura's (1963) work for distant amplitude from an arbitrary source in constant depth; adaptation to the real environment and solution of the integral equations involved are difficult for practical application, however; most recently Hwang and Divoky (1970, 1971) developed a numerical scheme for obtaining the tsunami generation based on the permanent displacement of ground motion. The results appear to be very encouraging. Some of these new results are reported here. The counterpart of the generation problem is the coastal response. Natural shorelines are seldom straight and uniform in slope; they usually contain small bays, curved beaches, canyons and shelves. Such features cause wave reflection, diffraction and refraction, resulting in large wave run-up at certain locations for some incident waves and none in others. In particular, when the period of incoming waves matches the fundamental period or harmonics of a bay and refraction effects concentrate the waves at a certain location, very large oscillation and run-up may be observed. The variation of wave run-up with the wave period at the same location can be as much as one order of magnitude. Such a variation may

significantly modify the nature of waves and enhance damage in some places.

In regard to the theoretical developments of coastal response to incoming waves, Leendertse (1967) and Loomis (1966) have developed numerical procedures to determine the response of basins to long waves. However, the time-dependent elevations at the opening of the basins must be specified. Since waves within a bay are dependent on both the incoming and outgoing wave conditions, entrance conditions prescribed a priori are undesirable. Theories taking account of this have been developed (Hwang and LeMehaute', 1968; Hwang and Tuck, 1970; Hwang and Olsen, 1971).

The problems of tsunami generation and coastal behavior appear to be rather different in nature. However, from a hydrodynamic point of view, they contain a great degree of similarity. We will discuss these two subjects and point out the similarities and differences in their mathematical treatments.

Tsunami Generation

Mathematical Formulation of Generation

The generation problem will be presented first. In a Cartesian coordinate system the equations of motion are

$$u_t + uu_x + vu_y + g\eta_x = 0 \qquad (22\text{-}1)$$

and

$$v_t + uv_x + vv_y + g\eta_y = 0. \qquad (22\text{-}2)$$

The continuity equation has an unusual form to account for the change in depth during ground displacement.

$$(\eta - \zeta)_t + [(h + \eta)u]_x + [(h + \eta)v]_y = 0$$

where

η	= wave elevation from mean level
u, v	= water particle velocities in the x, y-directions

t = time
x, y = coordinates in horizontal plane at mean water level
h = water depth (including ζ)
g = acceleration of gravity
ζ = ground displacement

A space-staggered multitime step procedure developed by Leendertse (1967) is used. During the first half of the time step η and u are computed explicitly and v implicitly; in the second half-time step η and v are computed implicitly and u explicitly. This yields a tridiagonal matrix which is solved by elimination procedures. No detailed discussion of the numerical scheme will be given here.

A problem arises in the treatment of boundaries. Coastal boundaries are not difficult to specify, but the seaward boundary is another matter. The seaward boundary is troublesome because it is not a physical boundary, but rather a ficticious boundary arising from the finite capacity of the computer. The problem is avoided by placing the boundary fairly far from the source. Then, even though it is difficult to numerically specify accurate wave transmission across this boundary, it is possible to study the waves of interest before they are affected by reflections from the inaccuracies developed by the ficticious boundary.

The problem now is to specify the initial condition that is the essence of the problem. The input is the nature of the forcing disturbance, the ground motion ζ. In general, this is quite complex, and its transient history may never be reliably known for a given tsunami. We have therefore adopted a somewhat idealized model of the disturbance although, of course, the subsequent numerical development could be as easily utilized with the real history, were it specifiable.

The bottom motion is assumed to result from changes in depth, whether by horizontal or vertical motion. The area over which the disturbance occurs is taken to be given by the distribution of aftershock epicenters (Benioff, 1951). It is envisioned that motion begins at the epicenter of

the main shock and that the leading tip of disturbance propagates radially from that point at a constant velocity V. This "rupture" velocity may be of the order of 3 to 4 km/second (Plafker and Savage, 1970). Vertical motion at a given point begins upon passage of this radial front. That is, if (a,b) and (i,j) denote Cartesian coordinates of main shock epicenter and an arbitrary point within the generation area, respectively, then vertical motion is assumed to commence at (i, j) at a time, t, given by $t_{ij} = (\Delta x/V)[(a-i)^2 + (b - j)^2]^{1/2}$ with Δx the coordinate spacing.

The vertical displacement is known to be quite irregular in time and may be somewhat as shown by the solid curve in Figure 22-1, evidencing overshoot and random oscillation ending in a permanent displacement value ξ_{ij}'. Since the time scale of the fine structure is small in comparison to that of tsunami period and the waves generated by the fine structure will vanish from the leading waves after a long distance of propagation, it is justified to introduce a simplified monotonic displacement history as shown by the dashed curve in Figure 22-1. In particular, we have chosen the following expressions for bottom displacement:

$$\xi_{ij}(t) = 0 \qquad\qquad t \leqslant t_{ij}$$

$$\xi_{ij}(t) = \xi_{ij}' \sin^2 [(\pi (t - t_{ij})/2\tau)]$$

$$t_{ij} \leqslant t \leqslant t_{ij} + \tau$$

$$\xi_{ij}(t) = \xi_{ij}' \qquad\qquad t \leqslant t_{ij} + \tau.$$

The parameter τ is a characteristic time of ground motion and is difficult to specify; records of horizontal displacement of western earthquakes (Berg and Housner, 1961) indicate that $\tau = 10$ seconds is a reasonable value. The controlling factor for tsunami height at a distance is the permanent displacement ξ_{ij}, which may be obtained by extensive survey of the region and a degree of inference assisted by the assumption of zero motion along the periphery of the aftershock re-

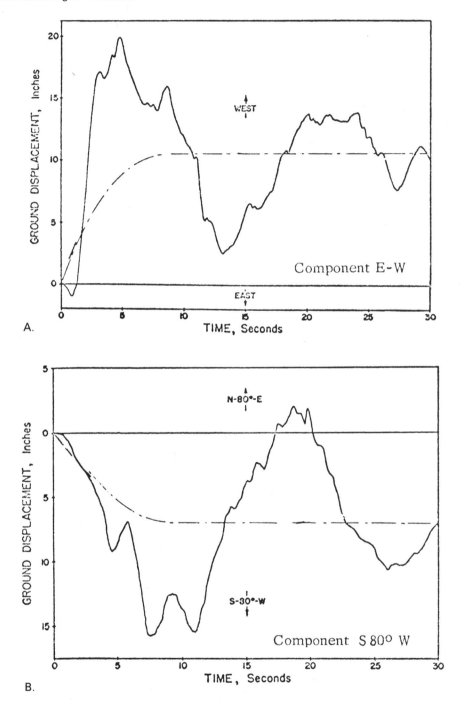

Figure 22-1. Typical recorded ground displacement histories for two western U.S. earthquakes (Berg and Housner, 1961). The average ground motion, as shown in broken lines, was adoped by Wilson (1969). *A*, El Centro, California, 1940; *B*, Olympia, Washington, 1949.

Figure 22-2. Aftershock distribution (Plafker, 1969); the dashed line separates uplift (below) and subsidence (above); the disturbed region is approximately 525 miles long and 225 miles wide.

gion. It is noted that horizontal displacement of a sloping bottom will also generate waves; to a first approximation, however, such a motion can be interpreted as an appropriate pattern of purely vertical displacement.

Application to the Alaskan Earthquake

Input Data. The primary method of determining the areal extent of the source disturbance is through the distribution of aftershock epicenters,

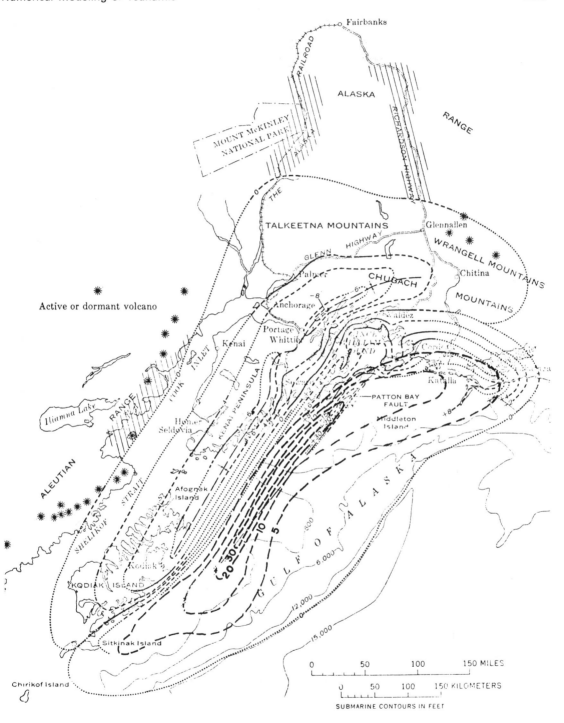

Figure 22-3. Assumed bottom displacement in feet (Plafker, 1969).

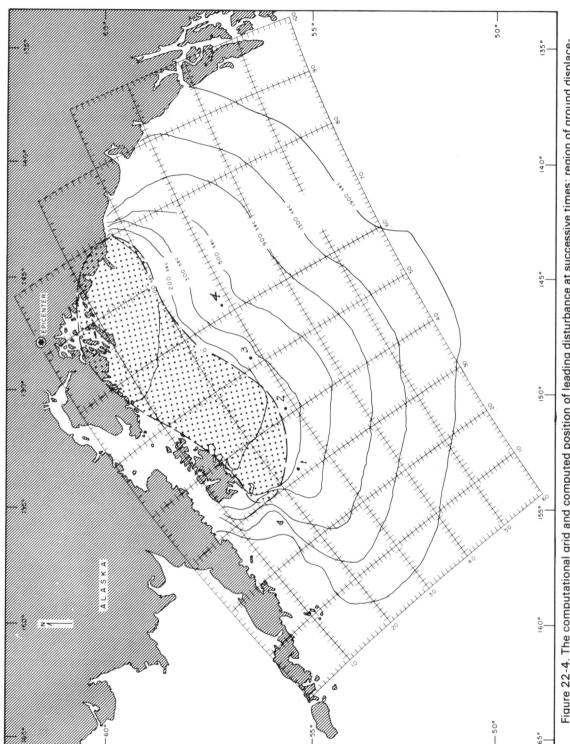

Figure 22-4. The computational grid and computed position of leading disturbance at successive times; region of ground displacement is shaded.

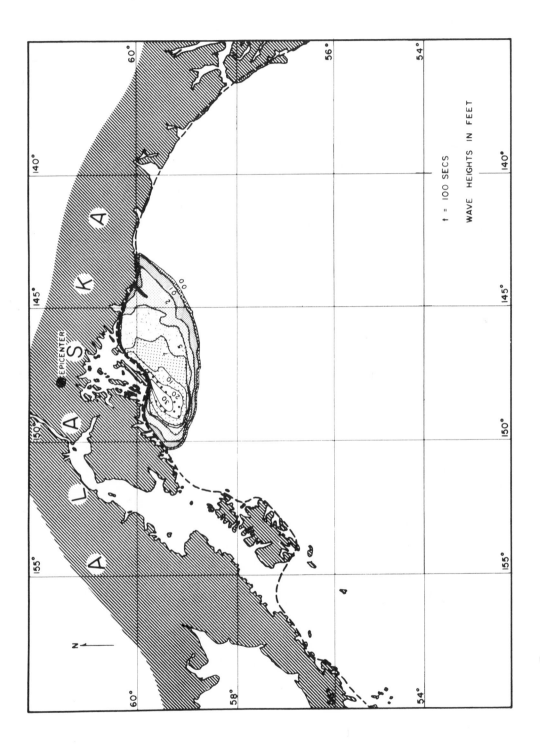

Figure 22-5. Computed water surface elevations 100 seconds after initiation of ground motion (Hwang and Divoky, 1970).

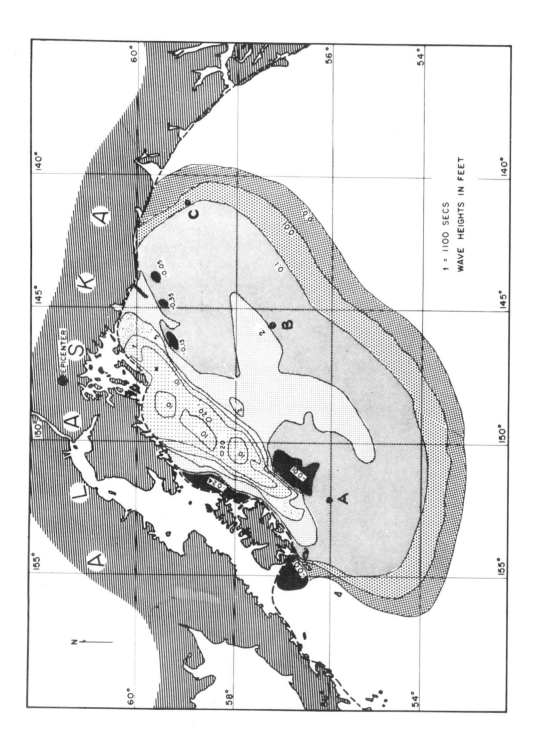

Figure 22-6. Computed water surface elevations 1100 seconds after initiation of ground motion (Hwang and Divoky, 1970).

the assumption being that the regions are congruent.

Figure 22-2, taken from Plafker (1969), shows the distribution of major aftershocks of magnitude greater than 4 occurring from March 27 to December 31, 1964. An enclosing curve would approximate the source region of the tsunami.

Within this region, the permanent displacement $\xi'(x,y)$ has also been summarized by Plafker (1969). Extensive surveys, barnacle growth, salt water damage, etc. allow estimates of the tectonic changes produced by an earthquake. Good prequake bathymetry and considerable commerce also help in reconstruction of the disturbance. In Figure 22-3 the pattern is shown. This is consistent with the data of Plafker and differs only in that contour lines have been interpolated where missing across submerged regions. This is somewhat arbitrary but, in any case, must be nearly correct.

The computational grid was established as shown in Figure 22-4; the grid spacing is 16.1 km. We have chosen $\tau = 10$ seconds and $V = 3.5$ km/second. The time step during ground motion was taken to be 2 seconds, and 100 seconds thereafter. Depths were read at each grid point from a nautical chart and were keypunched for computer; displacements were similarly prepared using Figure 22-4.

The Near-Source Wave System. Figures 22-5 and 22-6 show contours of equal water surface displacement from equilibrium at cited time intervals from 100 to 1100 seconds after the initial major earthquake (5:36 Alaskan time). The pattern grows rapidly toward the southwest, following the major rupture axis, which propagated from Prince William Sound towards Kodiak Island at a speed of about 3.5 km/second. At the same time, it spread seaward and shoreward—the latter motion being soon reflected by the coastline. An important feature is the complex topography of the sea surface even at early times.

These figures are most useful in showing the pattern of propagation of the leading disturbance. This is summarized in Figure 22-4 together with

the computational grid. Most interestingly, the shape of the source region is in general preserved.

To allow a better appreciation of the wave time history, results at the points indicated in Figure 22-4 are shown in Figures 22-7a–d. Again, considerable complexity is apparent. The wave forms shown are the superposition of many components traveling in various directions, including not only the fundamental wave due to the ground upthrust, but also components scattered randomly by reflection from the Alaskan coast. The underlying long wave seems to have a period on the order of 1.5 hours. This is in accord with reports of many observers (e.g., Wilson and Torum, 1968). In addition, there is a superposed higher frequency system due, as is noted, to geometric complexity; this is particularly evident in Figures 22-7b–d and appears to be dominately in the 10-minute period range. Interestingly, Van Dorn (1970) reports such higher frequency components in measurements he obtained at Wake Island, which were roughly of 12- or 13-minute period.

The results of this generation model have been checked with tide gages and location observations to be quite satisfactory (Hwang and Divoky, 1970).

Coastal Response

Mathematical Formulation for Coastal Response

Formulation of the coastal response is similar to that for the generation problem. The equations of motion remain the same but the continuity equation is simplified $\eta_t + [(h + \eta)u]_x + [(h + \eta)v]_y = 0$ since no impulsive volume change is present.

As posed, the ideal approach would require solution of nonlinear equations and would involve some iterative scheme to account for the multiple internal reflections within the coast (bay) and the reflected waves which travel outward from the coast to the open boundary. These waves would interfere with the incident waves and create problems for specifying the input conditions at the

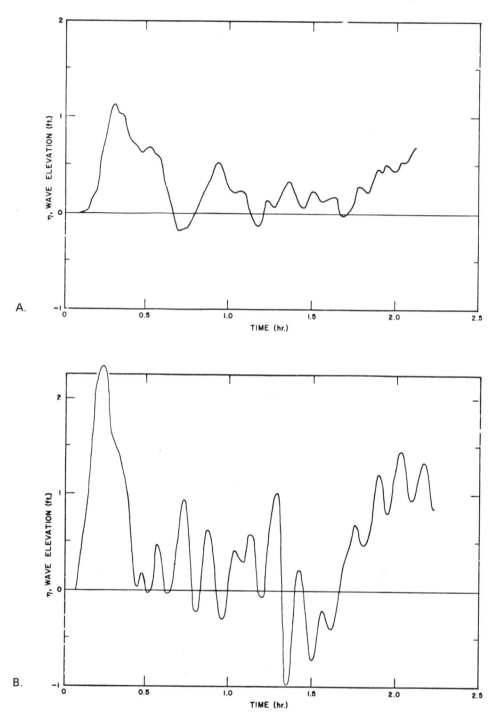

A.

B.

Figure 22-7 continued

Figure 22-7 continued

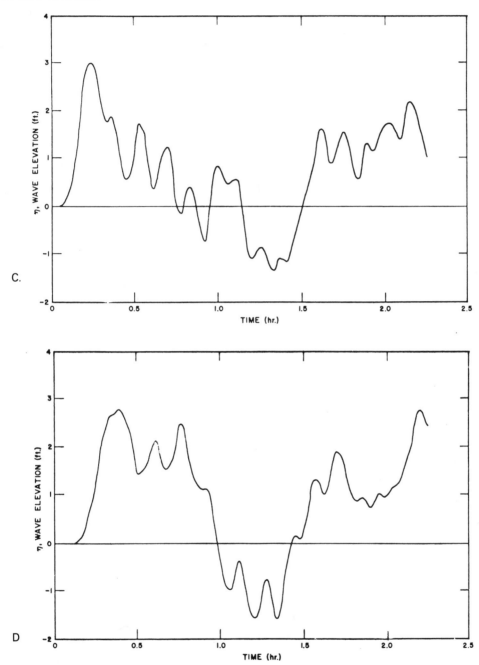

Figure 22-7. Computed wave history at points indicated in Figure 22-4. Notice that smaller period waves are superimposed on the larger period waves (Hwang and Divoky, 1970). *A*, point 1; *B*, point 2; *C*, point 3 and *D*, point 4.

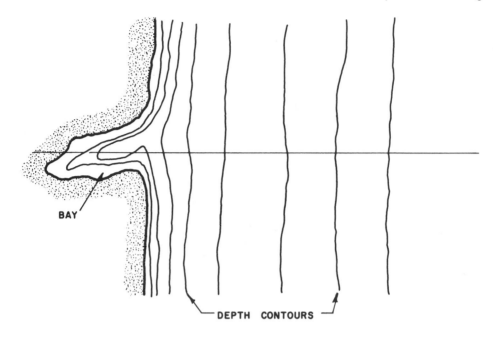

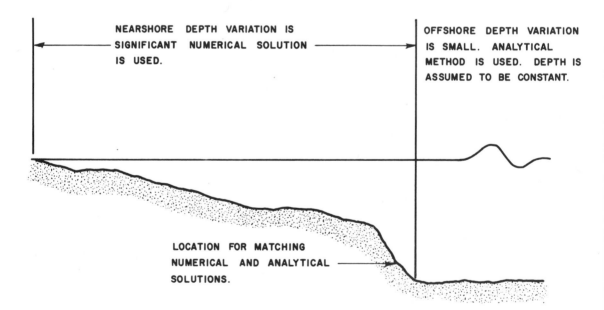

Figure 22-8. Schematic drawing of bottom topography. The topography near the bay is quite irregular but becomes two-dimensional in deeper water.

open boundary. The open boundary must be fairly close to the coast because of the finite capacity of the computer. As a result of complex interference patterns, the total transmission condition at the open boundary is rather difficult to simulate accurately. This makes the problem of coastal response exceedingly difficult to solve unless simplifications are made.

The first simplification is to linearize the equation of motion; then the principle of superposition can be used.

Upon linearization, equations (22-1) and (22-2) become

$$u_t + g \, \eta_x = 0$$

and

$$v_t + g \, \eta_y = 0. \tag{22-3}$$

Assuming the existence of a velocity potential Φ with $\Phi_x = u$, $\Phi_y = v$, integration of (22-3) gives the linearized Bernoulli equation, $\Phi_t + g\eta = 0$.

Assuming that $h \gg \eta$, the continuity condition becomes $\eta_t - [h\Phi_x]_x - [h\Phi_y]_y = 0$ so that $(1/g)\Phi_{tt} = [h\Phi_x]_x + [h\Phi_y]_y$ (22-4) through the substitution of equation (22-3).

The second simplification is to assume periodic incident waves $\Phi(x, y, t) = \varphi(x, y)e^{i\sigma t}$ where σ is the frequency of the excitation. Equation (22-4) now becomes

$$[h\varphi_x]_x + [h\varphi_y]_y + (\sigma^2/g)\varphi = 0. \tag{22-5}$$

The solution of equation (22-5) considering water depth, h, as a variable or a constant has been treated theoretically (Hwang and Tuck, 1970; LeMéhauté and Hwang, 1970; and Olsen and Hwang, 1971). For the sake of completeness an outline of the solution method is given.

The analytical solution of equation (22-5) in an arbitrary domain has eluded mathematicians for centuries. In this chapter, only a numerical solution is sought for the harbor oscillation. Because this differential equation is elliptic in nature and the domain of interest extends to infinity, it is

impossible for a direct numerical solution to include the entire domain. This is due to limited storage of a computer. The method to be developed here requires termination of the open region at some finite value. To assure the correct numerical solution, the wave form must be known at this outer boundary where the truncation is made. The boundary condition must not only satisfy the condition of the incoming wave, but also the scattered wave propagating away from the harbor.

In reality, the topography of a typical nearshore (including bay) and offshore area can be classified into two regions (see Figure 22-8): 1) inner region (nearshore), where variation of bottom geometry is large and usually has a significant effect on the wave characteristics and 2) outer region (offshore), where the water depth variation is relatively small. As the distance increases from shore, the bottom topography tends to be two-dimensional, the water depth variation decreases and the effect of bottom topography on wave behavior diminishes. Since the geometry within and near the bay can be very complicated, numerical calculations must be used to accurately represent the variations of bottom topography and wave behavior. A large number of mesh points is not desirable because the cost of computing time varies directly with the cube of the number of mesh points. Thus, the region must be small enough so the cost of calculation is feasible without affecting the accuracy of the results. However, it is possible to obtain an analytical solution in the outer region by considering the water depth to be constant and then to match the analytical and numerical solutions at the proper boundary. Through the use of this scheme, both incoming and reflected wave conditions may be properly considered.

To obtain the condition at the matching boundary, an analytical solution for the outer region is developed in the following equations. In the outer region, where the water depth is assumed to be constant, the governing equation reduces to the Helmholtz equation

$$\nabla^2 \varphi + k^2 \varphi = 0 \tag{22-6}$$

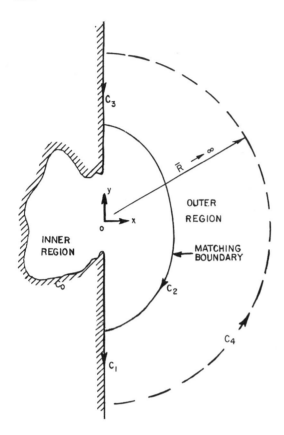

Figure 22-9. Schematic diagram used for obtaining boundary condition between inner and outer regions.

where $k^2 = \sigma^2/gh$.

The boundary conditions around the region (see Figure 22-9) are $\partial\varphi/\partial n = 0$ along the shoreline and $\varphi = \varphi_0$ at infinity where φ_0 is the prescribed boundary condition at infinity. This condition is valid if the condition at the matching boundary is such that the wave characteristics on both sides of the boundary are equal, and the equation describing this situation will be derived later in this section.

For solving equation (22-6), the velocity potential kernel φ will be considered in two parts: φ_0 resulting from incoming waves and φ' resulting from the presence of the boundary. Thus, the velocity kernel at any location in the region is $\varphi = \varphi_0$

$+ \varphi'$ where φ_0 and φ' both satisfy the Helmholtz equation (22-6). The boundary condition to be satisfied at is the shorelines C_1 and C_3

$$(\partial/\partial n)\,(\varphi' + \varphi_0) = 0. \tag{22-7}$$

At the matching boundary C_2, the velocity potential kernels $\varphi = \varphi_0 + \varphi'$ must be equal on both sides of the boundary. The perturbation velocity potential kernel φ' can be obtained by using Green's theorem

$$\varphi' = \int_S (\varphi'\,\partial G/\partial n - G\,\partial\varphi'/\partial n)\,dS$$

where G is the Green's function, which will be determined later, and S is the contour of integration. The integration is carried around the boundaries C_4, C_3, C_2 and C_1 with the integration directions as indicated in Figure 22-9).

The term G must 1) be a solution of the Helmholtz equation; 2) have a singularity at $R = 0$; and 3) have vanishing influence at infinity. The Hankel function of the first kind and zeroth order is a solution of the Helmholtz equation and has the following properties:

$$H_0^{(1)}(kR) \to i/\pi\; n(kR),\; R \to 0$$

$$\to -(i/\pi)[(2\pi/kR)\exp i(kR + \tfrac{1}{4}\pi)]^{1/2},\; R \to \infty.$$

so that it may be chosen as the Green's function. If the integration contour C_4 is extended to infinity $(R\to\infty)$;

Therefore, the value of φ on the matching boundary should be

$$\varphi = \varphi_0 + \varphi'(x, y)$$

$$= \varphi_0(x, y) + \int_{C_1+C_2+C_3} [(\varphi'\,\partial H_0^{(1)}/\partial n) - (H_0^{(1)}\,\partial\varphi'/\partial n)$$

where φ' is the solution of equation (22-5).

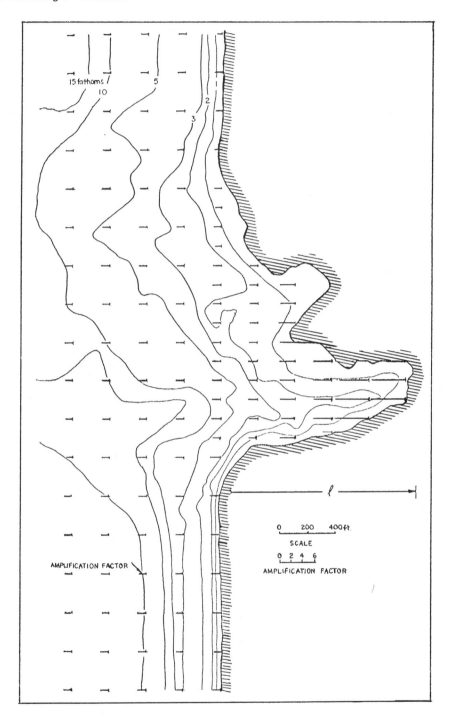

Figure 22-10. Wave amplification field at $(2\,\pi\,1)/L = 0.4$ (results were calculated using the variable water depth program) for Keauhou Bay, Hawaii.

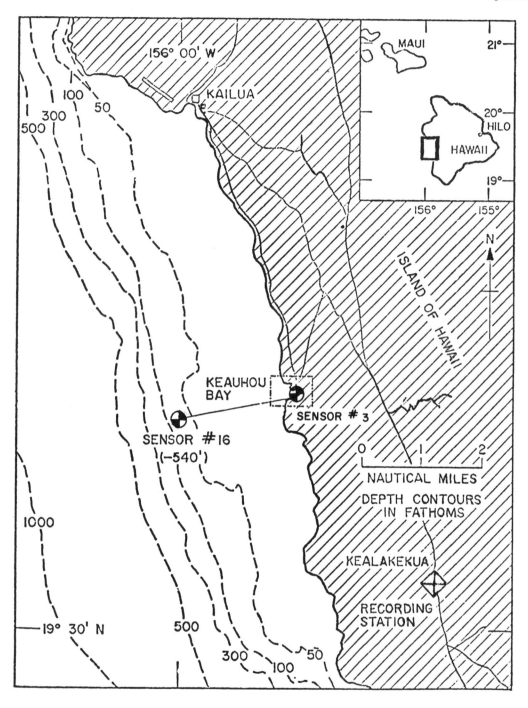

Figure 22-11. Deep water wave sensor locations off the Kona Coast of Hawaii, showing sensor sites (Olsen and Hwang, 1971).

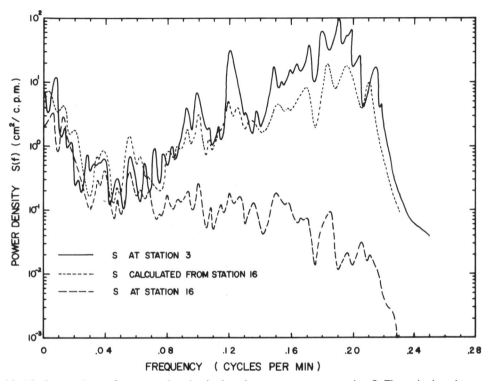

Figure 22-12. Comparison of measured and calculated wave spectra at station 3. The calculated spectrum is based on the measured (input spectrum) at station 16 (Olsen and Hwang, 1971).

Thus, the solution of the oscillations in a harbor of arbitrary shape and arbitrary water depth reduces to the solution of a boundary value problem in a finite region with one of the boundary conditions expressed as an integral equation (22-8) to be satisfied at the matching line as shown in Figure 22-9. Once the values of $\varphi(x, y)$ are known, the amplification factor A is simply equal to the modulus of $\varphi(x, y)$.

The solution of equation (22-5) with the boundary conditions (22-7) and (22-8) is rather involved. In fact, the solution for waves of higher frequencies is not yet available. For long waves with normal incidence numerical results have been obtained. Sample applications are shown in Figures 22-10 and 22-11.

Figure 22-10 shows the wave amplification field for Keauhou Bay (Hwang et al., 1969). These results were obtained using variable depth theory.

Hwang and Olsen (1971) have used these data to compare results obtained from field measurements. The measurements were taken at sensors 16 and 3 as shown in Figure 22-11.

Figure 22-12 is obtained by considering the spectrum measured at station 16 as an input spectrum and calculating the spectrum at location 3 through use of the bay response and shoaling. The results agree reasonably well between computed and actual measured values.

In a case where water depth is constant or nearly so, as in a harbor, the calculation of response behavior is much simpler. The method is to seek solutions of the Laplace equation $\nabla^2 \Phi = 0$ throughout the fluid contained within the boundary surfaces. If the wave is assumed to be of small amplitude, the velocity term in the Bernoulli equation may be neglected. Thus, the governing dynamic boundary condition on the free surface becomes $\eta = (1/g)(\partial\Phi/\partial t)$ at $z = 0$.

The linearized kinematic condition at the free surface, which follows from the fact that surface water particles stay on the surface, is expressed in the form $\partial\eta/\partial t = \partial\Phi/\partial z$ at $z = 0$. The condition on the fixed boundaries is that the normal velocity equals zero; that is $\partial\Phi/\partial n = 0$ on the boundary S.

Since we are dealing with uniform water depth h, the condition at the bottom is simply $\partial\Phi/\partial z = 0$ at $z = -h$.

Finally, the condition at infinity requires that $\Phi = \Phi_0 + \Phi'$, where $\Phi_0 = \cos(kx \cos\beta) \exp[-is, \omega t - ky \sin\beta)]$, and Φ' is an outgoing wave. Since the water depth is uniform, we may assume that velocity potential is a product of functions of x and y, z and t, such as $\Phi(x, y, z, t) = (1/\omega i)\varphi(x, y)Z(z)e^{-i\omega t}$ where ω is the angular frequency.

Substituting the above expression into the Laplace equation, separating the functions of x and y, and z and equating them to a constant, say k^2, we have

$$(\delta^2\varphi/\delta x^2) + (\delta^2\varphi/\delta y^2) + k^2\varphi = 0 \qquad (22\text{-}9)$$

and

$$(\partial^2 Z/\partial z^2) - k^2 Z = 0. \qquad (22\text{-}10)$$

The solution of (22-10) together with the bottom boundary condition and the kinematic surface boundary condition is simply $Z(z) = -Ag \cosh K(z + h)/\cosh kh$.

The problem now is to obtain the solution of (22-9) with the boundary condition $\partial\varphi/\partial n = 0$ on the solid boundary S. The condition at infinity can be determined as if the harbor were absent. This is due to the fact that the influence of radiated waves from the harbor tends to be zero at infinity. Thus, for a straight-crested standing wave at infinity with crest at an angle β to the shoreline, we have $\varphi_0 = \cos(kx \cos\beta) \exp[-iky \sin\beta]$ $(0 < \beta < \pi)$.

For a standing wave of unit amplitude at infinity, the solution of the equation (22-9) together with the boundary conditions can be found through the introduction of a source function $Q(\xi\eta)$ along the boundary S where $\overline{\xi}$ and $\overline{\eta}$ refer to coordinates on the boundary.

Thus, the value of $\varphi(x,y)$ at any point (x,y) is equal to the sum of two parts; one is the influence from infinity $\varphi_0(x, y)$ and the other is the contribution of the source distribution, that is, the scattered wave caused by the presence of the boundary. The latter will be given by $\int_S dSQ(\xi,\eta)$ $G(x,y;\xi,\eta)$ where $G(x,y\ \xi,\eta)$ is the Green's function and $Q(\xi,\eta)$ is the unknown source distribution, which can be determined from the boundary conditions.

The Green's function has to be chosen so that it is the solution of the equation (22-9), satisfies the radiation condition at infinity and has a singularity at the source point. Thus, we choose the Green's function to be a Hankel function of the first kind rather than of the second kind to guarantee that the disturbance, due to the harbor, at infinity takes the form of an outgoing wave rather than an incoming wave. $G(x,y;\ \xi, \eta) = -\tfrac{1}{4} iH_0(1)(kR)$

where $R = [(x - \xi)2 + (y - \eta)2]^{1\cdot2}$

so that the value of $\varphi(x, y)$ at any point (x, y) is

$$\varphi(x, y) = \varphi^0(x, y) + \int dS\, Q\,(\overline{\xi}, \overline{\eta})G(x, y; \overline{\xi}, \overline{\eta}).$$

The problem now is to determine the strength of the source distribution $Q(\overline{\xi}, \overline{\eta})$. This can be accomplished by applying the boundary condition which gives

$$\lim_{x,y \to \xi,\eta} \frac{\partial\varphi_0(x,y)}{\partial n} + \frac{\partial}{\partial n} \int_S dSQ\,(\xi,\eta)\, G\,(x,y)$$

$$(\xi,\eta) = 0.$$

Since the limit is singular inside the integral, it has to be treated with care. We evaluate the integral in (22-11) by using contour integration. The path of the integral is along the boundary except around the point $(\overline{\xi}', \overline{\eta}')$ where the contour is deformed into a small circle with a radius ϵ. Since the contribution around a large semicircle is zero, the integral in (22-11) may be evaluated as follows:

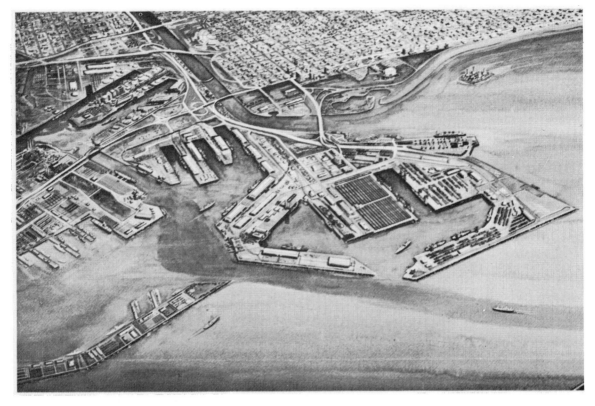

Figure 22-13. Aerial view of Long Beach Harbor, California.

$x, y \overset{\lim}{\to} \xi', \eta' (\partial/\partial n) \int_S dS\, Q(\xi, \eta)\, G(x, y; \xi, \eta)$

$= \oint_S dS\, Q(\xi, \eta)\, G_n (\xi', \eta'; \xi, \eta)$

$+ x, y \overset{\lim}{\to} \xi', \eta' \int_6 dS\, Q(\xi, \eta)\, G_n (x, y: \xi, \eta)$

where the sign $\oint_S$ refers to a principal value in the sense of Cauchy.

Since the Hankel function can be approximated by $-\tfrac{1}{4} i H_0^{(1)}(KR) \to (1/2\pi) \ln(kR)(R \to 0)$ the second integral of the right-hand side of (22-12) may be integrated analytically. We have

$x, y \overset{\lim}{\to} \xi', \eta' (\partial/\partial n) \int_\epsilon dS\, Q(\xi, \eta)\, G(x, y; \xi, \eta)$

$= R \overset{\lim}{\to} 0 (\tfrac{1}{2}\, \pi)\, Q(\xi', \eta')$

$\int_{-\pi}^{\pi} (\partial/\partial R) \ln k\, R\, R\, D\theta = \tfrac{1}{2} Q(\xi', \eta').$

Thus the integral equation becomes

$\tfrac{1}{2} Q(\xi', \eta') + \oint_S dS\, Q(\xi, \eta)\, G_n (kR)$

$= -(\partial/\partial n)\, \varphi_0 (\xi', \eta')$

where $G_n(kR) = -\tfrac{1}{4} i \partial(H_0^{(1)}(kR))/\partial n$ which must be evaluated numerically. This calculation is performed by dividing the boundary into short segments. For accuracy the lengths of the segments must be such that several, say eight, are contained in one wavelength. Within each segment, Q is assumed constant. The integral becomes a summation, and one arrives at an algebraic rather than an integral equation:

$\sum_{j=1}^{n} B_{ij}\, Q_j = b_i$

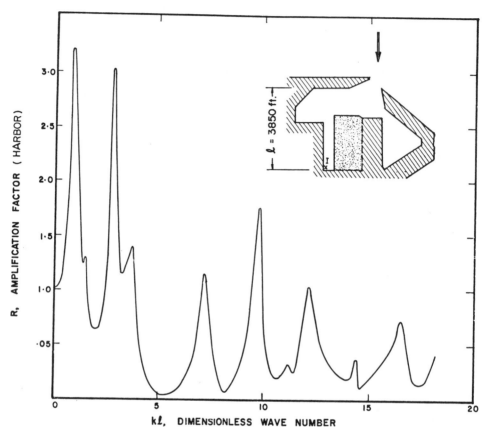

Figure 22-14. Response behavior at location *I*, Long Beach Harbor, California. Water depth is approximately 55 feet. Arrow indicates incoming wave direction.

where

$$B_{ij} = \tfrac{1}{2}\delta_{ij} + \int_{\Delta S_j} G_n(X_j, Y_j; \xi_i, \eta_i) dS$$

$$b_i = -(\partial/\partial n)\varphi_0(X_i, Y_i)$$

$$\delta_{ij} = \text{Kronecker delta}$$

and $(X_j \, Y_j)$ is the midpoint of segment S_j. Ultimate solution involves inversion of the B matrix, thus giving Q and hence φ and Φ.

Examples of Response Calculations

An example calculation of the response curve for Long Beach Harbor (Figure 22-13) is given in Figures 22-14 and 22-15. The response behavior of the harbor at locations *I* and *M* are shown. It is interesting to note that both locations have more or less the same fundamental amplification factors, while location *M* has much smaller harmonics. The reason for this is that the fundamental is a pumping mode. The amplification factor is the same everywhere in the harbor, while the second harmonics are the contribution of the long slip. The amplification tends to decrease at the mouth of the slip.

Figure 22-16 shows the instantaneous velocity field and amplification factors at $kl = 9.5$. It is interesting that the nodal line does not occur at the entrance, as is usually assumed in harbor resonance studies.

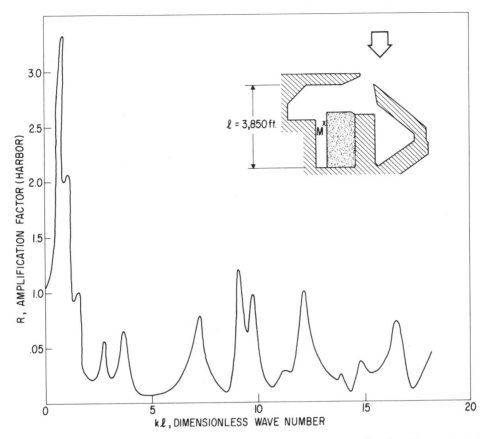

Figure 22-15. Response behavior at location *M*, Long Beach Harbor, California. Water depth is approximately 55 feet. Arrow indicates incoming wave direction.

The results in Figures 22-14 and 22-15 have been confirmed experimentally by field measurements.

Discussion and Recommendations

The link between the generation and the harbor response solutions presented here is propagation of the generated waves over oceanic distances. This problem must be solved using a spherical earth. The grid in a finite difference solution must be of sufficient density that the essential features of the waves are retained and yet sparse enough so that computer evaluation is rapid and accurate. The equations to be solved are $u_t = -(g/a)\eta_0$, $v_t = (-g/a \sin \theta)\eta_\varphi$

and

$$\eta_t = (1/a \sin \theta) (\partial/\partial_0)[(h + \eta) \sin \theta] + (\partial/\partial\varphi) [(h + \eta)v)] + \xi_t$$

where

a = radius of earth
θ = latitude
φ = longitude
w = wave velocity component in the θ-direction
o = wave velocity component in the φ-direction
ξ = ground motion.

The Coriolis force has been neglected.

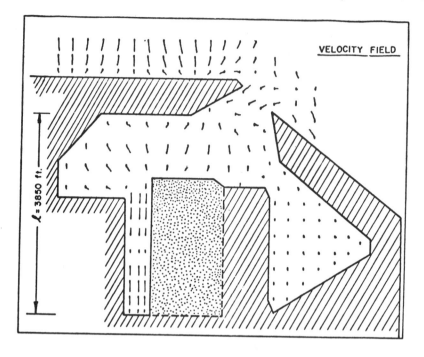

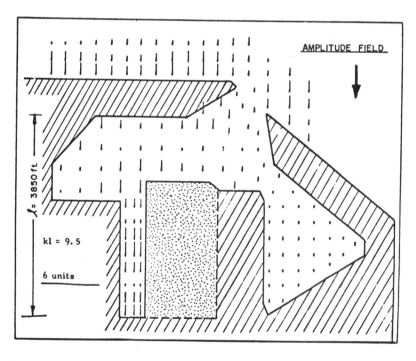

Figure 22-16. Velocity and amplitude field of Long Beach Harbor at wave period of 60 seconds. Water depth is approximately 55 feet. Arrow indicated incoming wave direction.

Work is currently underway to solve these equations for the Pacific basin with an arbitrary generation location.

The tsunami warning system is at present capable of predicting arrival times of waves once the epicenter, magnitude and time of an earthquake are known.

It is possible to study each earthquake-prone area of the Pacific basin and to at least probabilistically establish the most likely characteristics of each generation area.

With these data and the methods presented here it would then be possible to predict the probable wave height at various critical locations given only the earthquake characteristics now available to the warning system.

A further use of the methods discussed in this chapter is in harbor design and modification. Earthquake consequences could be studied a priori and protective measures incorporated into the engineering design phases of the work. In other words, the effects of tsunamis could be considered as design criteria of a far more specific nature than engineers have to work with today.

Problems for future research lie in the near field of the epicenter and in refinement of coastal effects. In the near field, for example, the high-frequency components of the earthquake motion are significant in wave generation; also, detailed analysis must be made of the horizontal ground motion consequences. In the far field these effects are negligible, but they cannot be ignored in the region where transients are significant.

Numerous coastal effects have not been included in the development described here. There is no provision for breaking waves or for wave instability, though tsunami waves are known to exhibit these characteristics and the run-up is affected by these phenomena. Madsen and Mei (1969) and Peregrine (1970) have studied the wave stability problem extensively. Another effect not included here is edge waves. These are significant because of their slow decay. Shelf resonance is an important effect which deserves more study. Finally, trailing waves are also important because they are of short period, which may cause mooring problems for ships in harbors.

Acknowledgments

The writer would like to express appreciation to B. LeMéhauté and W. Van Dorn for their valuable discussions and to D. Divoky and L. Butler and Miss L. Webb for extensive contributions to this chapter. Finally, I would like to express thanks to K. Olsen and A. Cox for their interest in the program which made this chapter possible.

References

1. Benioff, H. 1951. Earthquakes and rock creep. *Bull. Seismol. Soc. Amer.* 41(1): 31-62.
2. Berg, G.V. and Housner, G.W. 1961. integrated velocity and displacement of strong earthquake ground motion. *Bull. Seismol. Soc. Amer.* 51(2): 175-189.
3. Honda, H. and Nakamura, K., 1971. *The waves caused by one-dimensional deformation*
3. Honda, H. and Nakamura, K. 1971. *The depth.* Sci. Rep. Tohoku U., Sendai, Japan, vol. 3, pp. 133-137.
4. Hwang, Li-San and LeMéhauté, B. 1968. *On the oscillations of harbors of arbitrary shape.* Tetra Tech Rept. TC-123A, prepared under AEC Contract AT(26-1)-289(M002).
5. _____ et al. 1969. *Oscillations of harbors of arbitrary shape and variable depth and Amchitka field experiments.* Tetra Tech Rept. TC-148.
6. _____ and Divoky, D. 1970. Tsunami generation. *J. Geophys. Res.* 75(33): 6802-6817.
7. _____ and Tuck, Ernest O. 1970. On the oscillations of harbors of arbitrary shape. *J. of Fluid Mechanics* 42: 447-464.
8. _____ and Divoky, D. 1971. A numerical model of the Alaskan tsunami. In *The Great Alaska Earthquake of 1964.* National Academy of Sciences, in press.
9. Ichiye, T. 1958. A theory of the generation of tsunamis by an impulse at the sea bottom. *J. of the Oceanographical Soc. of Japan,* 14(3): 41-44.
10. Kajiura, K. 1963. The leading wave of a tsunami. *Bull. Earthquake Res. Inst.* vol. 41, pp. 535-571.
11. Leendertse, Jan J. 1967. Aspects of a computational model for long period water-wave propagation. Rand Co. Rept. RM-5294-PR.
12. LeMéhauté, B. and Hwang, Li-San. 1970. Harbor design: scale model or computer? In

Topics in Ocean Engineering, C.L. Bretschneider, ed. Houston: Gulf Publishing.

13. Loomis, H. 1966. *Some numerical hydrodynamics for Hilo Harbor, Hawaii.* Institute of Geophysics, University of Hawaii.

14. Madsen, O.S. and Mei, C.C. 1969. *Dispersive long waves of finite amplitude over an uneven bottom.* Dept. of Civil Enginnering, M.I.T. Rept. 117.

15. Momoi, T. 1964. Tsunami in the vicinity of a wave origin, 1, 2. *Bull. Earthquake Res. Inst.* vol. 42, pp. 133-146, 369-381.

16. Olsen K. and Hwang, Li-San. 1971. Oscillations in a bay of arbitrary shape and variable depth. *J. Geophys. Res.* Publication pending.

17. Peregrine, D.H. 1970. *Long waves in two and three dimensions.* Symposium on Long Waves, University of Delaware.

18. Plafker, G. 1969. *Tectonics of the March 27, 1964 Alaska earthquake.* Geophys. Surv. Prof.

Paper 543-1, pp. 4-25.

19. _____ and Savage, J.C. 1970. Mechanism of the Chilean earthquakes of May 21 and 22, 1960. *Bull. Geol. Soc. Amer.* 81:1001-1030.

20. Van Dorn, W.G. 1965. Tsunamis. In *Advances in Hydroscience* 3. New York: Academic Press, pp. 1-48.

21. _____ 1970. Tsunami response at Wake Island: a model study. *J. Mar. Res.* 28:336-344.

22. Webb, L.M. 1962. Theory of waves generated by surface and sea-bed disturbances. In *The nature of tsunamis,* by B. Wilson, Appendix, 1 National Eng. Sci. Co., Tech. Rept. SN 57-2.

23. Wilson, B. 1969. *Earthquake occurrences and effects in ocean areas.* U.S. Naval Civil Eng. Lab. TR. CR. 69.027, pp. 99-100.

24. _____ and Torum, A. 1968. *The tsunami of the Alaskan earthquake, 1964.* Coastal Eng. Res. Center Tech. Mem. 25, p. 50.

part 9: william g. van dorn

Tsunami Engineering

A Research Oceanographer in the Scripps Institution's Division of Oceanic Research, Dr. Van Dorn holds his B.S. in mechanical engineering from Stanford University, his M.S. and Ph.D. in oceanography from the University of California (1953).

Before coming to Scripps, Dr. Van Dorn worked with Northrop, Solar, Convair, and Aerojet Engineering Corporation; his experience with these organizations was in stress analysis, research, design, and the testing of airframes and rockets.

Dr. Van Dorn's background in oceanography includes participation in most of the nuclear tests in the Pacific. His fields of interest include air-sea interaction, waves produced by explosions and undersea earthquakes (tsunamis), circulation around islands, and the dynamics of breaking waves. He has developed a large number of special devices and instruments for the measurement of oceanographic phenomena, and acted as consultant and advisor to many Federal, State, and corporate agencies. He is currently a member of several National committees concerned with environmental effects and public safety.

23

tsunami engineering

Introduction

Although tsunamis have sporadically wrought destruction along oceanic coastlines for centuries, only within the last decade has a sufficient knowledge of their characteristics and behavior been obtained to provide some general guidance to engineers faced with the problem of designing coastal structures. While even today we lack adequate statistics to properly assess the long-term probability of tsunami occurrence, the theoretical and computational tools are in hand to make reasonable estimates of local effects from hypothetical tsunamis. Application of these methods should substantially reduce the incidence and extent of future damage.

Tsunami Phenomenology

Before discussing their engineering aspects, it is important to place tsunamis in a proper perspective as natural phenomena; to understand how they are generated, how they expand and propagate across the ocean and, lastly, how they are transformed in shallow water and rush up along the shore.

Tsunami Generation

Destructive tsunamis principally arise from vertical sea floor dislocations associated with shallow-focus earthquakes of Richter magnitude 7.5 or larger. Even in a particular source region only about one such earthquake in six produces a destructive tsunami, and there may be long intervals between occurrences. For example, the Aleutian Arc is a continuously active seismic belt, along which several tsunamis were generated in the early 1800s. These were followed by a lapse of 100 years before three more occurred in 1946, 1957 and 1964. The dislocated source area—as well as defined by the seismic aftershock pattern—is usually elliptical and may cover 100,000 square miles of the sea floor. Maximum vertical dislocations as high as 50 feet have been observed, tapering off toward the source margin. Judging from accelerometer records, local dislocations occur within 10-15 seconds, and the rupture propagates from the epicenter along the major fault axis at speeds of 2 to 3 miles per second. The entire disturbance is usually over within a few minutes. During this period the sea surface is deformed to fit the dislocation pattern, following which it spreads out as a system of free gravity waves in all directions.

Mid-Ocean Character

In the open sea the leading edge of the wave system propagates at a velocity $v = \sqrt{(gh)}$,* where

*Actually it travels slightly faster by an amount that increases with the cube root of travel distance (8, p. 14).

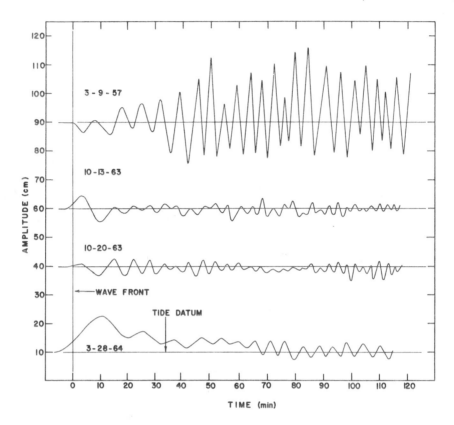

Figure 23-1. Initial portions of four actual tsunami records obtained at Wake Island. Records have same
amplitude and time scales but are uncorrected for instrumental prefiltering.

Table 23-1
Data for Tsunamis Recorded at Wake Island

Tsunami date	Time of quake (GMT)	Time of first arrival	Epicenter location Lat.	Long.	Richter magnitude	Angular distance (radians) Wake	Hawaii	Travel time, Wake t_1 (min)
3/9/57	1422	2054	51N	175W	8.3	0.613	0.593	272
10/13/63	0518	0911	44N	150E	8.1	0.518	0.837	233
10/20/63	0053	0443	44N	150E	7.7	0.515	0.834	250
3/28/64	0336	1023	60N[a]	146W[a]	8.4	0.907	0.690	407

[a]Coordinates are for point of tsunami origin.

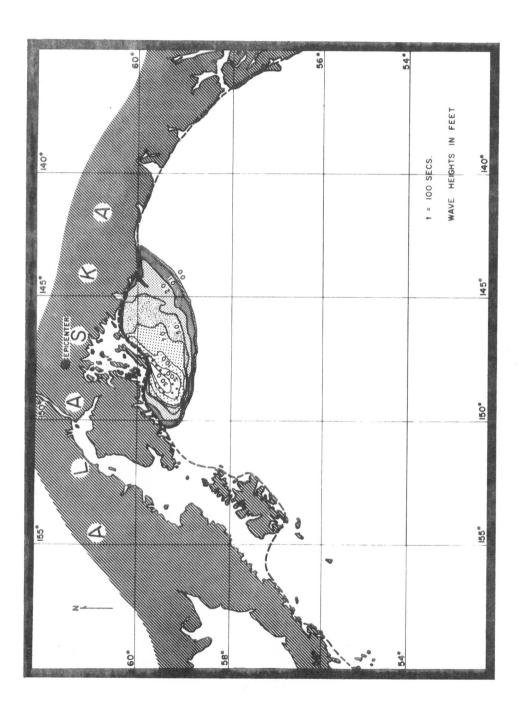

Figure 23-1. Surface elevation, Alaska, 1964, at 100 seconds (from Hwang, Ref. 2).

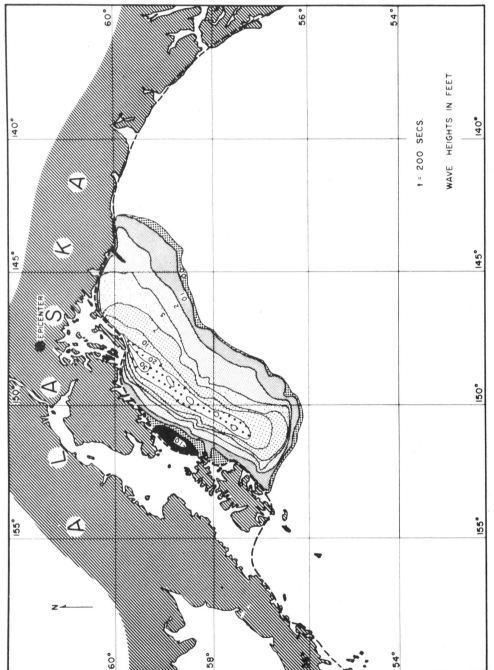

Figure 23-3. Surface elevation, Alaska, 1964, at 200 seconds (from Hwang, Ref. 2).

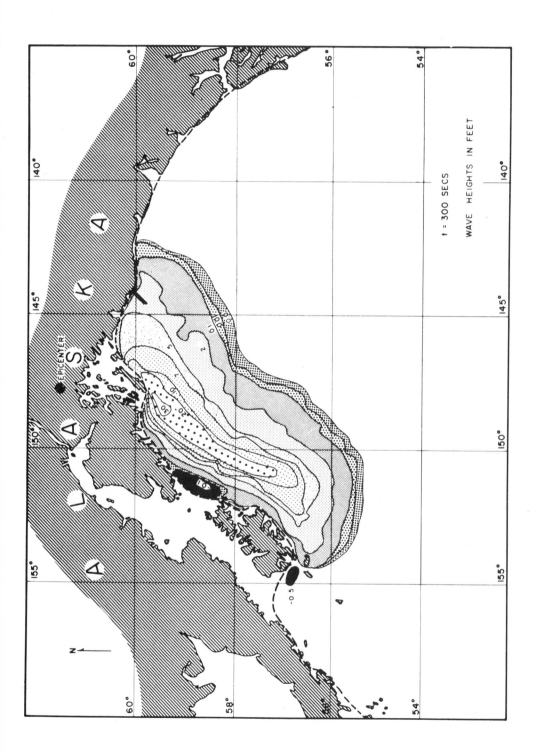

Figure 23-4. Surface elevation, Alaska, 1964, at 300 seconds (from Hwang, Ref. 2).

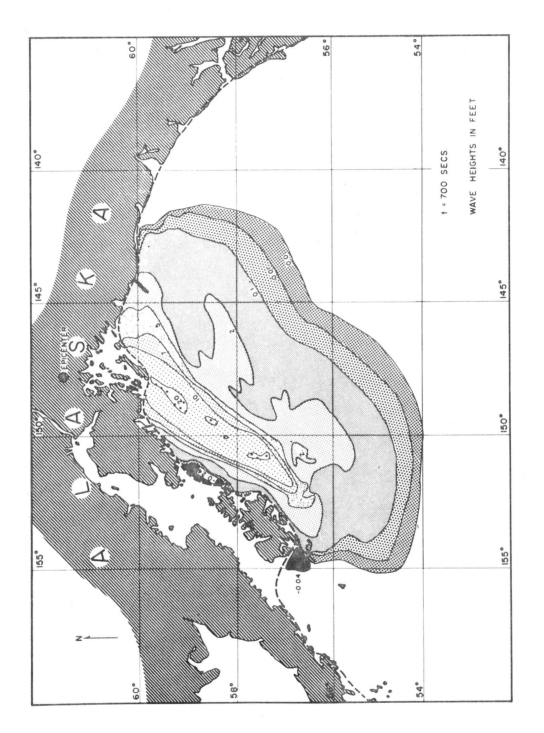

Figure 23-5. Surface elevation, Alaska, 1964, at 700 seconds (from Hwang. Ref. 2).

g is the gravitational acceleration and h is the local water depth. In the Pacific Ocean, this speed may be as high as 500 mph. As the wave pattern expands, its average amplitude is reduced because it contains a constant energy that is spread thinner because of lateral expansion, and also by radial stretching that somewhat resembles an accordion bellows.

Although deep-sea recording devices have been developed and are now being deployed aboard ocean weather ships, no tsunami has ever been recorded in mid-ocean. They have, however, been observed at special recording stations on small islands. While even a small island somewhat distorts the wave pattern (IX), such records provide some insight into the character and variability of tsunami sources. Figure 23-1 shows the first two hours of records for four tsunamis observed at Wake Island, for which the pertinent source data are given in Table 23-1. Each record consists of a train of dispersive waves of progressively increasing frequency superimposed on a broad, irregular series of oscillations. The important feature of these records is their extreme variability. The upper and lower traces are from large destructive tsunamis, whereas the two center traces are from very small tsunamis that were barely detectable in Hawaii, despite the fact that the dispersive waves in the lower three records are roughly the same size. The extreme difference in their respective coastal severity is attributed to the fact that the low-frequency phase is almost missing in the center records, indicating relatively smaller sources and dislocations. The maximum amplitudes of all records are only a few inches in mid-ocean, yet the large tsunamis resulted in run-up heights of 20-50 feet in coastal areas of Hawaii and North America.

Coastal Modification

As a tsunami approaches shallow water along the coasts of continents or large islands, the relatively systematic wave system is completely altered by complex interaction with the coastal shelf and irregular shoreline, so that it locally appears as a highly periodic series of low-frequency oscillations of much larger amplitude than those present in mid-ocean. In some sense the coastline might be considered as an array of high–Q filters subjected to a broadband impulse that selectively amplifies those elements of the incoming spectrum to which they are naturally tuned. As a result, the tide gage records at a particular station look very similar for different tsunamis, whereas the same tsunamis produce widely different signatures on different gages. Thus, the local response to any tsunami is a property of the environment, except that the gross amplitude still depends on the tsunami spectral intensity within the pass band of the filter response. This feature is of great engineering significance because one need only understand the local response at a given place in order to design for any remote tsunami—provided only that the maximum design intensity can be properly estimated.

Methods of Analysis

If the source motion can be specified, one can in principle determine the resulting wave system by solving the hydrodynamic equations of motion. If the source can be approximated by analytic functions in water of constant depth, it can be treated as an array of discrete source elements whose spectral contributions can be summed to give the local time history at any remote point (VIII). A number of special cases of this type have been considered by Kajiura (V), but without application to real tsunamis.

However, it is apparent from several independent lines of evidence that natural sources are quite complex and usually occur in areas of irregular topography where simple analytic methods cannot be used. In a very recent study, Hwang et al. (II) developed a computer program to numerically integrate the equations of motion for arbitrary, space-and time-dependent sources and have computed the wave development for several large tsunamis, using source models obtained from observational data. Figures 23-2 to 23-6 show five consecutive wave contour plots from their recon-

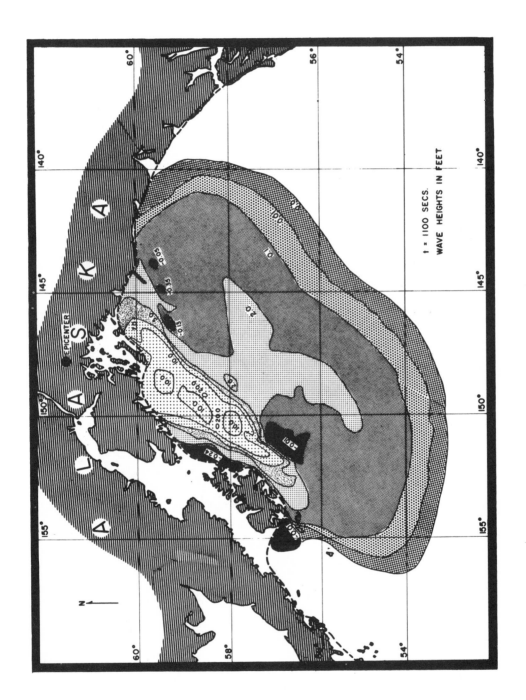

Figure 23-6. Surface elevation, Alaska, 1964, at 1100 seconds (from Hwang, Ref. 2).

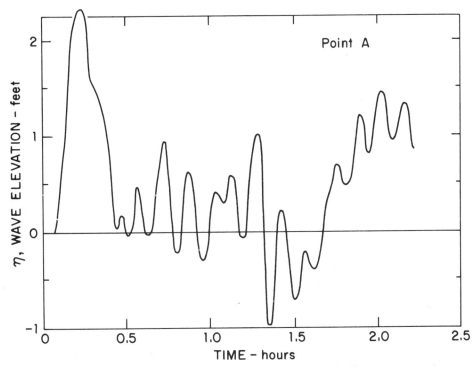

Figure 23-7. Wave history at point *A*, Figure 23-6, (from Hwang, Ref. 2).

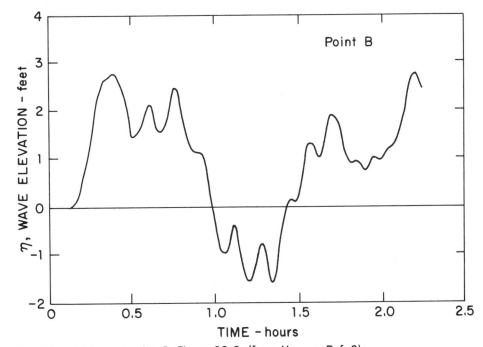

Figure 23-8. Wave history at point *B*, Figure 23-6, (from Hwang, Ref. 2).

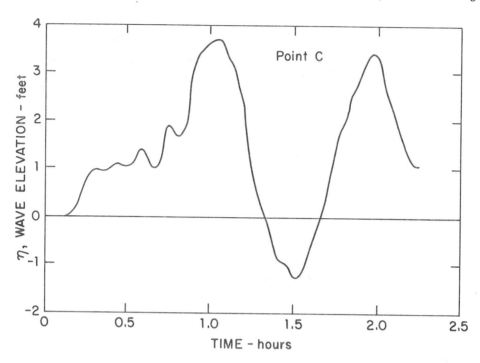

Figure 23-9. Wave history at point *C*, Figure 23-6, (from Hwang, Ref. 2).

struction of the large tsunami of March 28, 1964, in southeastern Alaska, in which the highly irregular nature of the wave pattern is clearly evident. Figures 23-7 to 23-9 show the time histories of waves propagating past the three observation points indicated in Figure 23-6, illustrating that there are pronounced differences in the amplitudes and phases of the wave systems propagating in different directions. This radial dependence leads to corresponding directional differences in the severity of local effects in remote coastal sectors.

Because of the close grid spacing required to give adequate resolution in the source region, the calculations were terminated at 1100 seconds, and the resulting wave systems were then propagated over the ocean by conventional ray refraction techniques. Table 23-2 compares the computed maximum wave amplitudes offshore to corresponding tide gage amplitudes at 23 selected stations around the Pacific Ocean and at several islands in mid-ocean. While there is a general corre-

spondence in diminution of wave intensity with distance between the computed and observed amplitudes along the coastline of North and South America, small discrepancies occur for the larger island stations and the computed values for Japan are two orders of magnitude too low.

These discrepancies can be attributed partly to the inaccuracy of the ray-optics approximation and partly to the highly selective local response factors at individual gage locations. Extension of the numerical grid calculation to oceanic propagation and to the local region near each gage station would probably resolve much of the present disagreement. Appropriate computer techniques for analyzing local response to arbitrary incoming wave systems have already been developed (3), but time and cost precluded application to the above study. The important point to be made here is that there appears to be no simple shortcut to predicting local behavior, even if the character of a remote tsunami source is well defined.

Table 23-2
Comparison of Computed Offshore Amplitudes
with Local Tide Gage Amplitudes
for Alaskan Tsunami of March 18, 1964

No.	Tide Station	Computed Amplitude, ft.	Tide Gage Height, ft.
1.	Unalaska, Alaska	0.5	2.6
2.	Sitka, Alaska	2.2	14.3
3.	Crescent City, California	2.2	22±2
4.	LaJolla, California	1.9	2.7
5.	Acapulco, Mexico	1.9	3.5
6.	Quepos, Costa Rica	1.9	1.5
7.	San Cristobal, Galapagos Island	1.9	3.8
8.	Talara, Peru	1.9	3.5
9.	Arica, Chile	1.9	7.0
10.	Valparaiso, Chile	1.9	6.7
11.	Honolulu, Hawaii	1.6	3.7
12.	Hilo, Hawaii	1.6	12.5+
13.	Johnston Island	1.5	1.0
14.	Midway Island	1.2	0.9
15.	Pago Pago, Samoa	1.5	1.3
16.	Wake Island 1.0	1.0	0.5
17.	Canton Island	1.4	0.2
18.	Apra Harbor, Guam	0.8	0.4
19.	Moen Island, Caroline Islands	0.8	0.6
20.	Kwajalein, Marshall Islands	1.1	1.0
21.	Aburatsu, Kyushu	0.8	7.8
22.	Chō shi, Honshu	0.8	6.9
23.	Hokodate, Hokkaido	0.8	9.0

Tsunami Engineering

Present Status

In common with earthquake engineering, tsunami engineering is in its infancy, despite a long history of destructive events. They differ in that earthquakes may occur sporadically over extensive areas, while severe tsunami effects are restricted to narrow coastal zones. Additionally, while a single large earthquake produces severe effects only within a relatively small region, a large tsunami affects the entire oceanic perimeter.

At present, tsunami engineering has been largely confined to limiting construction within obvious danger zones to those structures that require access to the sea and to provide protective measures, such as sea walls and breakwaters, where a sufficient economic justification exists. However, increasing pressure to extend our uses of available shoreline for recreation as well as commerce is forcing the engineer to seek new means of refining his judgment of tsunami effects and tsunami risk in many localities where ordinary protective measures will not work or cannot be justified.

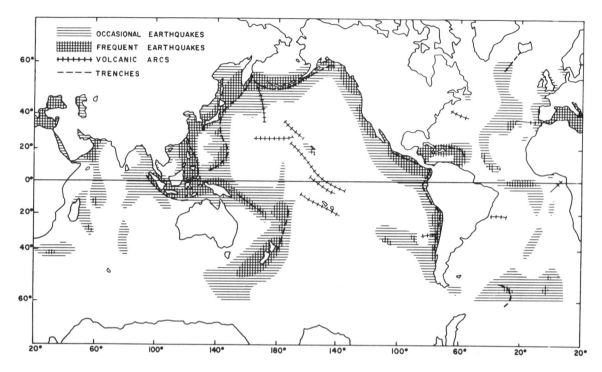

Figure 23-10. Oceanic zones of recent earthquake activity, showing association with trench systems and island arcs. Pacific preponderance is apparent.

Tsunami Risk

In the engineering sense, tsunami risk is the educated assessment of the probability and nature of extraordinary wave activity at a given location for any probable tsunami. Such an assessment involves consideration of probable tsunami sources—whether or not historically active; estimation or calculation of local wave effects from reasonable source models; and, estimation of the probability of such occurrences. We will consider these factors separately.

Potential tsunami sources have certain common characteristics. They are seismically active regions with a recent geologic history of large tectonic displacement. They are zones of past or present vulcanism and usually parallel great oceanic trenches. Figure 23-10 shows the world-wide distribution of these regions, which are principally confined to the Pacific Ocean, with

small centers of activity in the Caribbean and Mediterranean Seas and in the East Indies. Table 23-3 summarizes the historical distribution of large tsunamis, about 80% of which also have occurred in the Pacific. Currently, South America, Japan and the Aleutian Islands are sources for most destructive tsunamis, and the southwestern region from the Philippines to New Zealand is inactive. These regional differences in tsunami productivity are thought to be related to corresponding differences in the focal depths of large earthquakes, and to predisposition of crustal plate motions to horizontal versus vertical motions.

For use as design criteria, the best estimate of possible local tsunami effects can be obtained by calculations of the type described above, considering in sequence all potential source regions. In areas where recent tsunamis have occurred, suitable source models can be reconstructed from the aftershock patterns, fault rupture solutions and

Table 23-3
Number of Distribution of Large Tsunamis

Dates	Atlantic	Mediterranean	Pacific					
			Aleut.	Japan	S. Am.	E. Ind.	Other	Total Pacific
Before 1500	2	9		8		1		9
1500-1800	14	8	1	26	18	14	3	62
Since 1800	13	6	7	27	22	45	47	148
Total events	29	23						219

observed coastal displacements. In other areas, tectonic studies and evidence of old dislocations both above and below sea level can provide clues to the orientation, size and direction of motion of potential sources. Because these sources usually parallel the shoreline and are so long (~600 miles), only ten or fifteen such source models, staggered around the Pacific basin, would provide representative coverage in all sectors. This need only be done once to provide input for all future local design problems. The total cost of programming and calculation would probably not exceed $100,000, exclusive of the detailed response calculations at any particular location, which might cost $10,000 to $20,000 each.

Until such calculations become available, however, the only alternative is to estimate effects by analogy to recently observed events where wave effects are known and to modify or interpolate results for new sources and locations, allowing an ample safety margin for recognized uncertainties. These can be very large, as illustrated by the scatter of wave run-up observations for two large Aleutian tsunamis in the Hawaiian Islands (Figure 23-11). While the mean run-up heights—dashed curves—differ by a factor of three between the incident and lee sides of these islands, the local heights deviate from the mean by an equal factor.

Because of the similarity in local effects at a given place for different tsunamis, model experiments can sometimes provide guidance for specific problem areas. Such a model was recently constructed for Hilo Harbor (7) to study various protection schemes. Because of lack of suitable input criteria, the model input was arbitrarily adjusted so as to duplicate as nearly as possible the local inundation patterns observed from several previous tsunamis. Such methods will not work, of course, where prior data are not available. An equivalent computational model might have provided equally valid answers at much lower cost.

Engineering structures are usually designed for an expected lifetime over which the costs can be amortized. This requires an estimate of the probability of tsunami occurrences in areas where suitable protective measures are lacking or where some trade off exists between protective cost and expected risk. Although statistics are sometimes

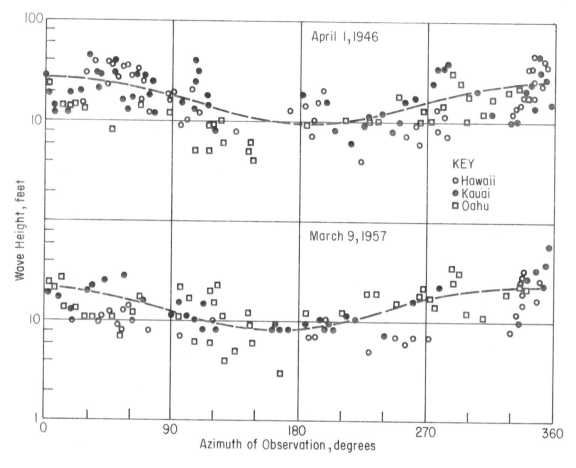

Figure 23-11. Azimuthal distribution of 254 run-up observations in Hawaiian Islands for two large Aleutian tsunamis. The 0° azimuth corresponds to incident wave direction from the respective epicenters.

misleading, they have been used in certain cases where tide records were available for a number of past tsunamis, and it was assumed that future events will follow the pattern of the past. Figure 23-12 shows plots of maximum wave height (tide gage) versus annual frequency of occurrence N for Hilo, Hawaii, and San Francisco and Crescent City, California (10). For each such curve, a set of curves can be computed from the Poisson distribution function $q = 1 - \exp(-ND)$, where q is the probability that a given height will not be exceeded in D years, as shown in Figure 23-13. The confidence value of such curves could be

greatly improved if a causative connection could be established between earthquake frequency, magnitude and the occurrence of tsunamis. Some attempts to correlate these factors for earthquakes near Japan have been reported by Iida(4).

Nature of Tsunami Damage

Tsunami damage falls into two general categories; i.e., that due to waves approaching from a remote source, as exemplified by Hawaii, and that arising within the epicentral area itself, as in Alaska in 1964. Because the latter event is

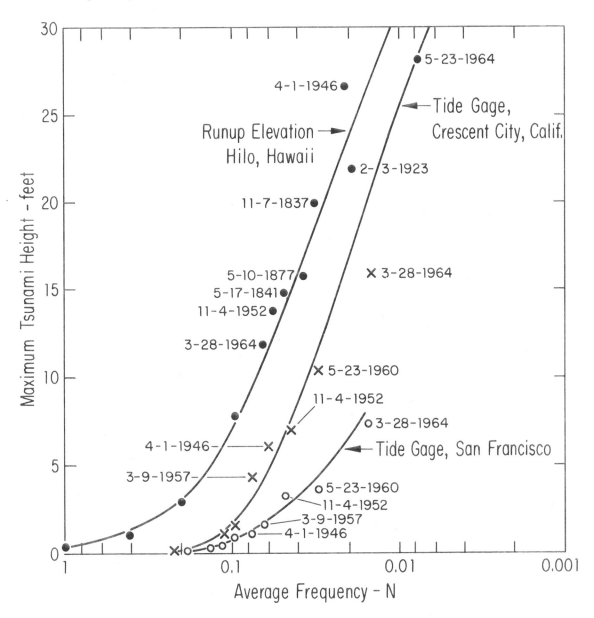

Figure 23-12. Distribution function for maximum tsunami waves (from Wiegel, Ref. 10).

exceptional and has been treated exhaustively (1), we will consider here only remote tsunami effects.

Most areas of engineering importance are low-lying coastal regions suitable for harbors or other shoreline facilities and habitations. In such areas tsunamis are manifested as a series of highly periodic surges or bores. The dominant wave period depends on the location and varies between several minutes to an hour or two. Wave activity may commence suddenly at peak amplitude and diminish slowly over several days, or build up gradually for six-eight hours before again subsiding.

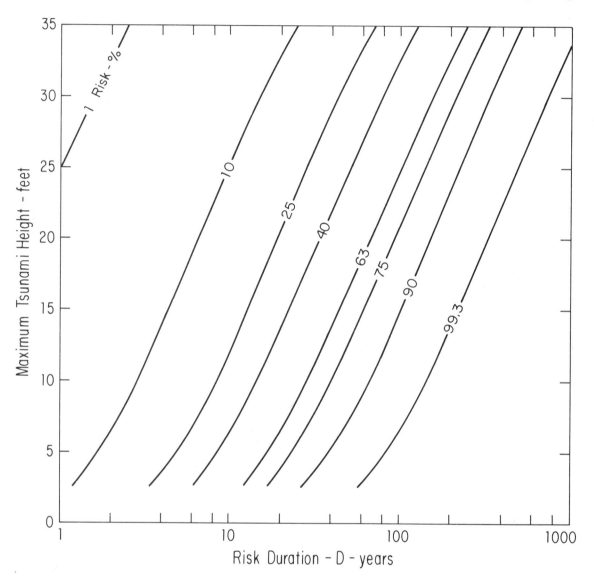

Figure 23-13. Probability of maximum wave height (tsunami) at Crescent City, California, exceeding a given value in D years (from Wiegel, Ref. 10).

These individual bores, of course, are superimposed upon the prevailing tide stage and surf conditions, which may greatly modify the damage interval and peak intensity.

In most cases the chief damage factors are extensive inundation, high-velocity flow, flotation of buoyant structures, transport of damaging debris and local erosion. Representative extreme bore heights of 10-20 feet and flow velocities of 30-50 feet/second have been observed. From analysis of structural failure and/or movement of heavy objects, dynamic pressures as high as 400-1800 pounds per square foot can occur (6). Damage studies after past tsunamis generally support the following categories of susceptibility:

1. Reinforced concrete buildings with secure foundations will withstand the above extreme conditions with little structural damage, although doors and windows are carried away.
2. Steel-frame structures will survive if not otherwise damaged by transported debris. Curtain walls are carried away.
3. Welded sheet metal-frame structures, such as gasoline stations, are usually dismembered and torn apart.
4. Trucks, autos and even heavy equipment such as tractors and railroad cars are overturned or transported considerable distances.
5. Piers and wharves of reinforced concrete have good record of survival unless collided by drifting or moored vessels.
6. Vessels of all classes are sunk or stranded.
7. Frame-stucco houses will withstand inundation up to about 3 feet before floating away—provided they have strong foundation connections. Higher water will carry away walls and collapse structure.
8. Alluvial areas not densely planted will suffer extensive erosion—roadways and railroad embankments are often washed out unless protected by paving.
9. Stone breakwaters and seawalls suffer extensive damage if overtopped.
10. Erosion and scour have caused collapse of bridge supports.

Designing for Tsunamis

The only specific effort to design a protective structure against tsunami attack in the United States has been the Hilo Harbor study of Palmer, et al. (7). Detailed model experiments were conducted to investigate a number of possible breakwater and seawall combinations, with the object of reducing wave activity within the harbor to tolerable limits. Two alternative designs were finally recommended, but both proved too expensive to execute in terms of cost-use benefits and neither promised complete waterfront protection.

Where protective measures cannot be justified, means should be sought to minimize damage from future events, given the best appraisal of local flow patterns during tsunamis. Such means include:

1. Consideration of levees or seawalls designed for overtopping, with possible catchment and drainage areas behind.
2. Limiting construction in potential inundation zones to structures designed to withstand expected forces or to those elevated so as to allow water to flow beneath them. Such areas can also be planted as parkland to provide some flow inhibition.
3. Pave, or otherwise protect potential erosion zones where critical to structural elements.
4. Provide adequate access and escape routes for vehicular and pedestrian traffic and instruct local agencies to establish and rehearse evacuation plans.
5. Provide for clearing harbors of vessels and prohibit other floating objects that might constitute collision hazards, such as water or fuel tanks, large mooring buoys, wooden wharves, etc.

Most of the foregoing merely represent common sense application of sound engineering practice, once the nature and severity of local effects can be anticipated.

Conclusions

While linear hydrodynamic theory in principle can describe all but the very local details of tsunami behavior, only numerical methods or hydraulic models can provide realistic predictions. Except for locally generated tsunamis, scaling problems and the earth's sphericity limit application of hydraulic models to terminal effects in areas where sufficient data exist from past tsunamis to calibrate the model input. Numerical studies of tsunami generation indicate that the resulting wave systems are quite complex and have substantial radial differences that are further modified as they propagate across the ocean. These latter modifications are not realistically treated by the conventional practice of ray-refraction, and there is need for extending the numerical grid method to a spherical ocean with real topography. Such methods are now under study.

Even with local predictions, numerical models offer substantial cost and time advantages over hydraulic models, and it seems likely that the future of ocean engineering will increasingly depend on application of computer techniques—not only to tsunamis, but to many other wave and flow phenomena. Such methods are already routinely used in forecasting storm tides and river flood propagation.

References

1. *The Great Alaska Earthquake of 1964.* Oceanography and Coastal Engineering. National Academy of Sciences. Washington, D.C. 1972.
2. Hwang, Li-San, David Divoky and Albert Yuen. 1970. *Amchitka tsunami studies.* Rept. TC-177, Tetra Tech, Inc., Pasadena, Calif. (Prepared for U.S. Atomic Energy Commission)
3. ____, and E.O. Tuck. 1970. On the oscillations of harbours of arbitrary shape. *J. Fluid Mech.* vol. 42, part 3, pp 447-464.
4. Iida, K. 1963. Magnitude, energy, and generation mechanisms of tsunamis and catalogue of earthquakes associated with tsunamis. *Proc. Tsunami Meetings Tenth Pac. Sci. Congress.* IUGG monograph no. 24, Paris.
5. Kajiura, K. 1963. The leading wave of a tsunami. *Bull. Earthquake Res. Inst.*, Tokyo Univ., vol. 41.
6. Matlock, Hudson, Lyman C. Reese and Robert B. Matlock. 1961. *Analysis of structural damage from the 1960 tsunami at Hilo, Hawaii.* Report for Defense Atomic Support Agency, Univ. of Texas.
7. Palmer, Robert Q. and Gerald T. Funasaki. 1966. The Hilo Harbor tsunami model. State of Hawaii, James Look Laboratory Tech. Rept. no. 1.
8. Van Dorn, W.G. 1965. Tsunamis. In *Advances in Hydroscience*, vol. 2. New York: Academic Press.
9. ____. 1970. Tsunami response at Wake Island. *J. Mar. Res.* vol. 28, no. 3, pp 336-344.
10. Wiegel, Robert L. 1970. Tsunamis. In *Earthquake Engineering*, Englewood Cliffs: Prentice Hall.

other volumes
in the

topics
in
ocean
engineering
series

contents
volume 1

contents
volume 2